Goodlands

THE WEST UNBOUND: SOCIAL AND CULTURAL STUDIES

Series editors: Alvin Finkel and Sarah Carter

Writing about the western regions of Canada and the United States once turned on the alienation of the peoples of West from East. The mythology of a homogenized West fighting bravely for its rightful place in the sun deflected interest from the lives of ordinary people and from the social struggles that pitted some groups in the West against others—often the elite groups who claimed to speak for the region as a whole on the national stage. Seeking to challenge simplistic interpretations of the West and its institutions, The West Unbound focuses instead on the ways in which particular groups of Westerners—among them women, workers, Aboriginal peoples, farmers, and people from a diverse array of ethnic backgrounds—attempted to shape the institutions and attitudes of the region. The series embraces a variety of disciplines and is intended for both university audiences and general readers interested in the American and Canadian Wests.

SERIES TITLES

Goodlands
A Meditation and History on the Great Plains

FRANCES W. KAYE

AU PRESS

Published by AU Press, Athabasca University

1200, 10011 – 109 Street, Edmonton, AB T5J 3S8

ISBN 978-1-897425-98-5 (print)   978-1-897425-99-2 (PDF)
978-1-926836-41-6 (epub)
A volume in The West Unbound: Social and Cultural Studies
ISSN 1915-8181 (print)   1915-819X (electronic)

Cover and book design by Marvin Harder, marvinharder.com.
Printed and bound in Canada by Marquis Book Printers.

LIBRARY AND ARCHIVES CANADA CATALOGUING IN PUBLICATION

Kaye, Frances W., 1949-
        Goodlands : a meditation and history on the Great Plains / Frances W. Kaye.

(West unbound, social and cultural studies series, ISSN 1915-8181)
Includes bibliographical references and index.
Also issued in electronic format.
ISBN 978-1-897425-98-5

1. Great Plains—History. 2. Indians of North America—Great Plains—History.
3. Agriculture—Great Plains—History. 4. Agriculture—Environmental aspects—
Great Plains—History. I. Title. II. Series: West unbound, social and cultural studies

F591.K39 2011   978   C2011-900923-4

We acknowledge the financial support of the Government of Canada through the
Canada Book Fund (CBF) for our publishing activities.

Canadian   Patrimoine
Heritage   canadien

To Howard

Contents

Acknowledgements

I would like to extend my thanks to the Council for International Exchange of Scholars for a Fulbright Fellowship to the University of Calgary, Alberta, during the 2001–2 academic year and to the Department of English and the University of Nebraska–Lincoln for a Faculty Development Fellowship during the same year. These fellowships made the research for this book possible. The libraries and staff of the University of Calgary and the University of Nebraska–Lincoln, and also the staff of the Local History collection at the Calgary Public Library, were invaluable. I am very grateful to my colleagues in the Department of History at the University of Calgary, particularly to Sarah Carter, Don Smith, and Gretchen Albers (who read several drafts), as well as to Jane Kelley of the Department of Archaeology. At the University of Nebraska–Lincoln, I thank the Department of English, especially Joy Ritchie and Sindu Sathiyaseelan, my very kind research assistant and computer whiz, the Center for Great Plains Studies, especially John Wunder, George Wolf, and Rick Edwards, and the Institute for Ethnic Studies. I am extremely grateful for the comments of two anonymous reviewers, who helped improve the manuscript immeasurably, and for the hard work of my editors, Walter Hildebrandt, Pamela MacFarland Holway, and Joyce Hildebrand. I would also like to thank my friend Chris Garza for his encouragement, my brother, Robert Grey Owl, for the reality checks, my son, Joel Kaye, for the distractions, the dogs who have accompanied me, especially Fireball and Autumn, and most of all, my husband, Howard Kaye.

Introduction

Wheat farmers hate gophers. The little critters cut the stalks and run off with the grain—particularly when the year is dry and the crop is light. Furthermore, to the disgust of ranchers, they dig holes in pastures, where large and commercially valuable animals like cows and horses can fall and break their large and commercially valuable legs. On the other hand, gophers are valuable members of midgrass prairie communities. In dry years, they strip the leaves and seed heads from the grasses, limiting above ground vegetation that would otherwise transfer limited moisture from the perennial underground forest of roots and rhizomes to the air through transpiration. Gophers—and their allies in drought, the grasshoppers—invented the summerfallow, but they use it selectively, fallowing the most land in the driest summers. The grasses themselves co-operate—on tallgrass prairie, the big bluestem, Indian grass, and switchgrass grow eight feet high in moist years, while in dry years they fade back and let the little bluestem, the stypas, and the other "bunchgrasses" take over and hold the soil; thus, less foliage is exposed to transpiration and little ground to evaporation, again conserving water for the perennial root forest.[1] Wheat, on the other hand, is an annual grass. The gophers' mowing may slow down transpiration, but there is no living root forest to benefit, only dead and shallow structures like frost-killed petunias in an urban flowerpot. Gophers' incessant burrowing aerates the land and separates the root forest, thinning it out so it can breathe and grow, just as the urban gardener separates the rhizomes of iris. Gophers store their seed underground, and in the event of a long drought, these storehouses become one source for grassland regeneration after the return of the rains. Gophers are the messengers of Gaia, small piping indicators of the complex biofeedback mechanisms that mark the whole blue-green Earth as a single living organism of interlocked living systems.

The Laramide orogeny of some 65 million years ago, the great collision of tectonic plates that raised the Rocky Mountains, set up the conditions for the grasslands ecosystem in the semi-arid rainshadow of the Rockies. The grasses and the gophers co-evolved with the buffalo and other even bigger ruminants, including something with a snout big enough to munch on Osage oranges. Badgers, ferrets, and hawks ate gophers, as did coyote, the trickster. Long cycles of glaciation and warming, drought and moisture, shaped the system. Rivers and wind lay down soil and stripped it away again. Dune systems grew and moved. Prairie pothole lakes formed in the remains of the glaciers, and waterfowl thrived. The long and short cycles of weather coiled past each other, and the gophers brought forth their young. When the first humans came onto the grasslands, whether emerging from the earth as the old stories tell or coming down from the north as more recent commentators would have it, they fit themselves into the cycles of the grasslands. It may be that they killed off the megafauna, or, more likely, that the cycles of cold and heat, moisture and drought no longer favoured the giant bison, the mastodons, and the others.[2] But the grass and the gophers continued their dance through the processions of the equinoxes and the tilts in the earth's orbit that change the name of the fixed star. At some point, the people began firing the grass, pushing the woody plants back to the verge of the creeks, and removing the overburden of dead plants. The young shoots showed improbably green on the scorched earth, and the buffalo and the gophers came to feast on bounty coming, like asparagus, from the deep and long-lived roots. Women with digging sticks foraged for prairie turnip, timpsila, and other roots, and joined the gophers in the work of aeration. The people lived well—they lived well indeed. Prairie is a diverse ecosystem, offering hundreds of plants and animals for food, medicine, inspiration, and co-management. But hunger, want, and warfare came too, as part of the cycle—and hard work and danger. Peoples moved. Newcomers came. And every year the gophers brought forth their young and the bison calves looked red in the sun and the grasses turned their tender faces to the sky.

For 65 million years or so, the ecosystem of the Great Plains at the heart of what is currently called the North American continent was exactly that—a heartland. The violent extremes of climate, the stunning expanse of earth tapered by glaciers and ancient seas to meet the sky, the grass-ful dance of above-ground[3]—all stretched across an invisible underground forest of roots; the gophers, the buffalo, the people, the hawks—all were at the centre of a universe that fitted them very well. When the horse—which had evolved precisely on that grassland—returned, it initially fit in as if it had never been away.

Every ecological system is necessary and sufficient for the plants and animals that have co-evolved with it and for those that have migrated slowly into it in response to the cycles of climate change that characterize Earth's history. The Arctic and the Kalahari are exactly home for the low-growth shrubs, the protectively coloured animals, and the people who hunt and forage there. Life may be hard because population is scarce and climate unforgiving, but neither the organisms nor the land is deficient. Far less deficient was the Great Plains for the first ten thousand, or forty thousand, or more years of its acquaintanceship with humans. Humans already occupied the land as it evolved from forest to grassland. Both archaeological and oral evidence agree that the Plains, away from the shelter of the mountains and the river valleys, was seldom traversed in the days when people walked and dogs carried their cargo on travois. (My travois dog sleeps beside me, her sturdy, big-chested body, which uses food so efficiently that she easily goes to fat in her latter-day idleness, bearing witness to the strength of her ancestors.) The oral history also tells us that emergence onto the Plains for the Lakota, the Blackfoot, the Kiowa, and the others was an emergence into a paradise, a garden that teemed with a diversity of prey animals, from buffalo to voles, and of vegetable treasures, from saskatoon berries to mouse beans. For people like the Mandans and Hidatsas, the Omahas and Pawnees, the Plains also provided space for riverine agriculture: corn, squash, beans, and sunflowers.[4]

Certainly, the hunters and gatherers and farmers could see ways to improve the Plains. Again, both archaeology and oral tradition agree that the people built buffalo pounds, especially ones that would hurtle the huge ruminants over cliffs so that they might easily be dispatched. People fired the prairies to repel bison with fire and to attract them with succulent new growth. Women knew the locations of all the berry, turnip, and other wild plant food grounds, though whether their practices actually enhanced the food grounds is not entirely clear. They did take berries, roots, and other products at a sustainable rate that left the grounds fruitful year after year. People cleared and planted riverine gardens and protected them from deer, birds, and other predators, including humans. They understood the land as part of a sacred tradition of earth and sky; they held sacraments such as the Sun Dance that expressed the courage and integrity of the people as worthy of the favour of the sun and the buffalo. Although many different groups of people lived on the Plains between their first emergence and some two hundred years ago, and although they understood various economic and sacred relationships to the region—including many that manipulated place, plants, weather, and animals for their own benefit—they worked from an ideology of sufficiency. What was there was what ought to be there. Droughts, severe winters, and even the deaths of individuals with superior skills in locating and securing food sources might bring about scarcities, even ones that lasted longer than a generation and required people to relocate in order to survive. But the human response to the Great Plains, until a few hundred years ago, was to use it, appreciate it, learn it, and manipulate it, but not to replace it or make drastic changes.[5]

For the Spanish who came with Coronado, the Great Plains were deficient in gold. The soft golden grass houses of the Wichitas were a mockery, not a marvel. A disappointed Coronado had his guide strangled. For the French and British fur trade explorers who came from the north and east, the Plains were deficient in fine furs and supported deficient people, like the Omahas, who demanded tolls of the traders coming through their territory, or like the Blackfoot Confederacy, who would not trap beaver and would neither trade with the Canadian traders nor allow the American mountain men to trap in their territory. But the true prophets of deficiency were the agricultural settlers and the people of their urban trade centres. They were prepped by theories of the Great American Desert and the Palliser Triangle to find deficiency. They also felt a strong sense of entitlement to something else, and they relied on theories about the "Manifest Destiny" of the "Anglo-Saxon race" to expand across the continent and to change the "desert" to the "Garden of the World," the theory that "rain follows the plough," and the idea that "free land," "virgin land," was just waiting for the touch of the "yeoman farmer" to "blossom like a rose" and bring forth wheat in the "Bread Basket of the World."[6] Tame grasses, tame water, tame cattle, land that was personal property, and a worldwide market system would end the deficiency and reclaim the empty land for civilization and Christianity, these newcomers believed.

The study that follows is a meditation about what happened when a mass of people hit a geographical and cultural region that they felt entitled to reclaim from deficiency. It is also about the intellectual resistance from groups of people, already weakened by disease and invasion, who nonetheless attempted to deal with vastly changed circumstances in both economic and sacred contexts; people who, unlike the settlers, began from the premise of sufficiency, not deficiency.

There is no single point at which the paradigm of deficiency replaced sufficiency; indeed, that shift is still not complete and might, perhaps, someday reverse. We might begin with Coronado's *entrada* in 1540–42, with the grant of Rupert's Land to the Hudson's Bay Company in 1670, with the Proclamation of 1763, or with the passage of the US *Homestead Act*, the Confederation of Canada, and the completion of the first transcontinental railroad in the 1860s. For the most part, it is this last decade that I have chosen for my starting point and that I have followed up to the present, with an outlook toward the future. My definition of the Great Plains follows that of my geographer colleagues at the Center for Great Plains Studies at the University of Nebraska (see map on next page). The region stretches roughly from the Missouri River to the Rocky Mountains and from the North Saskatchewan to the Rio Grande. It is the land that the governments

98th Meridian

Edmonton

ALBERTA
Calgary

SASKATCHEWAN

Regina

MANITOBA

Winnipeg

ONTARIO

MONTANA

Billings

NORTH DAKOTA

Fargo

MINNESOTA

SOUTH DAKOTA

Sioux Falls

WYOMING

IOWA

Cheyenne

NEBRASKA Omaha

Lincoln

Kansas City

Denver

COLORADO

KANSAS

MISSOURI

Wichita

Tulsa

NEW MEXICO

Oklahoma
City

ARKANSAS

Roswell

Lubbock

Fort Worth

TEXAS

LOUISIANA

MEXICO

THE GREAT PLAINS

0   200   400

miles

gave away as not quite good enough to be sold, unlike the land to the east, and not quite bad enough to be kept in the public domain, unlike the mountains, the deserts, and the arctic. Although the area is approximately two-thirds in the United States and one-third in Canada, I have tried to treat the two countries equally because the subtle (and sometimes not so subtle) differences in government policy and national narrative are useful for helping untangle environmental and cultural imperatives. Working with the paradigm shift from sufficiency to deficiency means that I have mostly omitted several narrative lines from earlier histories, such as the Wild West/Mild West dichotomy in many US-Canada comparisons, or the conflation of the Plains with the West Coast and Mountain West in one meta-region. As will become evident, I have been heavily influenced by many other writers, particularly Roger Epp, Sarah Carter, Barbara Belyea, Paul Voisey, Jim Pitsula, Angie Debo, James C. Malin, John Joseph Mathews, and Hamlin Garland.[7]

Except for my great-grandparents' adventure holding down a homestead in Colorado for a few years around 1880, my family has no farming traditions. My ancestors were coal miners and civil servants, merchants and soldiers, lawyers and teachers. Gardening, though, is a different story. My English grandparents grew bounteous vegetable and flower gardens in the long narrow lot behind their little house in South Calgary. Except for my student years, when I lived in dormitories or a co-op, I cannot remember living without a garden. True, we do not rely on our lettuce to feed us through the winter, and we know that if we don't bother to put it in, we can supply its lack from a farmers' market, but still, we follow the rhythm of planting and tending and harvesting at a level that, unlike the dirt under our fingernails, will not wash out. And I have lived nearly three-quarters of my life, as both child and adult, within sight and smell (I walked to school past cabbage fields in the Garden State of New Jersey) of farms. For thirty-two years, my family has lived on and with a ten-acre plot of land outside of Lincoln, Nebraska. It was a small but working dairy farm in the 1930s and a hobby farm with sheep and chickens from the 1950s to the early 1970s. We have planted vegetable gardens and fruit trees, and have watched the wide leaves of our rhubarb shrivel up after our neighbour sprayed herbicide on fields upwind from us. We have watched the tallgrass prairie regenerate in the front pasture, aided by fire and mowing, and we have watched red cedars take over the unmowed and unburnt back pasture. We mine mud from the creek to patch up the holes around the overflow that would otherwise drain our pond—the recharge for our domestic well. One year, my husband waged a war with a solitary bank beaver (we named him David Thompson) who ate up all the willows and insisted on trying to block the overflow and raise the pond up over our driveway. Great blue herons fish the pond, and green herons nest in the boggy area around the inflow. Red-tailed hawks still whistle and soar, even though the stars have disappeared from the north sky in the light pollution of the Wal-Mart and Menards that moved in across the highway about five years ago. None of this makes me a country girl, but I know farming and the land differently than I would had I always had people rather than grasses as neighbours. And so Hamlin Garland, the son of the Middle Border, John Joseph Mathews, and the others do not seem very far away to me.

Born in Wisconsin, raised in Iowa, and holding down a claim in Dakota Territory before becoming a successful author, Garland would seem to be the consummate American homesteader—and it is from him I first understood that the *Homestead Act* and its variations were most successful to the extent that they did *not* produce family farms. Mathews, the Oxonian Osage, showed me how *un*-inevitable—in fact, how freaky—it was that European ideologies replaced Osage ones, revelations underlined by Carter Revard and Leslie Silko. James Malin's cantankerous opposition to the theory behind New Deal agricultural practices, his stubborn insistence on the existence of Great Plains dust storms long before the plough, his scorn about theories of climax vegetation, and his incessant questioning of what prairie restoration would restore prairie *to* influenced my conviction that no ecosystem is ever deficient for the plants and animals with which it co-evolved. Angie Debo showed in great detail both how Indigenous political and economic systems worked in the context of an overlain free market system and how they were systematically destroyed, both legally and illegally, during the twentieth century. Paul Voisey, like Garland but much more exhaustively, showed me that homesteading was sometimes only incidentally about farms. Jim Pitsula showed me that a market economy, operating exactly as it was supposed to, would rob the Great Plains of people and resources. Barbara Belyea awakened me to the *contingency* of all systems of categorizing geography, including those as seemingly "obvious" as river systems. Sarah Carter teaches me many things, but especially how receptive the Plains Cree were to farming, how skilful and inventive they were, and how government policy systematically and repeatedly scuttled their successes. Most recently, I have been influenced by Roger Epp and his theories of the political de-skilling of the rural West. Other intellectual debts will become evident as this book unfolds. All errors of fact and interpretation are, of course, my own.

In chapter 1, "A Unified Field Theory of the Great Plains," I lay out an overview of how the region has transformed since the deficiency paradigm has become the norm and why I think deficiency is, indeed, a "deficient" theory. I also deal with institutions such as the railways, cattle ranching, and the grain trade, which have definitively shaped the region but which I do not study in individual chapters. Chapter 2, "Exploring the Explorers," looks at how the idea of deficiency was laid down by the various European and Euro–North American explorers of the Great Plains, their editors back in "civilized" locations, and subsequent historians of exploration.

The next two chapters parallel armed resistances to the paradigms of deficiency by pairing Riel's Red River resistance to the Cheyenne with Sioux resistance to Custer's Seventh Cavalry and then Riel's 1885 resistance in the North West with the Ghost Dance leading up to the 1890 massacre at Wounded Knee. At the Forks of the Red and Assiniboine Rivers in Manitoba, old fur trade families, crofters "cleared" from the Scottish Highlands, Swiss soldiers, and the Peguis Ojibway-Cree had coalesced into a successful commercial settlement.[8] Although Canadian expansionists from Ontario believed that the community would be happy to join the brand new Dominion of Canada on Canada's terms, Red River (today's Winnipeg and environs), under the leadership of Louis Riel, successfully resisted the extra-legal taking of the community and managed to secure some rights for the old settlers in the new province of Manitoba. A decade later, Lakota, Dakota, and Cheyenne warriors decisively defeated a certain show-off US colonel at the Battle of the Little Bighorn. Yet both of these successful resistances turned to pyrrhic victories, as they gave the federal governments of Canada and the United States graphic images of the "savagery" and hence deficiency of the inhabitants of the Great Plains; this gave intending settlers moral permission to displace, subdue, or even kill them. In the mid- to late 1880s, various religious revivals arose on the Great Plains, from the Exovedate established by Louis Riel at Batoche to the Ghost Dance among the Lakotas. Both of these movements were suppressed by the superior force of arms of the two federal governments, and both were used to extend the already coercive material and spiritual dispossession of Indigenous and mixed-blood groups in favour of European and Euro–North American settlers. The spiritual aspects of resistance survived, however, and helped mitigate the continuing attempts to "kill the Indian, and save the man," as Richard Henry Pratt, founder of Carlisle Indian School in Pennsylvania, put it.[9] Despite Anglo writer John G. Neihardt's contention that the people's dream died in the bloody mud at Wounded Knee Creek, South Dakota, just after Christmas 1890, resistance *never* failed.[10]

Two Indigenous historians, discussed in chapter 5, are Hehaka Sapa (Nicholas Black Elk), an Oglala Lakota, and John Joseph Mathews, a mixed-blood Osage. For them, there was no question about whether "Indians" had survived the "Indian wars." They had. In 1932, each of these men published a book—Black Elk through the interpretation of his son, Ben, and the rewriting of John Neihardt. The volumes each suggested ways in which specifically Siouan constructs of the universe—and particularly the intricate interconnection of material and spiritual life in the specific ecosphere of the Great Plains—could frame a sustainable way of living that was completely different from the linear and progressive model of the Amer-Europeans and their historians.

To Amer-Europeans—John Joseph Mathews' term for people of European descent who inhabited America but had failed to become naturalized to the land and its customs—the end of the nineteenth century seemed to mark the end of the frontier, the defeat of the "deficient" people who *had* peopled the Great Plains, and the triumph of a bicoastal Anglo-Saxon democracy, premised on a market economy and a particular definition of Christianity. Chapter 6 looks at how the saga of the "Closing of the West" was created for the United States by Frederic Jackson Turner and for Canada by Harold Innis, and how the saga has been tweaked and rewritten by our contemporary New West historians.

Yet "Indians" were not the sole history of the Great Plains during the period it was being transformed into commercial agriculture. The eastern, central, and West Coast areas of North America were never "free land" in the way that the Great Plains was purported to be. Quebec, Plymouth, Williamsburg, and other seventeenth-century settlements were sited on or near Indigenous settlements and were dependent upon Indigenous people for their survival. Land was granted to seigneurs or to compacts and parcelled out to settlers. Eighteenth- and early-nineteenth-century settlers or agents purchased land or were granted it for service in war. Oregon Territory featured an early *Homestead Act* designed to draw settlers west (ignoring the Great Plains) in order to hold the territory for the United States against British claims. The mountain and desert Wests and the North remain largely federal lands not "settled" by agrarians. The Great Plains, however, was "free land" to be made into farm homes by idealistic young families. Or so, at least, said the backers of the *Homestead Act* and the *Dominion Lands Act* and even the *Dawes Allotment Act*, which broke up the reservations into individual allotments for Indigenous people and "surplus" lands for Amer-European homesteaders. In chapter 7, however, we see that the great success of the Homestead Acts was in transforming "free land" into capital for the market development of the Great Plains, not in turning "virgin land" into "family farms."

Homestead laws both implicitly and explicitly, especially in Canada, excluded most women from homesteading in their own right. The *Indian Act* in Canada and other laws and treaties defined race in terms of gender. Only male persons were described as Indian—women's Indian status was dependent upon being fathered by or married to an Indian, and could be erased by marriage to a non-Indian. The destruction of the buffalo economy and of the definitions of the sacred year around the buffalo affected men more severely than it did women. Chapter 8 discusses the ways in which deficiency definitions affected women distinctively.

The deficiency definitions of the West did not disappear in the twentieth century, nor did the Amer-European belief that it was appropriate to continue to take Indian land, lives, and culture because they continued to be deficient by Amer-European standards. Chapter 9 looks at the de-Indianizing of the state of Oklahoma, the former "Indian Territory," from before statehood up through the 1930s; we also examine the "mixed economy" that had been re-created and rebuilt since the various "removals" of people to Oklahoma. And Indigenous peoples were not the only ones who resisted the imposition of Amer-European agriculture and farms on the Great Plains. The Dust Bowl of the 1930s—following on the depressed years of the 1920s on the Plains—forced Canadians and Americans to rethink the whole prospect of living on the Plains. The Dust Bowl reinforced the deficiency idea, of course, but it also forced people to reconsider the way they were doing things. Not all Amer-Europeans shared the belief in the deficiency of the Great Plains: there had always been people seeking to become native to the place, like Osage agent Laban Miles, of whom Mathews wrote.

Chapter 11 discusses how two unusual leaders, George Norris of Nebraska and Tommy Douglas of Saskatchewan, attempted to mitigate what was going wrong for the people who were living on the Plains. Both recognized that the extreme individualism preached by Manifest Destiny narratives simply was not working on the Plains, although their ways of mitigating both market forces and the particularities of the environment were fairly conventional. Douglas, particularly, recognized that market forces, working as they theoretically were supposed to work, would inevitably impoverish and depopulate the Great Plains. He believed that government development and a planned economy would mitigate the unforgiving hand of the market. Chapter 12 looks at how planning and growth theory can help us understand how the history of the Great Plains developed under an explicit model of deficiency that does not necessarily provide a blueprint for a better future—except for a planned depopulation of Buffalo Commons. In fact, the global blunders committed in the name of planning foreshadow a dark role for the Great Plains in terms of the global economy. "Mouse Beans and Drowned Rivers," chapter 13, shows how, again, the theories of the deficiency of the land and of its Indigenous inhabitants intersect, this time resulting in the string of dams built to "reclaim" the Missouri River for flood control, power generation, and navigation for Amer-European market agriculture and cities, all at the expense of the subsistence, convenience, tradition, and commercial livelihood of the tribal communities that were systematically flooded.

Although we have been looking primarily at an agricultural history of the Great Plains, resource extraction has also been a significant part of the story. While the region (except for a small section near the Black

Hills) has been spared consideration as a "National Sacrifice Area" (à la the uranium-producing Four Corners region of the United States), extraction of fossil fuels, and particularly oil and gas, has played a large part in the economic prosperity—and subsequent economic busts—of the region. Extraction comes with certain environmental degradations that emphasize the expendability of the place and its human and non-human residents. Alberta's oil sands are north of the Great Plains, but the vast expenditures of water, energy, and habitat in producing oil are resonant with the petroleum industry's history from Texas and Oklahoma up through Wyoming and the Dakotas to Alberta and Saskatchewan. Roger Epp's consideration of how they add to the "de-skilling" of the rural West is a twenty-first-century explication of the deficiency paradigm.

The final chapter suggests a way in which we might reconceptualize our whole understanding of this region within a paradigm that does not depend on deficiency. Among the "deficiencies" of Indigenous people that Amer-Europeans attempted to rectify was the "lack" of a justice system. As innumerable inquiries into the provision of justice (or lack thereof) to Aboriginal individuals and communities have repeatedly concluded, the vaunted, adversarial, rights-based Anglo justice system has been, especially in the Prairies, a travesty for Native people, who are, from birth, more likely than anyone else to be "victims" or "perpetrators" of crimes. Although things may be getting worse for actual Indigenous persons, society is no longer uniformly proclaiming that it is Native people who are deficient. Rather, it is the imposed "justice system" that has failed. Chapter 15 looks at how social justice might improve were it framed in an Indigenous intellectual context. It argues that a similar reframing might enable us to better understand how to create a thoroughly twenty-first-century form of sufficiency on the Great Plains that satisfies human beings without devastating the non-human Plains ecosystem.

Almost every summer morning, the dogs and I leave the little house in Calgary and walk past the neatly groomed fairways of the golf course to a few acres of "natural area" park. About three or four years ago, the neighbourhood community had the park declared pesticide-free and staged a raid on yellow goatsbeard or false salsify (*Tragopogon dubius*), a Eurasian plant that in Nebraska is content to be a minor forb in the tallgrass prairie, but here in the fescue shortgrass is a serious invasive. So each morning we stop, I put my right foot on the leashes, and I grasp the stem of the goatsbeard. Pull steadily, straight up, so as not to break off the stem at ground level. It is best to work two days after a rain, so that the water has penetrated deep enough to soften the ground. Pull out the taproot, which looks like a skinny parsnip or real salsify. If you are patient, you can boil the roots and scrape out the edible flesh between the woody core and the skin and root hairs to make a tasty porridge. Supposedly goatsbeard, like other salsifies, is a remedy for liver and gallbladder malfunctions, but it would take a very patient herbalist to work with it. Some of the other exotics in the natural area park—brome grass, dandelions, European vetches—were deliberately introduced to North America for their nutritive values, but goatsbeard probably just came along for the ride, mixed in with the seeds of those more prized Eurasian fodder plants. No one, not even I, bothers to cook up the yellow goatsbeard. The plants are allowed to dry out and disintegrate on the paths or are carefully bagged in plastic for the trash. I can see them now at any distance across the field, their shade of yellow entirely distinct from any other yellow, their silhouette of leaf and stalk standing out, now that I have hunted them for so long, from all the other grasses, forbs, and woody shrubs.

Except for Autumn, the strong travois dog (licensed as an Australian cattle dog because the City of Calgary has no categories for Indigenous North American dogs), the dogs and I are as much invasive exotics as the yellow goatsbeard, as the brome and dandelions, as the Hungarian partridges who fill the niche once claimed by prairie chickens, as the English sparrows and city pigeons. Yet sometimes we find coyote scat or surprise a jackrabbit hurrying off with stiff-legged bounds. I give thanks for the pin cherries and strawberries, the wolf willows and wild roses, the spruces and poplars, the fescue grasses and the spring crocuses, the ears of the prairie, and I wonder by what right I uproot my fellow invasive, the pretty yellow flower that, if left alone, produces a perfect hoary globe, like a giant dandelion plume, that might delight a small child; the flower that, if left alone, could be harvested in the fall as food and medicine.

On most fall and winter and spring mornings, the dogs and I leave the big old farmhouse in Nebraska and walk down our driveway, past the pond that feeds the well, along the abandoned railroad tracks, then up the gravel service road to the tower that sends 911 signals across the southern half of Lancaster County. Red-tailed hawks perch on the guy wires of the tower and launch off to search for the small mammals who make up most of their diet. On our own land, we walk through the regenerated tallgrass prairie, where the big and little bluestem, the switchgrass, the Indian grass, and the rest are slowly taking back these few acres from the brome that was planted there some ninety years ago. Already the yellow sweet clover that came up the year after we pastured the neighbours' horse is gone. We have wild roses and many forms of composite sunflowers and asters and daisies and iron weed, distant cousins of the yellow goatsbeard. We have the woody sumacs, whose "fire-fangled feathers" give fall colour to the field. We lack leadweed and sensitive plant and most of the other legumes of the prairie, who did not shelter a population along the creek beds sufficient to accompany the grasses back into the ploughed and seeded monoculture of the brome. The creek is the home of black willows of the kind one might cut to build a sweatlodge, and of one huge and symmetrical cottonwood tree. As we walk up to the tower, we walk between a fenceline of mulberries and Siberian elms, both deliberately introduced exotics who are now invasive, and a field that used to be sown in wheat or milo, both semi-arid plants, but that now is always given to the thirstier soybeans or corn. Only the corn is native to the Americas, but these commercial hybrids are a long way from the multi-coloured "Indian corn" of the Pawnee and Omaha and Oto corn villages that dotted southeastern Nebraska a few centuries ago. I am grateful for the properly named velvet leaf, which, exotic as it is, breaks up the monoculture. And I am grateful for the cattle who glean the fields after harvest, giving them shape and dimension. I know that on a late fall afternoon, coming home after dark, it is wise to be sure that the blacker shadow of the willow tree by the pond does not hide an Aberdeen-Angus heifer who has got through the fence. And I do not pull out the yellow goatsbeard that so modestly raises its head from the tall grass or the hedgerow. In Nebraska, it seems to have become naturalized, in equilibrium, not threatening to claim more than a sustainable niche in the floral ecosystem. Perhaps the dogs and I should aspire to the humility of Nebraska goatsbeard. Each morning I choose the highest point of our walk to face the four directions and salute the array of leafy beings against the great prairie sky.

According to Janine Brodie, "Regionalism structures *political conflict around the distribution of resources across geographic space.*"[1] All regions are imaginary—the sharp borderlines and different coloured spaces of the map are intellectual constructs, not physiographic ones—but, like most imaginary human constructs, maps control some of the ways human minds can conceptualize, in this case, place. If we look again at our particular map of the Great Plains, we can see that its outlines are an amalgam of physiographic and political features. On the land itself, elevation gradually rises and average annual precipitation gradually drops from east to west, while summer temperatures rise and summer daylight hours diminish from north to south across the Great Plains. Both native vegetation and contemporary cropping patterns spill over the edges of the region in all directions. Rural areas of the Great Plains share with the rest of rural North America, and indeed the world, problems such as depopulation, loss of political power, soil and water degradation, siting of material and human "waste" facilities, and low and uncertain commodity prices. Urban Great Plains centres are indistinguishable in their Wal-Marts, fast food franchises, and drug problems from other North American cities. Yet the variability across its area and the indistinct boundaries of the Great Plains do not negate the value of discussing it as "region" in order to structure our understanding of the political and economic tug-of-wars that have characterized this place and are rendering the rural areas—the vast majority of the land—socially and demographically unviable, except for the growing populations and high birth rates on reserves and reservations.

Let us look briefly at the geographical and human history that does unify the Great Plains and enables us to speak of it as a region that is more meaningful than either a single state or province or the larger and far more amorphous region designated "the West." As we noted, about 65 million years ago, the great tectonic plates on which the continents ride ground together in the Laramide orogeny, pushing up the Rocky Mountains. The soil of the plains is largely derived from the weathering away of the Rockies by wind, water, and frost, and the deposition of soil wherever the wind or water slows down enough to drop individual grains. Because the prevailing winds come from the west, they tend to shed most of their water on the west side of the mountains, since the air cools and condenses out moisture as the winds rise to pass over the obstruction. The resulting rain shadow east of the Rockies determines the semi-arid nature of the plains. Glaciation, the recession of the glaciers, and the concentration of meltwater in ancient Lake Agassiz (whose remnants are Lakes Winnipeg, Manitoba, and Winnipegosis) repeatedly flattened the region, but also decorated it with ancient shorelines, lateral and terminal glacial moraines, and prairie pothole lakes, formed where chunks of ice surrounded by glacial till melted, leaving holes in the till. The Black Hills and Cypress Hills became stone islands in the seas of ice, providing refuge for a variety of species: even today, they support different flora and fauna from the surrounding plains. Huge deep beds of gravel underlie the smooth surface of the prairies: like sponges, they collect water as it flows through the flat, braided rivers of the plains south of the Missouri and seeps down to aquifers, particularly the Ogallala Aquifer, which underlies the land from Nebraska's sandhills to the Llano Estacado of Texas. The deep, dark soils of the Great Plains, an annual average precipitation of nine to twenty inches (17 to 50 cm), and frequent lightning-caused fires allowed a characteristic vegetation of grasses and associated forbs to evolve, along with gallery forests along riverbanks and ravines.[2]

Before the nineteenth century, the Plains had supported human societies for millennia—longer in the south than in the north, for the most part. The people had blended horticulture in corn villages along the rivers with hunting and gathering. Their travois pulled only by their sturdy dogs, they could neither follow the buffalo nor ride them down, but they could predict where they would be and painstakingly herd them into pounds or over cliffs.[3] In the nineteenth century, it was the Great Plains that had the distinction of becoming, in a way not true for any other region, Amer-European "free land"—despite being the heartland of flourishing and expanding horse-bison-Sun Dance cultures. As early as the 1820s, the US federal government was eying the Plains as land too far west or too arid for Amer-European settlement and as a dumping ground for Indigenous peoples until such time as they either assimilated or died out. Before the *Kansas-Nebraska Act* of 1854, the whole eastern tier of the US Great Plains was Indian Territory, and Oklahoma and the Dakotas retained this distinction until statehood, though actual Indigenous occupancy was progressively more restricted. Starting with the passage of the *Homestead Act* in 1862 in the United States and continuing with the passage of the *Dominion Lands Act* in 1872 in Canada, the Great Plains in both countries was the main area opened to homesteading—in which the intending settler bet three (in Canada) or five (in the US) years against the government for 160 to 640 acres of land. In reality, more land on the Great Plains was purchased (through pre-emption, from railroads, from the Hudson's Bay Company, from government entities, or from other settlers) than was actually proved up in homesteads. In any case, this segment of national land policy—although it was also used in the southern United States, the upper Midwest, and the Pacific Northwest—overwhelmingly centred on the Plains, introducing the land to free market economies in a most incongruous way. The land and the terms of its incorporation into the current market system distinguish the Great Plains from the US Midwest and the St. Lawrence/Great Lakes lowlands, which were sold for money or in exchange for military service or which were granted or sold to seigneurs or other landlords who intended to tenant them for a profit. At the same time, the Plains differed from the Shield, the mountains, the desert, and the North, which never entered private ownership.

In *The Fur Trade in Canada*, Harold Innis laid out an enduring relationship between the European metropolitan centres, the Canadian entrepôt cities, and the fur-producing staples hinterlands. J.M.S. Careless developed the theories, Paul Voisey modified them for the Canadian Prairies, and William Cronon, in *Nature's Metropolis*, further modified them for the US Great Plains. But "hinterland" is a purely economic and relational status. The Great Plains of both Canada and the United States are now economic hinterlands—even if Calgary now boasts more corporate headquarters than any other Canadian city but Toronto. During Blackfoot and Lakota times—when many nations shared the culture marked by the bison, the Sun Dance, and eventually the horse—the Great Plains was the centre of the universe, the place where creation began. Full of sacred sites as well as both faunal and vegetal abundance, linked to trading routes that provided any wants the Prairies did not produce, this region was no hinterland until it was encountered by Europeans. But the connotation of "hinterland" is not simply relational—it implies some kind of deficiency, as in the title of the play "If You're So Great, Why Are You Still in Saskatoon?" It is important to reiterate that no region in the world is deficient—or excessive—in terms of the organisms that have co-evolved with it. The Great Plains is *grassful*, not treeless.[4] The Great Plains is semi-arid, its weather as variable as anywhere in the world, and often violent, but these are conditions that promote a complex grass and grazing ecology. Drought is a recurring condition on the Great Plains, a deficiency for a sedentary agrarian society but an advantage for a pastoral lifestyle in ways that contemporary whitestream plains society does not yet seem to have fathomed.

The movement of peoples onto the Great Plains between the 1860s and 1914 is an epic of one of the great migrations in human history. It is more (and less), however, than the valorized saga of "conquering" the land and establishing the breadbasket of the world and the home of millions of valiant family farmers where once had been a desert occupied by a few nomadic bands of Indians. "Desert" is, first of all, an unreliable term. Remember, no ecosystem is deficient in terms of the organisms that have co-evolved with it. The Great Plains is a complex and dynamic ecosystem with biofeedback mechanisms—such as gophers and grasshoppers—that keep it viable in the face of one of the most extreme and variable climates on earth. "Nomad," in the sense of a non-planning, erratic wanderer, is as suspect a term as "desert." Plains people before the advent of the horse visited various areas on a regular seasonal cycle, anticipating bison, elk, and other animal migrations as well as utilizing roots, tubers, berries, and other vegetable foodstuffs in season and preparing and storing them for winter. The return of the horse to the Great Plains increased the distance the people could cover and the materials they could carry with them, freeing them to use the whole Plains instead of being tied to the general vicinity of major rivers.

After the European "discovery" of North America but before any prairie schooners had crossed the Missouri or cart brigades had set out hopefully from Red River, numerous migrant peoples had entered the Great Plains from the east, south, and northwest. Although, as Vine Deloria suggests, the Siouan peoples may have come from the Black Hills area, by 1492 they seem to have been living in the Great Lakes/Ohio valley region, from whence they migrated west. The Osages settled in the southeast (Missouri and Oklahoma) and the Lakotas and Assiniboines in the northwest (Montana, Dakotas, Manitoba, Saskatchewan), with other groups strung out in between. Partly they responded to a push from the east, as European settlement and trade patterns started a train of displacement, and partly to the pull of the hunting and gathering opportunities of the Great Plains; perhaps they were merely returning to an ancestral homeland. By the 1810s, the southeastern peoples who had assimilated far too successfully for their Amer-European neighbours—the Cherokees, Creeks, Choctaws, Chickasaws, and Seminoles—began moving west, mostly to escape Amer-European encroachment. By the 1830s, the bulk of these people, including mixed-blood (with white and black) and African-Americans both enslaved and free, had been forcibly removed by the US government to Indian Territory (Oklahoma); in the following decade, numerous midwestern and eastern groups such as Shawnees, Miamis, Wendats, Senecas, Ottawas, Delawares and others were less violently but still forcibly removed to Indian Territory (Kansas and Oklahoma). When the *Kansas-Nebraska Act* of 1854 opened the land to "squatter sovereignty" and set off "Bleeding Kansas," a long-running prelude to the American Civil War, no land in Kansas was legally available for either free-state or slave-state settlers—it was all set aside in treaties or in trust for Native nations.[5]

Coming south from Hudson Bay and west along the St. Lawrence and Great Lakes, mixed-blood descendants of the fur trade settled at the confluence of the Red and Assiniboine Rivers. Their numbers were augmented by both the children and the retired workers of the fur trade, and eventually by Scots and Swiss immigrants. In the south, another mixed-blood community, of Spanish, Native, and Moorish descent, moved slowly into the Llano of West Texas and New Mexico, surrounding and to some extent blending with the long-settled Pueblo agriculturalists and the Athapascan-speaking pastoralists from the northwest. Similarly, the Kiowas moved, over several centuries, from the northwest down through the Black Hills to central Oklahoma.[6]

Euro/Afro/North American settlement of the Great Plains, then, came into a complex and diverse ecosystem that at many places may have been at or near the carrying capacity of the land. Mounted hunter-gatherer cultures both competed with and complemented older corn village/hunter-gatherer societies, such as the Pawnees and Mandans, who were established in many river valleys. The Canadian prairies, with a shorter ice-free history than the United States and a much shorter growing season, for the most part lacked the corn villages, but this ecological niche was accounted for by Scots and Métis horticulturalists in Red River and by Ojibwa (Anishinaabe) wild rice harvesters to their east. Euro/Afro/North American settlement did not introduce agriculture to the Great Plains, but it did introduce large-scale commercial monocultures in both field crops and animal husbandry. Eco-historians question the sustainability of bison herds even before commercialized bison hunting led to the collapse of the herds in the 1870s and theorize that even by the 1830s, the bison were both overstocked and over-hunted. Indeed, it is likely that bison numbers were never stable. Despite theories of climax vegetation (implying also climax fauna), the Great Plains is marked by variability—even instability. As James Malin has pointed out and contemporary ecologists such as Don Gayton have emphasized, the grasslands have developed symbiotically *with* crisis—dust storms, prairie fires, long droughts, floods, and population explosion and collapse. The sunflower-bordered roads of which Willa Cather writes so fondly are less examples of J.E. Weaver's progression to climax than of the *alternations* among various forbs and grasses in adaptation to changing conditions.[7]

Once Euro–North Americans encountered the Great Plains, they imaged it alternately as desert or garden, responding less to dry or wet conditions in the place itself than to ideas of what they wanted it to be. Thus, various reincarnations of the wishful thought that rain follows the plough lasted from the 1870s in Nebraska, through the localized dam building and irrigation era of the early twentieth century and the mainstem Missouri dams of the 1940s, until at least the early 1990s and the building of the Oldman Dam in Alberta. Global-warming denial is the most recent and least imaginative version of the mantra. Nonetheless, once the collective decision was somehow made that the Great Plains was to be a garden, settlement and transformation of the region from what Scott Momaday called a "lordly society" of "fighters and thieves, hunters and priests of the Sun" to a society of production agriculture linked to world markets happened extremely rapidly.[8] Unlike agricultural frontiers to the east and to some extent the west, which had undergone a period of subsistence agriculture as described by such pioneers as Ontarian Susannah Moodie or Michigander Caroline Kirkland, the Plains jumped into global competition in two generations. The process began in the 1860s with the original US *Homestead Act*, the end of the US Civil War, Canada's Confederation, the completion of the Union Pacific, the first transcontinental railroad, and the cession of Rupert's Land to the Dominion of Canada. The taking of the Great Plains was completed in the first decade of the twentieth century with Alberta and Saskatchewan becoming provinces in 1905 and Oklahoma achieving statehood in 1907.

There were a number of interlocking ideas involved in these two generations of taking of the land. One was the assumption, so basic as to be unstated, that had governed westward expansion from Europe since before the time of Columbus. An ideology still embraced by some neo-conservative thinkers, it held that a Christian society with an expanding population, an agricultural land-use ethic based on individualism and private property, and an increasing mastery of science and technology had an inherent right to land and natural resources, a right that naturally trumped the rights of anyone else with whom such a society might come in contact. By the 1860s, this basic belief had also evolved to require, quite explicitly in the United States and more hesitantly in Canada, a free market economy buttressed with an infrastructure—internal improvements and a banking system— provided or facilitated by the federal government. For Canada's first prime minister, John A. Macdonald, using "free" land to attract immigrants and building a railway to get them to the Prairies was part of what came to be called the National Policy. The railway would fulfill the promise to British Columbia that it would have a land link to the Dominion of Canada, would hold the newly acquired Rupert's Land territories against US expansionists looking north and Fenians looking to avenge Ireland by taking England's North American territories, and would transport the settlers to the "free" land. Once there, these pioneer farmers would create a market for machinery made in central Canada, thus developing an industrial base for the new country.

The square survey that enabled the hopeful homesteader to stake a quarter section is a perfect blend of federal infrastructure and individual enterprise. Thomas Jefferson had dreamed of an America based on yeoman farmers, each tending his own plot of ground and practising virtues that would lead to a settled and happy democracy. Alexander Hamilton, on the other hand, envisaged a commercial and urban America that would be a financial power in the world. Philosophically, the various Homestead Acts were purely Jeffersonian—contented families living in white houses with green trim and looking out on red barns with white trim would live happily ever after on their 160-acre farms. In practice, the Homestead Acts were a lot more Hamiltonian. As Alberta historian Paul Voisey has shown in his wonderful study *Vulcan*, some settlers were genuinely interested in putting together family farms of 160 acres or more. A few were seriously involved in constructing much larger farms. Many were engaged in speculation, perhaps holding down a homestead and dabbling in town lots, changing both profession and residence with bewildering speed and frequency. In fact, the Homestead Acts were the greatest instruments of middle-class capital formation ever invented. It was not wheat that made Alberta and Saskatchewan magnets for immigration of both capital and people, it was the land itself. The land became capital and wheat was the obvious, if temporary, mechanism. As a long-term wheat-producing asset, the land would never be worth what mortgage companies and intending buyers poured into it, a truth that continues to haunt the Plains in terms of perennial grain surpluses.

Turning land into capital was also crucial for the free-grass ranching that flourished on the Great Plains in the margin between Indigenous people assigned to reserves and the entrance of the homesteaders. As ranching historian Warren Elofson has shown, the inherently unsustainable open-range cattle operations could never be economically viable, but they siphoned large quantities of money into the country.[9] As they went belly up after the big Die-Ups, the money stayed with the sellers of cattle, the ranch hands who had been paid for bogus homestead and pre-emption entries to secure water rights for the ranchers (a practice more common in the US than in Canada, where leases provided a somewhat more rational basis of allotting land), and the various other payees and middlemen who handled the cash or started small viable herds of "orphaned" cattle. While the homesteaders focussed on the deficiency of the land, the open-range ranchers claimed the pastures as a paradise for cattle; they underestimated, though, the deficiency of both southern range cattle and the beefier British imports to sustain themselves on the northern prairies without shelter, protection from predators, or supplemental feeding.

The Homestead Acts were more successful at creating capital than at creating viable family farms; they were also extremely successful at moving land first from Native sovereignty to the public domain and then to private ownership. Like the Land Ordinance of 1785, which set out the form of the square survey, based on astronomical observations rather than on the lay of the land, the Homestead Acts commodified land, moving it from the commons to individual plots for individual ownership. An obsessive belief in the magic of fee simple ownership of land, including surface and usually mineral rights in perpetuity, fuelled the Homestead Acts as well as the *Dawes General Allotment Act* of 1887 and its variants from 1885 to 1906. Although allotment was supposed to make it easier for Native families to hold onto their land, in practice, it resulted in massive land losses: from 1887 to 1934 (when allotment was repealed), Indian land in the United States dropped from 135 million to 47 million acres.[10] The *Dominion Lands Act* was an inevitable result of the US *Homestead Act*. Canada could compete with settlers, particularly immigrants, only if it also offered them "free" land on even easier terms than in the United States. The gut-level commitment to owning land, particularly on the part of European peasants who had never been able to secure tenure on the soil they worked, as well as the ideological commitment to private property, especially in contradistinction to the radicals of 1848 in Europe, who extolled variations of communism, obscured the truth that the Great Plains was *not* free land in the nineteenth century.

Before the homesteaders, the Great Plains was purposefully occupied and used in ways that countered climate variability with geographic mobility. Indigenous people did not *follow* the buffalo herds—rather they anticipated buffalo movement and stationed themselves where experience told them the bison would be moving. Or, if their forecast was wrong, they moved toward alternate or supplementary food sources, such as deer, elk, berries, or prairie turnip. Allotment of specific small parcels of land, for both Indians and homesteaders, meant that modifying the effects of a variable climate and producing a uniform product for a world economy would substitute for modifying place of residence to sustain a plentiful subsistence living. And even if fee simple had been the key to a more prosperous life for humans on the Great Plains, the 160-acre homestead, laid out arbitrarily on the grid system, was by no means the most propitious choice. The river lots that the Métis had borrowed from the Laurentian valleys in Quebec granted each landholder access to water, wood, and transportation along the river; a kitchen garden; fields for grain; and finally, communal hay and pasture lands. Because they included both river frontage and uplands, the river lots allowed, on a small scale, the geographic mobility that had marked successful human adaptation to the Great Plains and took advantage of the micro-climates that affected everything from subsoil moisture to frost-free days.

To the extent that Indigenous people could control their allotments, they, too, chose river frontage mixed with upland to provide a source of indigenous food plants, access to hunting land, and access to pasture and crop land. The completely arbitrary nature of the range and township system that assumed all land was essentially interchangeable was singularly ill-equipped to help the Native family or the settlers adapt, within the confines of their own homestead, to extreme variability. Although the implied uniformity of the land was amenable to commercial agriculture for global export, it was not a particularly intelligent way to utilize the Great Plains. As James Scott has shown in *Seeing Like a State*, the square (or cadastral) survey was not really in keeping with peasant agricultures in Europe, where it was introduced for the benefit of state administrators and tax collectors, but it was even less suitable to the Great Plains. Yet neither lawmakers nor homesteaders questioned whether the allotment of the commons through the square survey was a superior form of land use—it was, by that time, self-evident. Indigenous people throughout the Great Plains did question the assumption during the whole allotment period, both by opposing allotment in general and by trying to secure plots that combined riverine and upland acreage and adjoined land held by other family members. Neither government nor incoming settlers, however, credited their arguments.

Canadian officials were not as obsessed with allotment as were Americans, probably because the original reserves were much smaller than American reservations, so there was less need to create the idea of "surplus" land that must be captured for intending homesteaders from the East. Prairie reserves, however, lost about half of their land between 1896 and 1928. Nonetheless, officials were equally deaf to the pleas from peoples such as the Dakotas of Saskatchewan for promised hay and pastureland. And, as Sarah Carter has shown, Indian Superintendent Hayter Reed's obsession with the idea of the inevitable progression from "savagery" through "barbarism" to "civilization" fatally hampered Cree adaptation to farming by equally foreclosing both communal and individualistic adaptations of sedentary farming techniques to northern Plains climate variability, thus eliminating Crees as economic rivals to neighbouring Canadian-European farmers.[11]

If the *Homestead Act* was the advertisement, the railroad was the vehicle for the intending settlers. The construction of the railroads was high drama, especially the first transcontinentals in each country, the Union Pacific Railroad and the Canadian Pacific Railway (CPR). Both lines gobbled human lives, money, timber, stone, and iron. Both were exercises in what economists call "premature enterprise"—too risky to build for the normal economic gains to be expected at the time. As Robert Fogel explains, in the

United States, a handful of entrepreneurs developed an elaborate kickback scheme centred around their Credit Mobilier company to raise the rate of return enough to attract investors. In Canada, John A. Macdonald's still unfinished CPR would almost certainly have fallen into bankruptcy and ruin had it not been for the Northwest Resistance in Saskatchewan (or the Northwest Rebellion, as MacDonald would have called and understood it) and both England's and Ontario's perceived need to ship soldiers out West to put down the "Indians."[12] Though it may be a metaphor to call the Euro–North American settlers of the Great Plains an army of occupation, the first passengers on the CPR were quite literally soldiers sent to affirm Canadian sovereignty over the Great Plains.

The saga of building the CPR across the muskeg of the Shield, over and through the mazy mountains of the Rockies and the Selkirks, and down the gorges of coastal ranges is definitely a national saga. Although the railway was originally built to tie British Columbia to the rest of the newly confederated Canada, in terms of loss of sovereignty, mono-crop settlement, and economic development (or exploitation), the effect of the railway was greatest on the Plains. The railway became one of the most vivid ideas—positive or negative—in the intellectual repertoire of the Plains. Settlement of the sort that occurred was impossible without the railroads, yet the railroads, grain elevators, and markets, and Macdonald's National Policy, were also the scapegoats for the prairie pioneers' indomitable sense of entitlement. Both countries paid for the railroad building partly through granting each rail company land in a checkerboard pattern on either side of each completed mile of track. Where the land did not seem to be of sufficient quality that its future sale would pay off the rail-building costs, the rail companies could select "lieu lands" in a more promising area. Settlers looking for free land who found that half the most valuable areas were railroad property, and for sale only, felt themselves to be victims of a "bait and switch" scheme. They were even more incensed when the railways were slow to choose their lieu lands, leaving settlers uncertain, sometimes for decades, about what lands would turn out to be "free."[13]

The gospel of the Homestead Acts is enormously appealing: by sweat equity, deserving families would create farm homes for themselves while helping to feed a hungry world. That such an enterprise was also a gamble, a lottery, was acknowledged both by the homesteaders, who frankly stated that they were betting Uncle Sam five years of their lives versus 160 of his acres, and by the actual lotteries that parcelled out the right to claim farm land or town lots after the seemingly endless "free land" ran out. The sense of entitlement that came from capturing and taming land—the belief that your five years on the land in the United States or three in Canada entitled you to a *farm*, including enough water to grow crops—plus that trust in gambling, in lotteries, has remained part of the idea bank of Great Plains thought today. Farming is always a gamble, but nowhere more than on the Great Plains, where the only constant in the climate is its variability and its "too-ness"—it is always too hot, too dry, too windy, too wet, too cold . . . It is not a place that breeds caution.

Homesteads, allotments, and the rigid assignment of farmland on Canadian reserves, then, succeeded in transforming the Great Plains from Buffalo Commons to fee simple agriculture in two generations. As Irene Spry has noted, it was the last chapter in the loss of the commons that had begun in medieval England. While the popular image of this human movement onto the Plains was, and is, that it civilized wild land and wild people and made the desert blossom like a rose, feeding a hungry world, the underlying economic interactions were somewhat more complex. As Hamlin Garland astutely pointed out in the 1880s and 1890s, the pseudo-homesteader who sold a relinquishment of a claim and moved to town or to an eastern city was more likely to prosper than those who stayed on the farm. Rates of return in agriculture are usually lower than those in commerce and manufacturing. The wet weather booms followed by the dry weather busts of the 1870s, 1890s, early 1900s, and 1920s and 1930s resulted in massive farm consolidations and a rise of tenantry—which, somewhat paradoxically, because it allowed for more flexibility than land ownership, was often more humanly and economically successful than hunkering down on the family homestead. Landowners, as Voisey points out, were likely to be cash poor because everything had gone for more land and equipment to work it.[14] As men and animals were replaced by machines, farming became more capital intensive and more tied to the commercial market.

In a market system, more risk is supposed to yield more potential gain, but as is the case in any kind of gambling, more risk always leads to more loss. Newspapers publish excited stories about the people who win the lottery—they do not publish long lists of people who lose the lottery. Farmers worked, in most cases, extremely hard. They kept clean fields and they planted and tilled and harvested. Writers and politicians praised them as the salt of the earth. As Patricia Limerick has shown, they felt entitled to succeed, to hit the pot of gold that had to be at the end of the double rainbow of hard work and high risk. Their chosen role became that of the "injured innocent."[15] Many Great Plains farmers were successful and prosperous, establishing farms that stayed in the family and sustained it for generations. Most were not. The economics of Great Plains farming have called for fewer people and more capital on larger and larger spreads of land. Since rates of return are lower on farming than manufacturing, commerce, and other kinds of extractive industries—such as petroleum—the consistent mismatch between expectations and return enticed farmers to keep on insisting that they ought to succeed. The old joke about the farmer who won the lottery and explained that he'd just keep farming until he lost it all shows a wry rural appreciation of the nature of the operation. Perhaps the problem was not with the land or the farmers but with the way society defined success.

Certainly, the Great Plains has been a hotbed of resistance to the commercial market. On both sides of the border, agrarian discontent has bubbled in waves. From the Grange, to the Populists, to the Nonpartisan League, to United Farmers of Alberta and the United Grain Growers, to the Co-operative Commonwealth Federation and Social Credit, to the Progressives, to the New Democratic Party, to Reform and Alliance, to the Saskatchewan and Wild Rose Parties, the lineage of western and agrarian discontent, on right and left, is strong. Yet in many ways, the Reform slogan "The West Wants In" may be the truest comment of them all. With the partial exception of the Non-Partisan League, the UGG, and the Progressives, and the almost total exception of the CCF in Saskatchewan under Tommy Douglas, agrarian protestors have been determined to make the market economy work for them. But the genius of free enterprise under the old pure Adam Smith definition of the elusive "free market" is to make decisions that exploit hinterlands for the sake of metropolises. The much-reviled National Policy, the CPR, the elevators, the grain merchants in Winnipeg and Chicago, the bankers and mortgage brokers, and all the other favourite targets of agrarian protest, along with Ottawa and Washington, are behaving exactly as they should—without even mentioning, as William Cronon has shown, that middlemen such as grain elevators actually provide a useful service to grain growers.[16]

The Great Plains should export its soil and water (in the form of grain and meat), its other natural resources, and the best and brightest of its children elsewhere: that is the way the free market is supposed to work. Someone like Tommy Douglas, who makes a reasoned plea for sustained sufficiency rather than the jackpots and busts promised by the free market, may succeed temporarily in constructing the Co-operative Commonwealth, but it is true that in the long run, Great Plains people have refused to accept governmental reforms that mute booms as well as busts and strive to abolish poverty before establishing wealth. And so Great Plains people support a market system that, *working as it should*, is bound to diminish the rate of return to the region as a whole. Furthermore, because capital is more mobile than labour, people are left behind in small towns of the Great Plains that dry up until there are more people in the nursing home, the only economic diversification in town, than in the school. The countryside produces more crops with fewer people and expects governments to find markets for them.

Except during wartime, the Great Plains has produced more wheat and corn than the market can absorb, at least at prices that return to farmers their cost of production plus a small profit—and certainly not at prices that would mitigate the mining of soil fertility and water for the production of grain. After World War II, the Great Plains supplied grain for the rest of the world and avoided both starvation in the war-torn countries and the collapse of grain prices that had so devastated farmers after World War I. But by 1948, there was a world surplus of grain that the US Department of Agriculture and the Canadian Wheat Board contrived to control by keeping their prices low enough to discourage other countries from going into the wheat export market. By 1963, however, this agreement had fallen apart, and the United States was gaining control of more of the world wheat market. In 1968, Prime Minister Pierre Trudeau asked President Richard Nixon to help salvage the year-old International Grains Agreement, which had set a minimum world price for wheat. Nixon declined, and Canadian wheat farmers, who had an even smaller domestic market than the United States, were in deep trouble.[17]

Not surprisingly, farmers believe deeply in the Great Plains as the breadbasket of the world and in the inherent nobility of producing beautiful, wholesome grain to feed hungry children everywhere. There is enormous joy in the "straight, dark rows" behind the ploughs of spring, in the intense green fields that follow, and in the steady stream of plump golden kernels the combine pours into the waiting grain hoppers.[18] Although USDA subsidies favour agribusiness at the expense of "family farms," American farm rhetoric from television herbicide commercials on up pay homage to the old Jeffersonian ideal of the yeoman farmer. Trudeau's 1969 federal task force on agriculture may have been realistic in its acceptance that family farms were being squeezed out and that Ottawa would not be able to sell all the wheat that farmers were producing, but it was arrogant and ham-handed in suggesting that the farm population should be reduced by two thirds and the grain acreage by half in about four years.[19] When Trudeau flippantly asked Prairie grain growers why he should sell their wheat, he badly misjudged not only the economic realities of wheat marketing since the beginning of Prairie grain farming, but also the importance of the entire breadbasket motif of western settlement. Ottawa should market the West's wheat because that is the basic premise of settlement and all the history of markets and the Great Plains since then.

Trudeau's arrogance hurt all the more because he had hit upon the real weakness of the Great Plains. The cheap food policies of both Canada and the United States and the encouragement and exportation of highly inefficient practices such as the transformation of multiple pounds of grain into single pounds of fatty meats, do not protect North American food safety or sufficiency and certainly do not protect the land, the water, or the human communities of the Great Plains. While it was impossible in 1969, and remains so now, to stop all food aid to the rest of the world, especially Africa, overproduction and export harms both the Great Plains and the areas that receive its grains. As Stan Rowe bluntly asks, why should Canadian or American governments or larger societies try "to save and maintain an exploitive, industrial, export-based agricultural system that has poorly served a large sector of the farming population, while at the same time running down the soils, diminishing surface and subsurface water, destroying natural landscapes and decimating native fauna and flora?" Nebraska, the Cornhusker State, grows far more irrigated corn than all the humans and animals in the state could possibly consume. Some of it ends up as high fructose corn syrup in carbonated beverages and thousands of other processed foods; some is distilled into ethanol—if oil prices are high enough. But under various farm plans, the US government markets it—or sometimes simply gives it away as food aid—all over the world. In Mexico and Central America, this cheap corn tends to displace subsistence farmers who had sold small corn surpluses in local markets.[20] And if they turn to high-priced coca crops, which then return to Nebraska as cocaine, a recreational drug for displaced farm boys and girls and others in Great Plains cities and towns—well, isn't that the way the market is supposed to work?

Traditions of agrarian discontent and western protest have settled down into a voting pattern in which the Great Plains states, which theoretically should never benefit from a free market, *always* support the Republican free market candidate for president, while the rural areas of Saskatchewan, which should theoretically benefit least from a free market, back the Ross Thatchers and Grant Devines and the Saskatchewan Party. Given the almost religious intensity of the belief in individualism and market forces that led to the commodification and settlement of Great Plains land, it is not surprising that this belief should remain so strong, especially as most economic diversification attempts on the Plains have failed. The US farm subsidy programs overwhelmingly support the largest and most capital-intensive farmers, while propositions from Liberal Ottawa setting out deliberate farm depopulation or a National Energy Policy—no matter how intelligent or defensible—have conditioned Plains people to distrust "government intervention." So American farmers "farm the mailbox," waiting for subsidy payments, and funnel more and more corn into ethanol despite the contention of some agronomists that it costs more petroleum to grow the corn than to buy gasoline.[21] The essentially conservative nature of agrarian discontent manifests in voting behaviours that can only reward farmers with subsidies of one sort or another—the exact opposite of a free market or of any rational system to protect the environment or the long-term economic viability of farmers or the rural West.

Although, as David Jones shows brilliantly in *Empire of Dust*, the twenties were economically disastrous for the Great Plains, it was the 1930s that exposed the failure of governments to deal with drought and depression, even though the New Deal and the *Prairie Farm Rehabilitation Act* permanently inserted the federal governments into agricultural decision-making. The American documentary film *The Plow That Broke the Plains* is an artistically effective condemnation of the farming practices that left delicate land uncovered by grass or crops and susceptible to the blowing of the Dust Bowl. The backlash against the film from Great Plains farmers, incensed that the government would censure or restrict their farming methods or "knock" the region, was so intense that the film was withdrawn from circulation until 1961, and then was shown only as art, not as history.[22] What farmers asked for, and eventually received, were crop supports and foreclosure moratoriums, which, however needed in the immediate crisis, were not alternatives to a free market system. While both Canadians and Americans reluctantly acquiesced to returning environmentally sensitive land to grass, including parks and leased and community pastures, Great Plains farmers and ranchers would not tolerate a wholesale assault on the fee simple cropping system, even though it did not work for the majority of them.

Diversification has become a watchword of western protest. To some it may mean simply diversification into different crops—canola instead of wheat, for instance. In Alberta, it usually means diversification from gas and oil. But most often, it means diversification into some form of manufacturing or "value-added" economic activity, not just the exploitation of natural resources. One of the West's great complaints about Macdonald's National Policy was that westerners were saddled with buying low-quality, high-priced farm machinery made in Ontario, while high tariffs kept out cheaper, better, American-made implements. American farmers of the same era complained of the banks and moneylenders who made them pay off their high-interest machinery loans in deflating currency.[23] One apparent answer would have been diversification into farm machinery fabrication in the Great Plains, but a dispersed market, a distance from materials, and especially the lack of the synergy of the "rust belt" industrial concentration doomed any such hopes. The 1960s and 1970s were the heydays of regional planning and economic development worldwide, as "more developed regions" attempted to stabilize their population growth by directing development to "less developed regions." The theory was simple enough—set up "growth poles" by importing industry and develop the economies around them. For the most part, such development failed because it never engaged the host economy. Economic enclaves flourished as long as they had development support and faded as soon as it was withdrawn. In fact, as we shall see later, economic development has failed in many ways in the Global South, and, although development theories are useful in explaining the Great Plains in the nineteenth century, they do not provide much sustenance for the twenty-first century.

Frank Popper, a planner from New Jersey who at first supported regional development but later became an astute critic, responded to the failure of most regional development projects by proposing Buffalo Commons. Originally an intellectual puzzle—how does one plan for de-development?—with the aura of Jonathan Swift's "A Modest Proposal," Buffalo Commons attracted so much attention that Popper and his geographer wife, Deborah, built a cottage industry around elaborating it. Briefly, Buffalo Commons is reverse development planning, an orderly program for the depopulation of the Great Plains, the clearing of the dying towns and the economically unviable farms, and the re-establishment of native grasses, buffalo, and Aboriginal people—with a few grizzled old homesteaders for their pictorial value for the new crop of eco-tourists who would be drawn to the new/old Great Plains. Like *The Plow That Broke the Plains* and Trudeau's plan for reducing farm populations, Buffalo Commons aroused a good deal of hostile interest in grass country. Perhaps the best response came from a pair of planners in Minot, North Dakota, who suggested working out the orderly depopulation of New Jersey so it could become a parking lot for New York City. Others, such as Maxine Moul, the former economic development director for the State of Nebraska, take Buffalo Commons very seriously as a useful instrument for rural planning. But like the regional development theories that it parodies, Buffalo Commons is firmly rooted in a free-market-with-government-tweaking model and bases its calculations on homogeneous space rather than distinctive places. Buffalo Commons is essentially conservative. Tyler Sutton and the Grassland Foundation have more recently proposed a variation on Buffalo Commons that would be developed by community groups, but the procedure is still in a hypothetical stage.[24]

In contrast to the underlying free market conservatism of most agrarian revolt on the Great Plains is a deeper and perhaps more valuable strain of resistance. The first resistance comes from the land itself. Except perhaps for truck gardens, land and growing things are resistant to the conformity and uniformity of production agriculture. Hydroponic greenhouses and hog confinement sheds grow more uniform "products" than crops grown in dirt and reliant on rainfall or than animals let out to "root, hog, or die." The Great Plains, with its enormous climatic variability and an evapotranspiration rate that usually exceeds natural precipitation, is a pretty chancy environment. Even seventeenth- and eighteenth-century peoples such as the Pawnees and Osages who combined riverine horticulture with large-scale bison kills and small-scale hunting and gathering could not always balance the various resources and environments available to them to avoid scarcity.

Although both rowcrop horticulture and cattle ranching (after the demise of the "free range" system) have produced much larger and more stable harvests in any given spot than did subsistence horticultural regimes based largely on hunting and gathering, their long-term sustainability, as Stan Rowe notes, is questionable. On the other hand, as Geoff Cunfer points out, Great Plains agriculture is as sustainable as any North American agricultural regime, which has tended toward being primarily a large-scale slash-and-burn regime. In fact, Rowe points out that there are no precedents elsewhere in the world for sustained agriculture in a semi-arid region. The row crops grown from Texas to Saskatchewan are the largest and longest experiment of their kind—yet perhaps ranching would be more sustainable. Monocropping mines the soil of both water and nutrients, and requires large inputs of chemical fertilizers, pesticides, and water. It exposes soil to blowing, contaminates surface and shallow well water with nitrates (my domestic well in Nebraska cannot be used for drinking or cooking), and severely reduces the flow of Plains rivers used directly or indirectly for irrigation. Fee simple land ownership and allotment tend to make migration a confession of personal failure rather than an expected and pragmatic response to the climate. As Robert Fletcher says, when Montana home-steaders quit, "they had nothing to take with them. They just spit on the fire and whistled for the dog."[25] This is quite a contrast to Hamlin Garland, leaving South Dakota with $200 for his relinquishment in his pocket, or even some enterprising homesteaders in western Nebraska who proved up, mortgaged the quarter section to the hilt, and took the money—leaving the land to the banks. Historians have not always distinguished between people who tried to make farms and failed, and the enterprising capitalists who were happy to turn land into money and then go on looking for the main chance. And perhaps those who, like the Montanans, left with nothing, should not be considered to have failed. It was only that the land won.

Stan Rowe, Wes Jackson, and, to some extent, Alan Guebert have been foremost and principled defenders of the land and of the Great Plains as a particular and highly desirable place. Rowe was an early proponent of what is now usually referred to as the Gaia hypothesis, the idea that as atoms are to molecules and cells are to organisms, so organisms, volca-noes, rivers, air, and every part of the ecosphere are to Earth itself. Thus the land is both the community and indeed the very self that we inhabit. We need, then, a mind that values the land—and in Rowe's case, especially the tallgrass prairie—for itself, and not just as a commodity to be mined. Wes Jackson, who accounts himself a student of Rowe's in his introduction to Rowe's *Home Place*, has also hosted Rowe for several of the annual festivals at the Land Institute outside Salina, Kansas, where Jackson has spent a lifetime sponsoring research on communities—be they human or grass—in the tallgrass prairie. Why, he asks, is that part of Kansas that was once Quivira now capable of supporting fewer people than in Coronado's time? And why does it export its soil and water in the form of field crops out-side the region instead of parlaying its resources of sunlight and rain into grass and then bison muscle produced within the region? Alan Guebert, a syndicated newspaper agriculture columnist from Illinois, though more sympathetic to production agriculture than the other two, also speaks for the land by looking, sometimes caustically, at US federal agricultural policy and its discontents. Despite the fallacies inherent in "thinking like a moun-tain" or talking like a prairie, these writers, especially Rowe and Jackson, have long and convincingly preached a gospel of the land. And, as Jackson says, at the Land Institute, he and his collaborators and interns are work-ing "to build an agriculture based on the way a natural ecosystem works . . . perennializing major crops to be placed in mixtures that mimic the vegeta-tive structure of that old prairie." Both Jackson and Rowe are particularly mindful of the thousands of years of land knowledge that European science and religion tried ruthlessly to eradicate.[26]

In addition to the land itself, resistance also comes from the First Nations of the Plains. Although putting "resistance" and "Indians" into the same sentence usually conjures up visions of "Custer's Last Stand" or Riel at Batoche, peaceful Indigenous resistance may actually be more meaningful. As Blair Stonechild and Bill Waiser point out, the usual conflation of Big Bear and Poundmaker with Riel obscures the fact that the Cree leaders were looking for a Cree solution that would include consolidated settlement of many bands in close proximity to one another with enough land for hunt-ing and agriculture, and a fully developed cultural life. The people of Indian Territory, what became eastern Oklahoma, similarly proposed that the ter-ritory become the state of Sequoia. Instead, Senator Dawes came back from retirement to preside over the allotment of Creek, Cherokee, Chickasaw, Choctaw, and Seminole land. These so-called Five Civilized Tribes (so spec-ified despite the fact that *all* North American Indigenous peoples were "civ-ilized" in the sense of being human cultures with well-defined, meaningful rules and value systems) had displayed amazing adaptability, rising from their Trails of Tears to independence and prosperity only to be torn apart by the US Civil War, then rising again from the punitive peace forced upon them to prosperity by the end of the nineteenth century. They had man-aged to accommodate the market within a system of communally owned land in which usufruct rights accommodated the control of anyone who consistently worked a plot of land. Creek Chief Pleasant Porter accurately pointed out that the Creek Nation had no paupers yet was being forced to adopt the land system of a mainstream United States that suffered pockets of horrendous poverty.[27]

Another centre of determined but severely compromised resistance developed in the 1940s and 1950s among the Missouri River Lakotas and Dakotas, Hidatsas, Mandans, and Arikaras. Many had survived allotment by claiming riverine land that would not support production agriculture but could be adapted to a combination of hunting and gathering with horticulture and small-scale ranching. Almost all their riverbanks were taken for the construction of the Pick-Sloan projects, a series of dams and reservoirs constructed along the mainstem of the Missouri to provide irrigation, hydro power, commercial navigation, recreation, and downstream flood control.[28] Both the Five Nations of Oklahoma and the Missouri River peoples had put together a hybrid society not unlike that of Appalachia or Newfoundland, in which traditional subsistence activities, low-wage occasional or permanent work, and, in some cases, welfare allowed people to create an existence that was humanly satisfying, if materially less than middle class. Although Indian policy in both Canada and the United States ostensibly aimed toward assimilation, in most cases it succeeded in producing only marginalization. The Crees, the Five Oklahoma Nations, and the Missouri River peoples, among others, all shared the capacity to wed their social marginality with land that the free market system also termed marginal—hilly eastern Oklahoma, the Missouri Breaks, the Cypress Hills. In so doing, they created a flexible way of life that adapted to Great Plains climate fluctuation by utilizing different altitudes and terrains at different times of the year and of the drought cycle. All three of these resistance communities were ruthlessly broken up, with no regard for treaty assurances, in the name of securing "better" free market conditions, but all three have shown us ways to adapt to the land in tandem with the market. Contemporary subsidy programs in Canada's North that enable people to continue traditional ways of living on the land while still participating in the cash economy demanded by settlements that provide formal schooling and health care demonstrate at least one alternative model to the deliberate destruction of subsistence communities.

Economic development in Europe and North America has, from the first, been synonymous with urban development, as we can still see in the concerted efforts of the Canadian government to move northern peoples to permanent urban centres. New Amsterdam became New York. A certain undistinguished swampy tract supporting small Indigenous villages became Washington DC, and Bytown became Ottawa with some Gothic buildings. Muddy moorings on the Great Lakes became Toronto and Chicago, and fur trade outposts on river confluences became St. Louis and

Winnipeg. With the exception of the national capitals, placed where they were for political and strategic reasons, these cities grew because they captured the trade of the surrounding areas and began to sell to settlers goods such as shoes that could more easily be manufactured than imported. Despite images of prairie schooners and white-hatted cowboys, the West has always been the most urban part of North America. As transportation hubs and population centres, cities feed on their own success. Growth promotes growth. But only up to a point. Internal improvements, first canals and then railways, expedited the growth of the Wests of both countries. But on the Great Plains, that growth slowed down some time in the 1910s or 1920s. The great population rush onto the Plains reversed as soon as it hit its peak, right around the time of World War I. The cities kept growing because the rural population was retreating to them as farms consolidated and fewer people were needed to work the land. But the cities of the Plains did not take off in the way that coastal cities, including Los Angeles and Vancouver, did.

The West Coast cities were the ones that boomed. Partly, they filled with Prairie people who had made good and moved to LA or Victoria. Regina, or even Calgary, in mid-January is a daunting proposition. Both Canada and the United States expended extraordinary amounts of money during World War I and even more in World War II. Some of it went to munitions plants in Montreal. Some went to aircraft factories in Wichita, Kansas. Much more of it went to the West Coast. San Diego, Los Angeles, San Francisco, and Seattle all developed aircraft manufacturing facilities during World War I so aviators could take advantage of the good weather. Between 1941 and 1945, the population of San Diego increased by 147 percent as the United States strove mightily to increase its military capabilities in the Pacific. Except for Los Angeles and San Francisco, already well established as traditional entrepôt cities, and Vancouver, a quintessential railway and port city, the urban West Coast was largely created by defence contracting. Except for Wichita, Great Plains cities never received a proportional share of this bounty and never saw the consequent economic takeoff. Empire aviators trained in the clear skies of the prairies, and Tinker Airforce Base in Oklahoma was named after an Osage general, but no one built thousands of planes in Saskatoon or Edmonton or Winnipeg. Because of the persistence and power of Senator George Norris, Omaha, Nebraska, did build the *Enola Gay* and other huge bombers, but after Norris's defeat and death, Nebraska lost all but the Strategic Air Command headquarters in Bellevue, near Omaha. The big contractors that had built the western federal dams during the Depression had done better at entrenching themselves in federal contracting than had the Great Plains states—and San Diego was a lot more convenient to the Pacific than were the Plains.[29]

If twentieth-century West Coast cities are—like so much in the twentieth-century West, particularly in the United States—the creations of the federal government, that is simply not the case on the Great Plains. There, the land seemed deficient compared to warm, sunny coastal cities with white beaches. We do have a few cities, such as Calgary or Tulsa, that are the gifts of the petroleum industry, but for the most part, if Plains cities are to grow, they must grow (as Calgary has discovered) in the context of sustainable growth of the whole region. In the absence of vast military spending, is there any way to fuel economic growth in prairie cities? That kind of spending is certainly not going to come—Winnipeg missed out on the F-18s with a bid that made market sense,[30] and none of Halliburton's booty is coming to Wyoming. If there is an answer, it might come, I think, from the third point of resistance, Tommy Douglas's Saskatchewan.

Unlike the bulk of agrarian reform, which, as already noted, has always been conservative in its insistence on making the market system work for this hinterland, Douglas recognized that there is no such thing as a free market economy—the pure Adam Smith is always tweaked. When the CCF called for the end of capitalism, Douglas explained that what they meant was the end of the dog-eat-dog capitalism that resulted in farm foreclosures and hobos and the enormous waste of human capital seen in the Great Depression. Douglas's solution to the tendency of the invisible hand of the traditional economists always to take from the Great Plains was to develop a tripartite provincial economy that combined entrepreneurialism and private enterprise with co-operatives and Crown corporations that ran effective monopolies such as utilities and auto insurance, as well as business ventures that could not gain backing from outside investors but that filled a niche in the Saskatchewan economy. Many of Douglas's ideas came from Europe, but they had Great Plains antecedents, too. In both the United States and Canada, as we have seen, the large-scale settlement of the Great Plains and its instant linkage to the market system were entirely dependent on the railways, and both of the initial transcontinental railways were adventures in premature enterprise. The United States coped with "creative" financing, and Canada gained British financing for its first transcontinental railway in response to the apparently opportune outbreak of armed resistance. Subsequent transcontinental railways displayed an alarming tendency to go belly up at inconvenient intervals, setting off panics or requiring more government intervention. Another Prairie-grown model for Douglas was the Rural Electric Administration and Nebraska Public Power, developed by Nebraska senator George W. Norris, and a model from Central Canada was the public development of hydro power.

Although many Albertans consider "Tommy Douglas" a bad word, the Conservative premier at the time, Peter Lougheed, agreed with the need for a modified free enterprise system to protect an energy-producing province. Even the Nobel Prize-winning economist of the Chicago School, Milton Friedman, hardly a socialist, points out the paradoxical relationship of energy and the market system: "Few . . . industries sing the praises of free enterprise more loudly than the oil industry. Yet few industries rely so heavily on special government favors."[31] Douglas had the advantage of being premier of Saskatchewan after World War II, when economies everywhere seemed invincible. Nevertheless, his government compiled an admirable record of balanced budgets and his Crown corporations provided jobs, research, windows on industry, and useful products and services to Saskatchewan while also returning modest dividends. Economic *growth* is easy in a staples economy—money comes in and products go out. Economic *development*, on the other hand, requires money to recycle within the community, supporting secondary and tertiary industry as well as the service sector. Douglas fostered development. Although his acceptance of dams as being without environmental cost and his assimilationist response to Indigenous peoples were not particularly oriented toward a Great Plains consciousness, his judicious use of government investment without creating white elephants (like some of those developed by Alberta out of its oil earnings) showed that sustainable economic development is feasible in even the purest and most isolated part of the Great Plains.

As Jim Pitsula and Ken Rasmussen show, the privatization of Saskatchewan in the name of the free market following the Douglas years was economically counterproductive, functioning, as it is designed to do, to return Saskatchewan to its hinterland status. And it may be that Saskatchewan, like the Creeks in Oklahoma a century ago, has lost that successful moment of resistance that might have been the seed of a new destiny for the Great Plains. Margaret Laurence, the region's great novelist, believed that the small towns of the world would protect a vital kind of human knowledge that would not survive in the big cities, similar to the medieval monasteries preserving knowledge that would otherwise have been lost in the world of the Dark Ages. Although free enterprise and western democracies may be the best systems the world has yet seen, they are certainly not flawless. Francis Fukayama was wrong—we are not at the end of history.[32] State-run, tribal, and theocratic societies are enormously important in our world, and some experimentation with other ways of governing economies would be wise, if only for self-protection. More than almost any other place in the world, here on the Great Plains, where free enterprise, working as it is intended to work, *should* make us a hinterland, we have both the incentive to experiment and the tradition of successful resistances—from the land, from Native peoples, and from Tommy Douglas (and Peter Lougheed).

John Richards and Larry Pratt, writing in the 1970s, identified a rentier mentality in Alberta and eventually in Saskatchewan that allowed provincial citizens to turn over the development of their substantial mineral treasures to large foreign companies. Although Richards and Pratt contrast this to Texas populists who got out their guns to fight for their mineral rights, in truth, the Texans for the most part are just as much of the rentier mindset except that they are dominated by their own domestic companies. More recently, Roger Epp has shown that the increasing de-skilling and de-politicization of the rural West has resulted in an aimless and even hopeless clientism, which sees the provincial government as both inevitable and useless: "The development initiatives of community-minded people founder on the difficulties of speaking about, and for, a community interest in a world that increasingly presents only individual choices." Similarly, Thomas Frank shows how Kansans, especially from rural and suburban areas, have become willing to substitute narrowly focussed "moral" issues regarding abortion and gay marriage for more universal matters of social and economic justice, which they passively cede to large corporations that render them just as powerless as the foreign corporations did Albertans.[33]

Like Epp, and like Sharon Butala, I believe that there is a Prairie consciousness that exists where Indigenous and newcomer peoples have merged their land and economic knowledge. I believe that this consciousness has to do with place, with finding the best symbiosis of land, plants, and animals, including humans. I believe that we must study the models of Native resistance that have been repeatedly demonstrated on the Great Plains. I believe that a certain attention to the rhetorics of deficiency of this land and a close attention to where they have broken down and where they have been most pervasive will allow us, as contemporary residents of the Plains, to articulate a Great Plains consciousness that will allow us not only to live with, rather than against, this demanding land, but also to suggest how peoples of all regions can live better upon this earth, which, despite the musings of scientists such as Stephen Hawking, is still our only home.

More than forty years ago, William Goetzmann wrote in his magisterial study *Exploration and Empire*, "It is the thesis of this book that explorers, as they go out into the unknown, are 'programmed' by the knowledge, values, and objectives of the civilized centers from which they depart. They are alert to discover evidence of the things they have been sent to find."[1] Goetzmann's thesis is still valid, and his 1966 book is still the gold standard on exploration of the US West, but readers' reactions to the idea of the "civilized center" and the "knowledge, values, and objectives" that such places promulgated has shifted somewhat. Goetzmann's own programming meant that he accepted both the inevitability and beneficence of the "Winning of the West," to use Theodore Roosevelt's term, by the civilized centres. Since 1966, other scholars have complicated these issues. Post-colonial theory worldwide has focussed on the ways explorers used "the other" in contrast to both their own cultures and familiar landforms and weather terms to explore the mysteries in their own psyches and societies. More important, it has returned subjectivity to those who are being explored, be they Pashtuns in Afghanistan or Nahathaways in Rupert's Land. The homogenizing views of "savagery," at least partially developed from the idea of the "infidel" during crusading days, tinged mainstream European explorers' perception of non-Europeans during the ages of exploration and empire, but those views are crumbling under challenges from turn-of-the-millennium critics supported with both the exploreds' own reactions and the messages of individuality and surprise noted by the explorers themselves.

Mary Louise Pratt is best known for her term "contact zone," but perhaps a more useful concept that she has developed is that of the "anti-conquest"—the narrative of exploration that pitches itself not as conquest but as innocent scientific or commercial exploring. The anti-conquest seems objective and neutral and is perhaps even couched in terms of universal benefit to humankind. Perhaps rather than an *anti*-conquest narrative, Pratt should have called this the *covert* conquest because its rhetoric leads inescapably to the incorporation of land and people into empire. Dean Neu and Richard Therrien carry the study of these procedures into the treaty period and up to the present, showing how something as seemingly neutral as accounting can be used, for instance, to justify the taking of 115,000 acres of Blackfoot land without actually handing over any money to the dispossessed Blackfoot people.[2] Mapping, likewise, suggests that an area was, previous to the explorers, "unknown" and thus underutilized and in need of liberation into its full potential—obviously a job for the explorers, or at least the European or Euro–North American civilizations that they represented. Naming and classifying the flora and fauna are similarly acts of covert conquest. Mapping, naming, and bureaucratic manoeuvres all suggested that the land was empty and unused, ripe for the picking, without any untoward suggestions of violence or coercion.

Both Lewis and Clark and their successors in the United States, and the North West Company and Hudson's Bay Company explorers and later surveys in Canada were part of the imperial scheme that Pratt describes, but there were variations. Before the Palliser and Hind expeditions, Canadian explorers were working for fur trade companies, and for them the land could not be empty. Because the Canadians were traders rather than trappers, they needed to know which people were where, what resources they traditionally commanded or could command, and what could induce them to trade with the company represented by the explorer rather than with a rival company or not at all. None of the explorers on the US side was as consistently concerned with the fur trade, not even Jedediah Smith. As Goetzmann points out, Lewis and Clark were instructed to look for the broadest possible uses of the West to keep it from being the preserve of any special interests—including the fur trade. Goetzmann saw American mountain men as aspiring entrepreneurs, interested in any kind of main chance.[3] Because they were themselves trappers, for the most part, they did not need to worry about trade relationships with Indigenous peoples, though they did need personal relationships because they were still primarily dependent upon Indigenous women to prepare furs and to provide them with meals and clothing, a relationship that Goetzmann does not touch upon.[4]

Because marriage in the custom of the country entailed certain reciprocal responsibilities that many mountain men either did not understand or chose to ignore, marriages, instead of establishing kinship relations, often led to hostilities. For the mountain men, Native people became obstacles rather than trading partners. Except to John Wesley Powell, who would eventually found the American Bureau of Ethnology, and to army officers surveying for potential railroad routes and collecting information about the war-making abilities of various Indians, Indigenous people were not of interest to US explorers or mountain men because the Americans did not visualize them as having any place in the future of the area. Goetzmann argues that when the mountain men turned to being Oregon Trail guides at the end of the fur trade era, they demonstrated the pro-settler bias that had been theirs all along. The fur trade expeditions found the passes for what became the Oregon Trail and provided guides for the "great and inevitable folk movement" that passed along it and other trails to the Pacific, bringing "the two chief forces of contemporary civilization, science and organized Protestant Christianity."[5]

Goetzmann's programming led him to miss the social implications in his own account—he even speaks with an alarming lack of irony about the cavalry's "'final solution' to the Indian problem"—and to portray Americans as winners in the contest for the West while Spaniards (Mexicans) and British (Canadians) are the losers because their primary focus was not on settlement or complete exploitation of territory.[6] The Canadian Prairies and British Columbia, one might conclude from reading Goetzmann, were merely the leftovers after the Americans had taken everything worth having. Not surprisingly, few Canadians, even the expansionist Ontarians, who pushed for the annexation and settlement of Rupert's Land, shared Goetzmann's conclusions—the Canadian climate was healthier with no stagnant ponds and the soil was better in the northern fertile belt than in the Great American Desert.[7]

Goetzmann's programming also led him to underestimate Canadian achievements—Alexander Mackenzie beat Lewis and Clark to the Pacific by a decade, which Goetzmann acknowledges only backhandedly—and especially to underestimate the systematic and scientific documentation carried out by fur trade explorers. Germaine Warkentin explains that the scientific work of Hudson's Bay Company factor Andrew Graham has only recently been credited to him instead of to his early collaborator Thomas Hutchins, but the voluminous notes of many HBC men have long been available.[8] They contain plentiful observations on soil, climate, terrain, and the other things that Goetzmann values. As Barbara Belyea points out in her edition of Anthony Henday's travels across the northern Plains, the HBC furnished all its explorers with detailed instructions for entering information of this sort. Henday was to use his compass and dead reckoning to figure out where and how far he had gone; he was to note the depths of the water, the rivers and lakes he navigated or passed, whether or not they joined the fabled Ocean of the West; the names of the "Nation" of people whom he passed; the nature of the land and vegetation; and any indication of minerals. Clearly, he was not just looking for information that would be useful to the fur trade.[9] Like Lewis and Clark, Henday was deficient in ways of measuring longitude and proceeded primarily by dead reckoning. Although he is generally credited with being the first white man to see the Canadian Rockies, Belyea suggests that his daily estimates of distance travelled were far too optimistic and that he was never as far West as historians have suggested. But it is his own voluminous documentation that allows her to come to this conclusion.

Belyea, more than any other scholar, has pointed out the enormous difficulty of squaring explorers' reports of where they were and whom they had contacted with the names on contemporary maps and the ascribed tribal names of contemporary peoples. Even landforms themselves change in their representation. European conventions of establishing watersheds divide streams into mainstems and tributaries. "Fall Indian" (probably the ancestors of today's Gros Ventre and Hidatsa and Arapaho people) conventions instead represented rivers as equal paths through the mountains. Lewis and Clark, following their conventions and the theory that mainstems in adjoining watersheds mirrored each other, asked the wrong questions of Indigenous people and of their maps, and ended up taking a long detour to the headwaters (the assumptions embedded in the language are almost impossible to avoid) of the Columbia.

Which current names on the map correspond with the streams that Henday travelled is perhaps unknowable. Although contemporary scholars such as Belyea, Malcolm Lewis, James Ronda, and Mark Warhus are looking carefully at maps produced for explorers by Indigenous people, they have not usually been able to gain access to the detailed oral traditions that the maps accompanied.[10] Certainly Indigenous people followed pragmatic routes such as those featuring easily discerned landmarks that may not have had any particular geological significance. As Malcolm Lewis points out, however, the Pawnees, for instance, used star charts that mapped villages on earth in accordance with the star patterns that were associated with the ancestors and founders of each of the Pawnee villages.[11] Lewis and Clark, Henday, and many other early explorers were relatively amateurish in their use of the scientific instruments available to them, but they did add detail to the ways that European maps, particularly Arrowsmith's, laid claim to North America by depicting her rivers and mountains and enormous breadth from sea to sea on a particular grid that mimicked the landforms not as one would ordinarily visualize them but as one might see them if the globe were both flattened and miniaturized. But in addition to producing these covert conquests, the explorers were also erasing a different way of knowing the land that was subjective and based on how to get from point to point following already known food sources, trade routes, and sacred places.

The grid map made homesteading possible but also hampered settlers in taking land by breaking it up in ways that contradicted the actual lay of the land, and it also hampered the preservation of mental maps that focussed on the everyday uses of the land. One can, for instance, contrast the explorers' maps with Amos Bad Heart Bull's remarkable series of maps of the Black Hills.[12] Scientific map-making is enormously useful because it reduces everything to the same scale and provides a method of linking all parts of the world, but it is not always the most useful way to produce a map for every-day chores, as anyone knows who has ever prepared a map showing the way to one's house marked by such landmarks as railway and grain elevator in the country or traffic light and neighbours' kitschy lawn ornaments in town. Because scientific map-making virtually drove out Indigenous map-making, it extinguished a way of seeing and conceptualizing the West.

The two sets of captains from the age of exploration who have arguably had the most influence on how we visualize and depict the West, including the Great Plains, are Lewis and Clark and David Thompson. The importance of Lewis and Clark comes more from their nationalistic significance than from their skill as either explorers or cartographers. States mark highways as part of the Lewis and Clark Trail, and buffs and re-enactors follow parts of their route every summer. The Missouri itself, despite the dams and lakes of the Pick-Sloan projects, still testifies to their passage in the names given its tributaries, and their inland voyaging has been painstakingly documented. Scholars and buffs have produced whole libraries of editions of their journals and of writings about them. One of their Indigenous guides, the teenaged Bird Woman, Sacagawea, is on the obverse of a gold-coloured US one dollar coin. Yet despite the emphasis on Sacagawea, there is little indication in popular accounts of the Lewis and Clark expedition that theirs was primarily a guided tour. Indigenous people could almost always answer their questions—their biggest difficulty may have been in deciding which questions to ask.

David Thompson is a less nationalistic hero than Lewis and Clark, though as poor-immigrant-makes-good-in-new-world, he perhaps could be more of one. Rather, Thompson is acclaimed as the apotheosis of the scientific explorer. He is the only Canadian that Goetzmann allows in his pantheon. Thompson, a lad of fourteen, arrived in Churchill on Hudson Bay in 1784. According to Goetzmann, Thompson's explorations and the North American maps of Aaron Arrowsmith, which recorded many of the discoveries of Thompson and other fur trade explorers, were the main inspiration of Lewis and Clark. According to Germaine Warkentin, Thompson was "the most outstanding of Canadian exploration writers in

English, possessing the most reflective cast of mind and the greatest powers of synthesis." She credits him with both scrupulous attention to detail and a synthesizing intelligence that allowed him to find the system of what he called "the Great Plains." She notes his "courteous inquiry" into the lives of Indigenous people but does not point out how dependent he, like other fur trade explorers, was on his mixed-blood wife and on the knowledge and skills of many other Indigenous companions and even rivals.[13]

Although European records of the Great Plains begin with Coronado (1541) in the south and Henry Kelsey (1690) in the north, Thompson's systematization is important. As Lewis and Clark would do later, Thompson thought in terms of rivers, but unlike them, he did not think in terms of watersheds. Instead, he saw the Plains from the Gulf of Mexico to Hudson Bay as a whole, traversed from west to east by rivers that ran eventually into Hudson Bay, the Gulf of Mexico, or the Arctic Ocean. Understanding correctly that the southern rivers were flat and wide, forming characteristic braided channels, he also observed that from the Missouri north, the rivers ran in deep valleys. Thompson accounted for this by proposing that a flood from the Gulf of Mexico had washed all the deep soil up to the north and left the south a cactus-covered desert. He also noted that the rivers of the plains did not form lakes. He had already remarked, "These fine plains will in time become the abode of mankind, probably some civilized leading pastoral life tending Cattle and Sheep." Farmers, he believed, would have to stay at the northern verge of the plains, where wood was available.[14]

Thompson's explorations were carried out between 1784 and 1812, but he only began to write his narrative in 1846, and it remained incomplete at his death in 1857.[15] Thus, the rhetoric of his narrative probably owes more to the 1850s, a time when Canadian expansionists were already laying claim to the West, than to his actual years of exploration. Thompson devotes much of his narrative to describing the Indigenous people he had met and lived among, particularly the Peeagans [*sic*] and Nahathaways (Crees), both of whom he describes in the present tense, though recollecting events from a half century earlier. Still, he prophesies that these Plains, whose people he describes so carefully, will "in time" become the dwelling of "mankind," a phrasing that erases those members of "mankind" he had known and replaces them with herdsmen who may be either Euro–North

American or partially assimilated mixed-blood or Indigenous peoples. Thompson was obviously writing for a Euro–North American audience, since he hoped to sell his narrative to help support his family, and he had become accustomed to Euro–North American assumptions about land and progress—if, indeed, he had ever shed or questioned them in his years as explorer and surveyor. Goetzmann argues that Canadian explorers were only representing the mercantile interests of the Canadian fur companies. Thompson's narrative, however, shows that he, at least, had come to think in terms of agricultural settlement.

While Thompson and others, such as George Nelson, wrote intelligently about Indigenous beliefs and about everyday life in Indigenous societies, Canadian exploration narratives were still shaped by ideas of savagery and civilization.[16] Thompson, for instance, doubted whether the Nahathaway language, which he found easy for a European to learn and useful for trading and which he accurately described as similar to other languages as far east as the Delawares, was complex enough to "clearly express the doctrines of Christianity in their full force."[17]

In marked contrast to Thompson's accounts of Indigenous peoples is Samuel Hearne's vivid, horrifying, and often anthologized description of the violent raid by his "Northern Indian" companions on a small camp of Esquimaux, published in 1795. That the raid took place and that it was highly unusual are both demonstrated by the fact that it "is still recalled with horror by the Inuit today" and that Franklin's expedition members later visited the spot and contemplated Hearne's veracity. But Hearne's text is suspect—it was, in the manner of the time, edited and probably rewritten by his literary friends. The details and the sensibility recorded in his witness to the killing of a young Inuit woman may be a legitimate eye-witness account, sensibility fictionalized to match eighteenth-century notions of savagery and civilization, or some combination of the two. At any rate, the most memorable and bloody of the descriptions does not appear in Hearne's field notes. Nor does Hearne—or his editor—attempt to understand the motivation for this encounter.[18] Although the massacre takes place considerably north of the Great Plains, it represents an *entrada* to the continent through Hudson Bay, the same entrance used by the traders onto the Plains. As Owram points out, this northern gateway resulted in a popular image of the North West as more north than west, and thus not hospitable to agrarian settlers.[19] Hearne's narrative at Bloody Fall provides an example of the Indigenous North American as bloodthirsty savage, though it may well have been largely the European editor's idea of what a bloodthirsty savage should be. Many of Goetzmann's narratives make the same point. It is so inculcated in the genre that Goetzmann himself, as we have seen, cannot entirely escape a sort of unconscious identification with it, just as he does not escape the assumption that Canadian and Mexican explorers (and the country they explored) were inferior to American explorers and American terrain.

By the 1840s, the Great Plains was tolerably well filled in on the European maps. The rivers were accurately delineated, and the general idea of the Rockies, Selkirks, Sierras, and coastal ranges was clear enough to show that the Plains lay completely in their rainshadow and did not communicate with the Pacific. The imputed savagery of the Plains peoples was both an impediment to Euro–North American settlement and a rationalization for Euro–North Americans to dispossess the Aboriginal inhabitants and to occupy the Plains. The Aboriginal inhabitants were self-evidently, according to spokesmen of European origin, not worthy of a land so rich and wide. This was the message passed on to the East and to Great Britain, though it was, until the 1860s, more resounding in the United States than in Canada.

The intellectual traditions that were being overwritten during the age of exploration and the response of Indigenous peoples to Euro–North Americans and their varied agendas are much harder to gauge. Scholars, particularly James Ronda, have begun to practice a species of North American subaltern studies to remove the overburden of written documentation and try to determine how Indigenous people understood the invasion. The Indigenous response to the explorers is important in its own right as a part of the intellectual history of the Great Plains and also because it is impossible to evaluate how Indigenous concepts have influenced mainstream formulations if we do not first recognize the Indigenous ideas. Ronda focusses on using Lewis and Clark's texts to discover Indigenous responses to the Corps of Discovery, particularly the response of the Mandans, with whom the expedition spent the winter of 1804–5 and who thus had the opportunity for observation.[20] More recently, Matthew Jones began an exhaustive survey of both written and oral historical records of the interactions between the Oto-Missouria people and Lewis and Clark.[21]

Like many of the African groups that Pratt discusses, who had to make sense of explorers, the Mandans had pegged Europeans and Euro–North Americans as traders interested mainly in furs. As such, they were expected to assume the mutual obligations of kinship, either through marriage or through ceremonial or personal recognition as fictive kin on the part of Indigenous community members. Thus, in the Indigenous world view, commercial relations were governed, first of all, by kinship relations. Likewise, relationships to the non-human world were a balance of practical observations of the habits of prey animals—such as bison—and ceremonial relations with the *manito* or spirit of each animal. Thompson provides a particularly clear observation of the balance.

> When we related the scarcity of the Bison and Deer [the Peeagans] were pleased at it and said it would be to them a plentiful winter. Their argument was; [*sic*] the Bison and Deer have passed the latter part of the summer and the fall of the leaves upon the Missisouri [*sic*], and have made the ground bare of grass and can no longer live there; they must come to us for grass to live on in our country. . . . The winter proved that they reasoned right.

In addition to such practiced observation, "the religious hunter, at the death of each animal, says, or does, something, as thanks to the Manito of the species for being permitted to kill it."[22] For the most part, Euro–North American tradition has been to amplify the use of observation and to minimize or eliminate the relationship to the *manito*. Thus, we see radio collaring as a major source of information about the movements of animals, especially large predators. The introduction of non-native food animals, especially cattle and hogs, has further emphasized scientific management, especially in feedlots and hog-confinement operations, and virtually eliminated the sacral, especially as none of the European earth-centred religions of animal guardians seem to have been transplanted along with the cattle and hogs.

Oral traditions also emphasize the importance of both secular and sacred observations as well as expectations of a certain fluidity among humans, animals, plants, and elements such as rivers, stars, and rocks, considered animate in many Indigenous cultures but not in most transplanted European ones. Among Euro–North American scholars, it is mostly anthropologists, folklorists, linguists, and literary scholars who have studied these relationships, while a new generation of Indigenous scholars is once again examining these from within the cultures. Because this cosmology has, unfortunately, become only a background to the hinterland intellectual interpretation of the Plains that I am pursuing here, I will not discuss it in detail.[23] The treaties—particularly on the Canadian Plains, where Indigenous negotiators had more hand in crafting some of the conditions—demonstrate how Indigenous traditions show through and contrast with the mainstream Euro–North American attitudes expressed both by settlers and in the "scientific" explorations that formed the proximate basis of settlement. Terms like "as long as the grass shall grow and the waters run" were sacred and specific parameters, not just figures of speech, though Euro–North American explorers and legislators, as well as the general public, viewed them as such.[24]

The number and magnitude of the surveys of the American West carried out by the railroads and United States Geological Survey are simply staggering. Mountains and canyons, flat lands and sloping ones, dry lands and wet, prairie, forest, and desert: all were traversed, measured, and mapped. Fossils and rock strata, plants and animals, Indigenous people and their languages all were grist for the often competing surveys. According to Goetzmann, scientists were to survey everything from "ancient Silurian mollusks" to "sun-bleached Comanche skulls."[25] This inclusion of people as part of the environment showed the underlying ideology of the great scientific and railroad surveys: this mapping and classification was solely for the purpose of Euro/Afro/North American utilization and culture. True, individual Comanches might continue to exist, but Comanche civilization, to Goetzmann as well as to the surveyors themselves, was as firmly past as the ancient Silurian mollusks themselves.

The railroad surveys were obviously for the purpose of opening commerce and settlement across the Plains, linking the coasts and firmly cementing the hinterland status of the Plains in their tributary position, while the mountains and deserts would be transformed from the sublime to the merely picturesque. According to Goetzmann, Gouvenor Warren's 1857 map for the railroad surveys was the culminating achievement of what Goetzmann calls the "Great Reconnaissance," the first accurate, instrument-based outline, a master map of the West. On the other hand, some of the geologists were too esoteric for practical westerners, failing to map simple, practical occurrences—such as coal mines or salt licks. Nor, despite their careful studies, did the botanists look for agricultural potential outside California and Oregon, assuming that the rest of the area, including the Plains, was really desert. *Pacific Railroad Reports* included observation on Indians, with an emphasis on their war-making capacities. This was no longer a narrative of covert conquest but a plan of warfare. Later, Wheeler would provide maps for soldiers to use in their campaigns against Apache and Paiute peoples. The pragmatic Hayden, who found ways to praise everything he mapped as either useful or picturesque, also surveyed the Great Plains in the firm belief that rain follows the plough.[26] For westerners and intending western entrepreneurs, science for its own sake or reforms like John Wesley Powell's that would limit individual exploitation were worthless. The surveys were blueprints for building the land into the market economy.

In Canada, the railway surveys came only after Confederation. The first non–fur trade surveyors to cross the Prairies were the Hind and Palliser expeditions of the late 1850s. The Hind expedition, sent by the Canadian government to see what the North West held, mingled a sentimental look at the supposedly remnant Crees with an eager anticipation of Euro–North American settlement. Henry Youle Hind had his own version of rain following the plough. It was prairie fires, he believed, that caused soil sterility and had wiped out trees "south of the Qu'Appelle and Assiniboine." Fire suppression, he believed, would allow willows and aspens to develop and humus would render the soil both more fertile and more moist. Fire does regenerate prairie grassland and clear it of woody shrubs and trees, but grassland soil does not lack fertility, and trees do not bring moisture except to the extent that shelter belts capture snow and slow down wind evaporation. Fast-growing trees, however, are thirsty and their large leaf areas transpire more water than the narrow leaves of grass. Salt cedar, for instance, is an introduced invasive tree that can actually empty rivers and reservoirs. Hind was similarly confused about the decline in numbers of the buffalo—though to what extent the Canadian buffalo had already declined by 1858 is by no means clear. He blamed both the "careless thriftless" Crees who took only the "tongue and hump" of the buffalo, and what he believed was the sheerly destructive use of impoundment. Shortly afterward, however, he wrote that "buffalo were fast disappearing before the encroachments of white men"—the reason that the Crees demanded presents before allowing Hind's expedition to cross their land, already beset by too many whites and Métis.[27]

Englishman John Palliser's appointment to head a survey expedition resulted from his own enthusiasm for the Wild West of the buffalo hunts that he had participated in and written about in 1847. He was not for the most part optimistic about agricultural settlement in the prairies, describing the area that is still called the Palliser Triangle today as an extension of the "more or less arid desert" separating the eastern states of the United States from the Pacific Coast. "This central desert extends, however, but a short way into the British territory, forming a triangle, having for its base the 49th parallel from longitude 100° to 114° W., with its apex reaching to the 52nd parallel of latitude." Palliser did, however, see a fairly good possibility of settlement in what the expansionists would call the "fertile belt" between the northern forests and the southern triangle. Hind's and Palliser's observations that there was a large arable belt of land in the prairies were bolstered by the theories of American geographer Lorin Blodgett, who wrote in 1857 that climate did not depend on latitude but in fact got warmer toward the west as well as toward the south. Thus, Canadian expansionists in the 1850s and 1860s perceived that settlement would come to Rupert's Land along the valleys of the Saskatchewan River, even though they were not like the wide alluvial lands along the upper St. Lawrence and lower Great Lakes.[28]

The idea of the West was inescapably part of Confederation. It would provide an outlet for Ontario, Quebec, and the Maritimes, and would allow the new Dominion to flourish rather than to stagnate, blocked from expansion by cold in the north and by the dynamic United States in the south.

The acquisition of Rupert's Land and of a Pacific port in British Columbia were necessary to the dream, and British Columbia was induced to join by the promise of a transcontinental railway, which was originally planned to follow the expected settlement path along the Fertile Belt. The 1874 Dominion Land Survey initiated the same kind of instrumental grid survey that was being undertaken across the US West. While the 1850s and 1860s had been relatively dry, the 1870s and 1880s found the prairies in a wet cycle that evoked an optimistic response. John Macoun, the botanist of the survey, was delighted by the prairie flowers and proclaimed the prairies—including the Palliser Triangle and the lands that Hind had found too rocky or alkaline—the best place in the world. His theory of wind patterns suggested that the southern prairies would furnish both enough water and a long-enough frost-free season to ripen crops. The cooler climate, according to Macoun, guaranteed that Canadian settlers would be healthier than those of the United States, while the American desert actually helped build up rains to fall in Canada. Macoun's creative and wishful thinking, and his enthusiasm for the southern prairies allowed railway builders to consider running the first Canadian transcontinental railway further south, near the US border, rather than north, through the parkland belt.[29]

One major advantage of the southern route was that it would establish an indisputable Canadian presence in the area and hold the region against American expansion. Canadian fears of American annexation were considerably more realistic than Macoun's accounts of climate. Prime Minister John A. Macdonald himself, as a young lawyer in Kingston, had witnessed one of the more quixotic of the Fenian raids in 1837 and had served as a lawyer for the doomed Polish dreamer who had led the abortive raid.[30] The American cry of "54°40' or fight!" had threatened to merge British Columbia with Oregon Territory and gain the entire West Coast for the United States, destroying the possibility of a British polity from sea to sea. The Minnesota Uprising of 1862, when Santee and Yankton Dakota indignation at persistent treaty violations boiled over into a widespread and bloody but uncoordinated attack on non-Native settlers, left between two hundred and eight hundred settlers and soldiers dead, and many times that number demoralized and fleeing from the frontier.[31] Despite the concentration of the US Army on the Civil War in 1862, American soldiers quickly arrested hundreds of Dakota and mixed-blood men, while many others, with or without their families, fled north to safety. More than three hundred Dakotas were convicted in summary trials, and while Abraham Lincoln did manage to arrange clemency for nearly 90 percent of them, thirty-eight were hanged in the largest mass execution in US history. Ironically, Lincoln, who had begun his public career as an unsuccessful Indian fighter in the Black Hawk War, signed their death warrants the month before he signed the Emancipation Proclamation. The recoil of the US settlement frontier following the Uprising and its aftermath undoubtedly dampened annexationist pressure to go North, but it did not stop it. The Dakota people who claimed and were granted refuge in Canada might have pointed out their ancestors' part in fighting back the Americans during the War of 1812 and their continuing role as guardians of British territory through scaring away intending American settlers, but they do not seem to have made anything of the issue. Sitting Bull, however, a decade and a half later, would unsuccessfully remind the Crown of its obligations to its 1812 allies.[32]

Canadian Confederation itself was partly a response to US pressure on Britain to grant its North American colonies to the victorious North in payment of the US government's claims against Britain for allowing the Confederate ship *Alabama* to raid Union shipping from British ports.[33] American Fenians took great interest in the Provisional Government at Red River, formed after the new Dominion government attempted to enter Rupert's Land before it had been completely ceded by the Hudson's Bay Company and without the consent of the actual residents of the area. Although Louis Riel, the leader of the Provisional Government, maintained a distance from the Fenians, his Catholicism inclined Ontario annexationists, overwhelmingly Protestant and frequently anti-Catholic, to assume that he was plotting with them. Despite Riel's steadfast discouragement of Fenian ambitions, there were still ambitious Fenians, and Confederation had not ended US government interest in Canada.[34] The US Senate held hearings in 1874 on the possibility of admitting a state to be called Pembina, north of Minnesota.[35] An all-Canadian railway route that crossed the Plains near the American border seemed like a prudent anchor against future Pembinas and possibly a permanent barrier to future threats of annexation. Regina's placement on the bald Saskatchewan prairie at Pile

O' Bones Creek marked the symbolic demise of the idea that the southern prairies were wasteland and announced that Canada would hold the south for Queen and country.[36]

While the CPR survey was not complete until shortly before the tracks were laid—quite improbably—through Rogers Pass, the Dominion Land Survey completed the basic mapping, naming, and intellectual incorporation of the Great Plains into a Euro–North American mindset. The lines of longitude and latitude on the maps and the section roads, correction lines ("correcting" the deficient globe for not being square),[37] and fields apportioned by the square survey on the land itself rendered the curves of rivers and landforms obsolete for describing and judging the land and for proceeding across it. Geography had changed. Euro–North Americans had never doubted that they had inherent rights to settle their people, their animals, and their plants upon the Plains, but they had debated whether or not it was worthy of them. Samuel Aughey, with his theory that rain follows the plough, Lorin Blodgett with his theory that west was just as good as south for providing a long growing season, and John Macoun with his enthusiastic descriptions of and belief in the moderating behaviour of winds and clouds established an argument for settlement.

Native people's thoughts were irrelevant to both Euro–North American prospective settlers and to their governments. Before the War of 1812, Native peoples had maintained political power by playing off European powers one against the other. The Louisiana Purchase and the Mexican-American War, by removing the French and Spanish from contention for lands the United States wanted to claim, dissolved this strategy on the southern Plains. The War of 1812, by removing the British from the Ohio Valley and all the lands south of the Great Lakes, similarly destroyed the divide-and-conquer (or at least divide-and-resist) strategy in the north. Henceforward, Euro–North American settlers would divide the Great Plains into two co-operating settler nations, and they would neither raise Aboriginal allies against their rival settlers nor invade their rivals' territories across international borders.

Imperial powers gain their power at least in part by their willingness to spend or risk the lives and welfare of their subjects for the gain of the empire. The subjects, especially where the government has guaranteed their individual rights, are naturally unwilling to risk their lives and livelihoods in battles in which they stand to lose much and gain little, though appeals to courage, patriotism, and love of adventure will lead individuals and nations to move beyond what prudence would dictate. Americans, it seems, were generally willing to move onto lands granted by treaty to one or more Native nations and then to demand federal help when the invadees very sensibly resented them, but they were not likely to raise warfare against neighbouring whites. Besides, such a course would undercut the implied argument that Euro–North Americans were entitled to the land because they were Christians and because European-style commercial agriculture was a "higher" use of the land than hunting and gathering and riverine horticulture. An agreed-upon boundary line between the United States and Canada, fairly permeable in both directions for Euro–North Americans, was easier to sustain and more satisfactory than Fenian raids or warfare between Euro–North American powers, each with its own Native allies. The massacre of the buffalo, the commissary of Indigenous Plains peoples, coupled with the ease of obtaining, by rail, commercial foodstuffs to support an army in the field meant that Euro–North American settlers and their governments could either defeat Indigenous people militarily or conclude treaties with them that ceded control of the land to the newcomer.

Except for those northern areas that still do not quite fit into the market economy, the European invasion of what we call Canada and the United States of America was completed on the Great Plains and in the American Southwest in the first quarter of the twentieth century. The disappearance of a frontier line and the passing of the frontier that Turner mourned in 1893 marked not only the disappearance of "free land" but also the unfreedom of those whose land it had been. The American frontier myth, from the Pequot War and King Phillip's War on, had defined itself by violence to Indigenous peoples, while the Canadian frontier myth that began with Champlain and Des Ormeaux and Brébeuf had modulated into a repudiation of violence as American and an exaltation of the Canadian West as being more British, civilized, and fair.[1] Americans killed Indians. Canadians, far more "humanely," forced them into starvation. Indigenous peoples, however, did not see themselves as doomed and fought back in every way that they could, with physical warfare, with spiritual reawakening, and with reformulation of Indigenous philosophy to be efficacious in a new age. Two narratives came out of this period of seeming conquest—a whitestream narrative of dominance and an Indigenous narrative of what Gerald Vizenor has called "survivance."[2] Curiously, however, each nation's whitestream narrative involves a martyr, and each Indigenous nation's survivance implies rebirth. For the United States, the martyr was George Armstrong Custer; for Canada, it was Louis David Riel.

Both Custer and Riel had early successful careers that prepared them for their later martyrdom, Custer as the Boy General of the US Civil War and Riel as the successful leader of the Provisional Government at Red River. Custer, especially as he constructed himself in his own letters and other writings, and as his wife, Libby, reconstructed him in her memoirs after his death, was undoubtedly a fascinating character—ambitious, charismatic, and surrounded by a menagerie of dogs, horses, and tamed beasts such as antelope. Libby Custer's books canonized her husband and demonized the Sioux. Custer as an Indian fighter, both as the perpetrator of the Washita Cheyenne massacre of 1868 and the victim of the Little Bighorn battle of June 1876, was essentially a national figure, not an avatar of the West. The Great Plains, as the site of many of the Indian Wars between the 1850s and 1890s, is the location of both his triumph and his defeat, but it is not central to the Custer story. *Fort Apache*, a 1948 film that both debunks and glorifies Custer, is set in Arizona, and the substitution of desert for Plains and Apaches for Lakotas and Cheyennes is immaterial to the national myth of Custer.[3]

Riel, on the other hand, is firmly set in the West. His role in the national myth is as the first western rebel. Thus, Preston Manning can—with all seriousness, though aware of the potential humour—portray the rise of the Reform Party (a conservative federal party intended to gain federal influence for the West) as "the third Riel Rebellion."[4] Since Red River and the North West were the only sites of military campaigns between Aboriginal or Métis and settler forces in Canada after the early days of New France, it would not be possible to transplant Riel within Canadian history, but even had there been other battles, Riel's identification with the New Nation of the Métis, with the English "half-breeds," and with the old stock English-speaking settlers of the North West makes him explicitly a man of the Plains. On the one hand, his interest in representing First Nations peoples and his seemingly contradictory plans to bring more European immigrants to the West link him to all colonization efforts. On the other, his identification first with Manitoba's entering Confederation, his own election to Parliament (and his surrendering his seat to Georges-Étienne Cartier, Macdonald's Quebec lieutenant), and finally his wholly involuntary connection to the completion of the CPR make him a national figure. Quebec's identification with Riel but not with the West also complicates his consideration as national, regional, or international figure.[5] Yet despite their differences, Riel and Custer both served as symbol and synecdoche for their respective federal governments, giving them permission to abrogate treaties and to exert ruthless pressure on Indigenous peoples through outright warfare, starvation, and massive, systematic human rights violations designed to stamp out not only effective Native resistance to the wholesale Euro–North American settlement of the Great Plains but any cultural continuity whatsoever for Indigenous people. The events surrounding Custer and Riel "proved" that the people of the Great Plains were terminally deficient and gave federal governments and popular culture in both countries "permission" to decree complete assimilation or extermination for Plains Indigenous peoples and their philosophies of life.

Both Riel and Custer have spawned scholarly industries trying to establish What Really Happened, primarily at the Battle of the Little Bighorn and at the trial of Louis Riel, as well as a whole historical context for each man. Poets, novelists, painters, composers, and film and video makers have been no less active. Huge, epic "Last Stand" paintings seem to have been the favourite visual medium for Custer, while Riel has been portrayed in plays and sculptures. Other figures from the Little Bighorn and the North West have also been valorized, including Sitting Bull, Crazy Horse, and Rain-in-the-Face (Ista Magazu); the "lone survivor" of Custer's command, the horse Comanche; and Gabriel Dumont, Big Bear, Poundmaker, and the captives Theresa Gowanlock and Theresa Delaney. We even have scholarly interpretations of interpretations of the Custer myth, including those by Canadian Brian Dippie and Blackfoot/Gros Ventre writer James Welch.[6] What I want to do here is to look at the ways in which the idea of Custer and the idea of Riel have shaped the intellectual history of the Great Plains. Although Custer's demise in 1876 predates Riel's hanging by slightly more than nine years, I will discuss Riel first, simply because the narratives associated with him are more complex and more directly associated with the Great Plains than those of Custer. Then when we return to Custer, we can see what is different and what is simply left out. The next chapter will look in much more detail at Riel in comparison to the Ghost Dance movement on the Great Plains and will examine spiritual and cultural revival rather than political and military imagery.

A 1979 CBC television movie called simply *Riel* is a treasure trove of images and provides a good place to start, more because of its blithe rewriting of history than because of its historical accuracy or insightful probing of myth and symbol. The film is apparently at least one of the targets of Rex Deverell's 1985 play *Beyond Batoche*, which problematizes a Euro–North American television play about Batoche. The young and impatient writer discovers that although he had always seen himself to be Riel, in a pinch, he identifies with Macdonald.[7] The CBC's *Riel* begins in the context of Buffalo Bill's Wild West, an artificially American and literally "Wild West" setting. A journalist approaches Gabriel Dumont, who actually did join Buffalo Bill after Batoche, for his reminiscences about Riel, and most of the rest of the film is a flashback to the sharpshooter's days with Riel, although it also departs to show Macdonald (played by Christopher Plummer), usually in the company of Donald Smith, a Hudson's Bay man and later a CPR baron. In fact, the film suggests that Riel and Dumont, Macdonald and Smith actually shared the same dream of a united Canada from sea to sea that was home to all—Indians, Métis, British, and French.[8] But the sectarianism and religious fanaticism of the French priests in Montreal and the Orangemen at Red River cause a conflict that ultimately makes enemies of Macdonald and Riel. Macdonald's grand obsession is the railway, and the film accurately shows that the North West became the crisis that allowed Macdonald to use the railway to ship troops west—and enabled Smith to borrow the money to finish it. Red River is a prelude to the North West, and Dumont is portrayed as having been a participant at The Forks as well as at Batoche. The time between 1870 and 1884 seems to be telescoped into two or three years. Perhaps in order to introduce a strong female character into the movie, Mrs. Schultz, wife of the leader of the Canadian annexationist element, is seen as having an affair with Thomas Scott. The historical

Scott appears to have been a rowdy, perhaps mentally deficient in some way, whose execution at the behest of the Provisional Government would eventually furnish the motive for hanging Riel, but in this rendition of the story, he appears to have walked off some set for *The Great Gatsby* and is one of the leading plotters against the Provisional Government and all it stands for. The film shows the British commander of the forces against Riel in 1885 suitably as Colonel Blimpish, and his victory at Batoche is (accurately) won by Ontario soldiers who mutiny against his vacillation and take the Métis lines. Métis women have little role in the film, and Indian characters are more or less indistinguishable from the Métis, except for their names and occasionally their braids or "medicine man" role.

Although the movie is unintentionally funny some decades later, it provides a lens for examining what Riel has meant to Canadians. For one thing, it is not at all a Prairie view. Riel and Macdonald have the same dream of the Nation, although Riel's symbol is the cross and Macdonald's the locomotive. Riel is played as deeply religious but not unorthodox, and his "insanity," for which he was hospitalized, seems to be a temporary nervous breakdown in response to his harassment by Schulz and the other Protestant settlers and his being barred from taking his seat as an MP. (Macdonald is portrayed as showing Riel a back door to the Parliament buildings so he can escape the vicious Orange [Protestant] mobs who are howling for his death in retaliation for the execution of Thomas Scott.) Riel dies a martyr's death, a sacrifice to the powers that be, in thanksgiving for the completion of the railway, the salvation of the nation. The television movie also accepts Indian complicity in the uprising—as had Riel and virtually all scholars until nearly the present.[9] It ignores the problems of the West at the time: lack of secure land titles for the Métis, crop failure and land claim problems among non-Métis settlers, and starvation and abrogation of treaty rights among the Crees. It also undervalues the Provisional Government at Red River.

None of this is necessarily the fault of the film or its writers and researchers. The story of Louis Riel and all the lives and stories that have come to be associated with him is enormously complex. Despite many superficial similarities, the differences between Red River in 1869–70 and the North West in 1884–85 are important. The idea that Métis and Indians were natural allies against Euro–North American settlers was unexamined by all but the Indians for at least a century. And Riel's use as both a positive and a negative symbol by Quebecois, westerners, and even various groups of mixed-blood and Indigenous peoples further complicates the issue. Custer, by contrast, is a relatively one-dimensional character whose meaning changes along with public perceptions of the federal government and of Indians.

Louis Riel, Sr., had been one of the Métis leaders in the Guillaume Sayer case of 1849, in which the Métis broke the Hudson's Bay Company monopoly on trade in Red River and gained free access to the markets in Minnesota. As one of the few European-educated Métis in Red River in 1869, as well as the son of his father, Louis, Jr., was a logical leader of the Red River people as they met the Canadians after the easterners' completely unilateral annexation of Rupert's Land. As Doug Owram has shown, the Ontario expansionists believed their own rhetoric about the inhabitants of Red River calling out to be relieved of the feudal yoke of the Hudson's Bay Company and being eager to join the world of the up-and-coming Ontario merchants and settlers. John Christian Schultz's paper, *The Nor' Wester*, had been started to promulgate such opinions, and at least some children of the fur trades, such as Alexander Kennedy Isbister, shared them.[10] Under other circumstances, bourgeois Métis such as the Riel family might well have been willing to join the Ontarians, but that was never, in fact, an option. Such a union ignored both the role of religion and the social structure of Red River: it is hard to make cause with someone who blackguards both your racial ancestry and your religion, and the annexationists were blunt about "half-breeds" and "Papists." As Sylvia Van Kirk and other scholars have shown, the gendered mobility of Métis society was putting enormous stress on family relations and overall social structure at this time. Young Métis women could marry "up" into Canadian society, but their brothers were regarded by the Canadians as not only unworthy of marrying into Canadian society but even unworthy of maintaining control of economic power in the territory.[11] Much the same pressure existed in other mixed-blood peoples of the Americas, and toward the end of his life, Riel himself envisioned Canadian Métis children being educated in Latin American countries, leading to a saving relationship for the Métis.[12]

Métis society was an outgrowth of the fur trade. While the Hudson's Bay Company traders originally tried to wall themselves off from relationships with Native women, the Montreal traders realized that such separation was completely unworkable. The traders needed Native women to help them survive, to prepare furs, and to establish the kinship networks that were the foundation for Indigenous trading patterns. Hudson's Bay Company traders could not remain aloof and succeed, so they, too, took wives according to the custom of the country. The children of the fur trade for the most part stayed in the fur trade. Women became the wives of new European factors and men married other mixed-blood women or Indigenous women. Only a few children were assimilated or even chose to be assimilated into actual European society, while a larger proportion chose or happened to live with their mother's people and to be accepted as Indigenous. George Simpson, long-serving governor of the Hudson's Bay Company, however, changed the pattern by putting aside his "country wife" and marrying his young cousin, Frances, who was soon followed by other English wives, drastically reducing the social status of mixed-blood women.[13] Julie Lagimodiere, on the other hand, the first full-blood European girl born in Red River, married Louis Riel, Sr., himself mixed-blood, and she and her sisters and other French-Canadian women fit smoothly into the French, Catholic, Métis society.[14] Nonetheless, the influx of large numbers of young, single Anglo-Canadian men into Red River in the 1860s once again provided potential husbands for young mixed-blood women but marginalized mixed-blood men both socially and economically, especially as an agrarian society rapidly replaced a fur trade and hunting society.

In 1869, Louis Riel was in a precarious position. His father had died. It seems likely that Louis, Jr., had been rejected by the family of his Montreal sweetheart because he had no particular prospects for making a living. The annexation crisis represented an opportunity. He threw himself into the leadership of the Provisional Government and managed both to maintain control of the situation in Manitoba and to negotiate with Ottawa the status of the Red River settlements as the province of Manitoba. Had Riel and the Provisional Government not succumbed to the imprecations of Thomas Scott and ordered his execution, Riel would not have provided a martyr to rally the Orange annexationists from Ontario, and he might have had a successful political career in Manitoba. This seems unlikely, however. Although the Canadians had attempted to enter Rupert's Land before they had legal authority, that authority was coming, and few central or eastern Canadians appreciated the niceties of international law that made the Provisional Government legal. They simply saw ungrateful half-breeds turning down the gracious offer of annexation. Macdonald had popular Canadian backing and the full co-operation of the British government in sending to the West an expeditionary force to defeat the Métis. And the Métis prudently left Fort Garry before Colonel Garnet Wolseley and his troops arrived. From founder of Manitoba and Member of Parliament, Riel went to being a fugitive, banned from Canada for five years, more as a sop to the angry Orangemen than for any actual crime or act of rebellion.[15]

Combined with his political career throughout his life was Riel's attempt to obtain from the Canadian government some kind of reparation for his economic losses or compensation recognizing the truly masterly way, except for the execution of Scott, that he had maintained order and avoided violence in a very precarious situation in Red River. Riel's detractors have claimed that his requests for some kind of pension or indemnity to help support himself and his family prove that he was in the Métis cause only for the money and that he would have been willing to sell out his allies at any juncture had Macdonald only been ready to buy.[16] This not only oversimplifies Riel, it oversimplifies the whole class basis of Métis society at the time of Confederation. During Riel's years in Montana in the early 1880s, he was able to marry and to support his beginning family on a very meagre schoolteacher's salary. Given his education, his experiences, and his belief in his own prophetic calling, it is hard to picture him as a homesteader or even as a miller like his father. Although he may have lived briefly with buffalo hunters in the Dakotas before moving to Montana, he had neither the training nor the ambition to be a hunter.[17] He had had the experience of being elected to Parliament and then prohibited from taking his seat. He was far too controversial to have been a John Norquay, the first Métis politician to gain considerable success in Manitoba as provincial premier from 1878 to 1887;[18] by 1890 and the Manitoba School Question, a divisive national fight over the use of French-language instruction, it was clear that no one else would follow the late John Norquay's career pattern, either.

In some ways, both the Red River Provisional Government and the Exovedate, the religiously based governing body Riel established at Batoche, were formalized resistance on the part of the young men who had been the elite of mixed-blood fur trade society and who had become or would become marginalized in the new settler society. In US terms, they would have been similar to Andrew Jackson's followers—frontiersmen challenging the Virginia and New England elites. But Jackson was elected to the presidency, and his followers were known as Indian fighters; the Jacksonians almost certainly included a few people of mixed-blood descent, but they did not identify as anything other than European-descended. The Provisional Government's execution of Thomas Scott was not unusual at a time when capital punishment was common, accepted, and swift in both Great Britain and the United States as well as Canada, but the idea that Riel, as the leader of a group of Catholic "half-breeds" had "murdered" a man portrayed as an up-and-coming Protestant from Ontario ensured broad Canadian support for both the persecution of Métis individuals after the arrival of Wolseley's army and for the general supplantation of the Métis in Manitoba. The execution of Scott may have been factually and procedurally justified, but it was a political disaster, an excuse for throwing out the Provisional Government with its alliance of Métis, English "half-breeds" and other fur trade peoples, and the descendants of the Scots Selkirk and Swiss Demeurons settlers, and for replacing them over the next two decades, after the defeat and death of Norquay, with the annexationist elite led by Schultz and the incoming Ontarians. Métis scrip, issued to resolve Métis land claims, allowed some Métis to locate and develop farms, but like most of the scrip programs in the US upon which it was based, its main beneficiaries were speculators who were able to scoop up concentrated areas of first-rate land on the cheap by exploiting the need of cash-poor Métis who expected to be able to re-establish themselves further west.[19]

Native people, as opposed to the Métis, played relatively small roles in both the Red River and Northwest Resistance movements. In most accounts of Red River, their roles are virtually invisible. The annexationists, apparently blithely unaware of the fear that the idea of an "Indian uprising" evoked in settlers—especially when some of the Dakota people involved in the Minnesota Uprising of less than a decade earlier had taken refuge in

Manitoba—tried briefly to raise Indian allies against the Métis, a scheme that was quickly squelched by more politic members of the movement. That the Indians considered the suggestion, however, indicates that they did not see their cause as identical to that of the Métis and the old settlers.[20] By 1869, Native peoples throughout the Great Plains had only to look east, south, or west to recognize that their autonomy was severely threatened and that migration was at best only a temporary option that would place them in the homelands of another people who were also severely threatened. Resistance was sometimes counterproductive (as in Minnesota) and sometimes temporarily successful (as in the Red Cloud War of the late 1860s in Montana and the Dakotas), but there were no extant examples of anything that looked like a satisfactory long-term solution. Indigenous peoples on the Canadian Prairies were anxious to protect their lands and buffalo as much as possible from the new Canadian expansionists. While the Canadians wanted peace and land cessions, Aboriginal people sought treaties that would give them the best chance of protecting a land base and obtaining assistance in converting their economies to a subsistence- and commercial-based form of agriculture.[21] Although, as events would prove, no faction would be a worthwhile ally to the Prairie Indigenous people, the annexationist Ontarians were a better bet simply because they would be the winners and wield the power. Indigenous leaders with centuries of diplomatic history dealing with other tribes, mixed-blood groups, Europeans, and Euro–North Americans were not naive in their choices.

Except for small-scale reprisals against the Métis after the arrival of Wolesley, the Red River Resistance seemingly ended peacefully and almost hopefully, though the promises given to the Provisional Government were never fully implemented by the Canadians. At first sight, the resolution of the conflict represented the triumph of the "civilized" British way of handling "Natives" as opposed to the Wild West formula of the Americans. As Owram has shown, Canadians since at least the 1850s had rather smugly contrasted their system of Indian affairs, based on treaties and courts, with the constant frontier warfare of the United States. The whole ethos of the North West Mounted Police as embellished by both historians and fiction writers contrasted the straight-dealing, scarlet-coated Mounties, relying on a personal manliness animated by the weight and civilizing justice of empire, with the two-gun craziness and military might of the Americans, enforcing the newest edicts of treaty-breaking governments in Washington. The public reason for the formation of the NWMP, after all, was to protect Native Canadians and Canadian territory from the violent lawlessness of American whiskey traders.[22]

The Northwest Resistance of 1885, discussed in detail in the next chapter, finished with a British-led army, as at Red River, investing Riel's home settlement at Batoche. Like the Minnesota Uprising, it was followed by trials, hangings, and imprisonments, but also by a vengeful and counterproductive agricultural policy and religious suppression forced on the Plains Crees, Assiniboines, and, to a lesser extent, Blackfoot in the late 1880s and 1890s. All of this was thinkable only because the resistance allowed Canadians to accept the demonization of Indians. Later, in the popular Euro-Canadian mind, the Northwest Resistance became Canada's fling at having a Wild West, as the 1979 film and the continuing summer dramatizations of "The Trial of Louis Riel" in Regina every year demonstrate.[23] Not nearly as dramatic as the Custer battle, perhaps—a poor thing, but our own. And Sitting Bull had toured with Buffalo Bill's Wild West only a few years before Gabriel Dumont, Riel's Saskatchewan lieutenant, joined the show.

The Battle of the Little Bighorn, coming between the two resistances led by Riel, is valuable for its symbolism, not for its military significance. It was by no means the most costly battle of the US Indian Wars—when Little Turtle, leading his Miamis and allied Shawnees, defeated Arthur St. Clair's troops in 1791, they killed more than six hundred US soldiers, three times the American losses at the Battle of the Little Bighorn. Like Custer's defeat, St. Clair's defeat rallied public opinion against Indigenous peoples—this time of the Ohio Valley—and the fledgling Republic poured its money into the military, enabling Mad Anthony Wayne to defeat Little Turtle three years later. Little Turtle, like the Cree leaders Big Bear and Poundmaker nearly a century later, turned to accommodation as the best way to gain living room for the Miamis and Shawnees. His opposition to the charismatic Shawnee brothers Tecumseh and Tenkswatawa prior to the War of 1812 and Tecumseh's death with General Brock at the Battle of the Thames in 1813 ensured that neither the British nor the Shawnees and

Miamis would ever again regain control of the Ohio Valley.[24] St. Clair's defeat came at the beginning of the United States of America's wars with the Indigenous peoples, when the outcome of both the wars and the new Republic itself were genuinely in doubt. Americans celebrated Wayne's victory, not St. Clair's defeat, in order to encourage themselves in their conquest of the continent. Custer's defeat also encouraged Euro-Americans to concentrate men and money in a battle against Indigenous people, but this time the defeat represented no actual threat—just an occasion for a rededication to the American project of what by then was known as Manifest Destiny. The battle was fought on 25–26 June 1876. Perhaps Custer, who had characteristically disobeyed orders in going ahead and engaging what turned out to be an overwhelmingly large Lakota, Dakota, and Cheyenne encampment, anticipated news of his great victory being announced at the Centennial celebrations upcoming in Philadelphia on the Fourth of July, Independence Day. The news, of course, turned out to be rather different.

The year 1876 was not only the centennial year, but also a presidential election year. Ulysses S. Grant, the general who had saved the Union, would be stepping down. The dashing Custer, who had made brevet general during the Civil War, would not have been an implausible candidate—certainly he was better known than another Civil War general from Ohio who *did* become president after a disputed election, Rutherford B. Hayes. Despite his Union background, Custer would have been a far better candidate for his own party, the Democrats, than Samuel Tilden of New York, who still battled Hayes to a dead heat. Part of the Custer legend is that he deliberately entered Sitting Bull's encampment to secure a great victory and a nomination by acclamation, though his own letters to his wife give no indication that he was hankering to be president.[25] Whatever Custer wanted, he certainly achieved posthumous fame. And, like the 1885 Northwest Resistance in Saskatchewan and its enormous value to Ottawa, if Custer's Last Stand had not existed, someone in Washington would have had to invent it in order to justify the wholesale abrogation of treaties (already underway) and the scorched earth policy that had all Native people confined to reservations, and then had the reservations halved and halved again and finally alienated through allotment.

In 1871, the US Senate abolished the treaty-making process between the United States and the groups Chief Justice John Marshall had called "domestic dependent nations" in his famous *Worcester v. Georgia* decision. Grant's Peace Policy had distributed the various reservations to the Quakers, Catholics, Episcopalians, and Methodists to administer, hoping to get rid of the graft in the Indian service. Unfortunately, it turned out that men of the cloth could be as venal and corrupt as anyone else and perhaps even more self-righteous. The Peace Policy was collapsing under its own contradictions as well as the inherent contradictions of the musical-chairs nature of US Indian policy, which kept concentrating Indigenous peoples and moving them away from land desired by Euro-Americans.[26] Custer's own 1874 reconnaissance of the Black Hills and his publicizing of the gold discovered there initiated the breakup of the Great Sioux Reservation and propelled the Lakotas west in 1876, in violation of the newly disseminated rules for where they were to stay. The whole Seventh Cavalry was dispatched to bring them back, and that campaign plus the subsequent campaigns of Generals Crook, Miles, Terry, and others were far more important than Custer's contributions in confining the Sioux. The victorious villages scattered after the Little Bighorn, with many going north to Canada in search of the same refuge that the Dakotas had found after 1862. Sitting Bull argued that Lakota aid to the British and Shawnees during the War of 1812 entitled his people to refuge, but the Canadians and the British Crown turned a deaf ear. The Lakotas could stay, but they could not claim any land, and there was eventually nothing to eat, no more buffalo and not even the meagre rations available to Canadian Indians. And so Sitting Bull and his remaining people came back to the United States and to Standing Rock Reservation, where Sitting Bull would eventually meet his death at the hands of Indian police.[27]

A century after the Little Bighorn, more had been written about that battle than any other in America except for Gettysburg.[28] Since the bicentennial year, Custer and his battle have attracted even more ink, especially during President Ronald Reagan's belligerently patriotic "Morning in America" years and in response to the significant archaeological finds that emerged after a fire cleared the battlefield area in 1983.[29] Just as important, AIM (American Indian Movement) and Vine Deloria, Jr.'s book *Custer Died for Your Sins* (1966) have made Custer as vital a symbol of American Indian resistance and revitalization as he had a century earlier been a symbol for American Indian extinction and assimilation. He has become the symbol of everything arrogant and bloodthirsty that Euro-Americans have ever done to American Indians—a rather heavy weight to bear.

If Canadian television producers in 1979 felt the need to set Louis Riel in the context of the Wild West, Custer's Last Stand *was* the Wild West, its re-enactment the penultimate act of Buffalo Bill's Wild West.[30] Not only did Buffalo Bill manage to woo Sitting Bull and other actual Lakota and Cheyenne fighters to his entertainment, he had himself, he proclaimed, taken the "First Scalp for Custer," killing a Cheyenne man named Yellow Hair (not Yellow Hand, as early scholars stated) in hand-to-hand combat during a "minor skirmish" on 17 July 1876. While revenge killing of those who had killed one's kinsmen or friends was sometimes part of Cheyenne and Lakota warfare, it was not supposed to be part of the US Army's method of operation, even for a somewhat freelance scout. Since Cody almost immediately returned to Chicago to use Yellow Hair's scalp in his own stage re-enactment of Custer's Last Stand, one could argue that he murdered the Cheyenne man solely to obtain a unique theatrical prop. Controversy relating to Riel is, generally speaking, confined to the factual, though the theatrical is certainly a part of all the artistic representations of the man and his cause. With Custer, the real became an artifact of the dramatic, and the participants in the act, at least the survivors, became participants in the re-enactment. Yellow Hair's scalp became a ghoulish trope for the way Custer's history had completely mixed artifact and symbol. The only really clear result of the Battle of the Little Bighorn was permission to kill Indians—and Cody the scout took the scalp for Cody the actor. Meanwhile, Miles, Crook, and company had begun a war of attrition against the American Indians that ended its active phase at Wounded Knee in December 1890 but has never hung up its symbolic rifle. The colour and pageantry of Custer's Last Stand has effectively drawn public attention and historians' interest away from the systematic bilking of American Indians in the near century and a half since.

By 1876, the United States already had a transcontinental railroad and others were furiously being built. America was aggressively bringing the "Indian territory" it had established in the Great Plains into the settler nation. But Custer's martyrdom not only justified the future repression and dispossession of Native people, it justified "the land of the free and the home of the brave" upon the occasion of its centennial. It also fed the belief that all the previous Indian wars, from the Pequot War through the Battle of Fallen Timbers and onward, had been just. Indians were bad guys who deserved to be conquered. Throughout the rest of the nineteenth century and all through the twentieth and into the twenty-first centuries, historians and creators of popular culture have focussed on Custer—and Riel. Indians fall out of the history books in both countries, and continuing bureaucratic dispossession, particularly in the United States, has proceeded merrily on.

The final act of Custer's Last Stand did not take place until December 1890 at Wounded Knee. Crazy Horse's people came back from Canada in 1877 and Crazy Horse died in the guardhouse at Fort Robinson in September at the hands of the soldiers—assisted by Crazy Horse's former friends.[31] Sitting Bull's people gradually drifted back to the United States, managed to find sanctuary as individuals with Canadian groups, or finally accompanied their chief back in 1881. Little of the land set aside for Lakota reservations was suitable for agriculture, but by 1890, Lakota people were beginning to accept allotted land that allowed subsistence hunting in riverine forests, especially along the Missouri, and even to begin raising cattle along with their horse herds. Food was often scarce. Social breakdowns came from the outlawing and repression of Lakota religious practices including the Sun Dance, from dividing tiyospayes (extended family units) into nuclear families on individual land, and from taking children from their parents to go to boarding schools and returning the survivors without the skills relevant to either reservation life or off-reservation success. As was true in Red River in 1869, the stresses were partially gender specific. Women's roles in child care, cooking, clothes making, and gathering were certainly changed—particularly by the boarding schools and the instruction given by woman missionaries—but were still reasonably intact. Unlike Mandan and Hidatsa women, Lakota women had largely given up horticulture when they came onto the Plains and devoted themselves more to the preparation of buffalo meat and hides. In the reservation era, to be sure, they cooked beef instead of bison and tanned and worked cowhides, but they still tended children, picked berries, and sold such things as beaded moccasins to traders and tourists for cash or basic staples. Men, on the other hand, were displaced. Warfare against either Euro–North Americans or other Indigenous peoples was prohibited. There were no more buffalo to hunt, and off-reservation trips in search of elk or other large game were also prohibited. By 1890, the Sun Dance had been outlawed, and a generation of consistent missionizing and religious persecution had forced most traditional ceremonies and healings underground, seriously undercutting the role and livelihood of doctors and priests. If farming had been viable, it undoubtedly would have become a very popular pursuit for Lakota men, since the traditional way for Lakota men to gain esteem was to distribute food to the poor, but the reservations combined semi-arid land with a lack of agricultural implements, draft animals, and seed. Herding and hauling provided an occupation for some men, and a position on the Indian police afforded authority, but most men found a meagre subsistence and a strong sense of redundancy. Thus, it is not surprising that the Ghost Dance spread widely among the Lakotas when messengers bought it back from the Paiutes in 1889.

Spiritual and Intellectual Resistance to Conquest, Part 2:   *4*
Messianism, the 1885 Northwest Resistance, and the 1890
Lakota Ghost Dance

The quarter century between 1865 and 1890 saw the completion of trans-continental railways in both the United States and Canada, the slaughter of the buffalo herds, and the nearly complete disruption of the golden age of the horse-buffalo-Sun Dance culture that had begun only two centuries before. In response to this Armageddon, messianic movements developed, flourished briefly, and were put down in blood and bullets. Although historians are familiar with this general framework and have studied in detail both the two resistance movements led by Louis Riel (mostly on the Canadian side) and the Ghost Dance (mostly on the US side), no one has attempted to compare and contrast these two movements. In addition, although James Mooney, Michael Hittman, Thomas Flanagan, Manfred Mossman, Gilles Martel and others have looked at the various Christian antecedents and analogues of the Great Plains messiah religions, no one has examined the parallels between these two movements and first- and second-century Christianity.[1] This last may seem a strange comparison, but it is probably misleading to try to discuss Christian influences on the

Ghost Dance without first exploring the ways in which early Christianity, as it sought to define itself in an imperial world of rapidly changing material conditions, was itself a Ghost Dance religion. Messianic religions have arisen all over the world where small tribal groups face larger, technologically or militarily superior groups. Similarities, then, may result from similar human responses to similar human situations, and not from borrowing.

After Manitoba became a province, many Métis left Red River for the Saskatchewan River country. When Canadian settlers also moved north and west, the Métis were once again embroiled in land claims, and recalled Riel, who by then was in Montana, to negotiate for them. This time, however, the conflict ended in 1885 in bloodshed, with battles between the Métis and a few Indians on one side and the North West Mounted Police, regular troops, and volunteer soldiers from Ontario and Quebec on the other. The troops were able to take the Métis village of Batoche, and Riel was tried and convicted of treason (although he was an American citizen) and hanged in November 1885. Riel had tried to foment a general Indian uprising, but most Indians stayed true to their treaties and kept the peace. Nonetheless, some young men who had used the unrest to settle old scores with individual whites and some leaders whose men had been involved in hostilities were hanged or imprisoned.

The 1890 Ghost Dance began in Nevada with the Paiute prophet Wovoka and soon spread widely throughout the US West. Wovoka told followers that if they danced a certain round dance, they would be able to visit dead relatives and the present world would pass away, to be succeeded by the world of old-time Indians and plentiful game. The Lakota people of North and South Dakota were among the most avid Ghost Dancers, but the prophetic movement turned to tragedy there. Like the Métis, the Lakotas had serious land-rights concerns with the federal government, and the combination of political agitation and a messianic dance movement frightened Indian agents into provoking violence. When Indian police came to arrest Standing Rock leader Sitting Bull (of Little Bighorn fame) in December of 1890, a scuffle ensued in which Sitting Bull, some of his family and followers, and several Indian police were shot and killed. Frightened dancers from Sitting Bull's and other bands were pursued by the army into the South Dakota badlands. When they came in to surrender to the troops at Wounded Knee Creek on Pine Ridge agency, another scuffle ensued between the troops and some men reluctant to surrender their guns. The troops, who surrounded the Indians, fired and continued to fire as the Lakotas fled. The bloodshed ended the widespread following of the Ghost Dance religion.[2]

North America has certainly had its share of revivalistic religious movements, probably far more than have been recorded. The first of which we have a record seems to have been what resulted in the League of the Iroquois some time in the sixteenth century. According to Alice Kehoe, "A saintly stranger, Dekanawidah, came among the Iroquois fervently seeking to create peace among their constantly warring communities."[3] Teamed with Hiawatha and later the war chief Thadodaho, Dekanawidah established a strong league of peace that was, nevertheless, fearsome to its enemies. After the French and Indian War and the American Revolution, when the Iroquois no longer held a commanding position among European or Euro–North American powers, another prophet arose among the Iroquois, Handsome Lake, who also preached peace and revitalization. He may well have been influenced by the Christian eighteenth-century Great Awakening and may in turn have influenced the nineteenth-century revivalism of the nearby "burned-over district" of New York State and its most famous prophetic movement, Mormonism. In the eighteenth century, Pontiac's Rebellion was to some extent a revitalization movement, while in the nineteenth century, the Shawnee Prophet Tenkswatawa and his brother, the war chief Tecumseh, led one of the most far-reaching and successful Native American revitalization movements.

Nineteenth-century white revivalism included the Shakers, whose worship ceremonies featured dance, and other more localized groups. Nor were false messiahs absent. Although William Dean Howells's novel *The Leatherwood God* is fiction, it is an astute psychological study of early nineteenth-century American messianic yearning.[4] In this case, the messiah is clearly depicted as self-deceived, but the focus is on the enormous desire of a small group of supposedly self-reliant men and women of the frontier to see themselves as being in the centre of the world, rather than in the middle of nowhere. Although these settlers are the dispossessors rather than the dispossessed, their hunger for meaning and stability are a gauge for the needs that would be expressed in the messianism at Batoche and Wounded Knee.

One could, of course, catalogue many more revivalistic movements among both Native and white peoples during the eighteenth and nineteenth centuries in North America, but let us move on to look at the specific intellectual histories of Louis Riel and the Lakota Ghost Dancers. The young Louis was raised as a Catholic and was particularly close to his mother and to his sister Sara, each in her own way an exceptionally pious woman. Sara would become a missionary nun, and Louis was picked as one of four promising boys from Red River to be sent to Montreal to study for the priesthood. Although he would leave the seminary before attaining holy orders, he was thoroughly schooled in the ultramontane style of Montreal's Bishop Ignace Bourget. As George Stanley points out, Bourget's particular brand of nationalistic ultramontanism is crucial to understanding Riel's later interpretation of what he took to be his own sacred mission. Ultramontanism was simply an authoritarian form of Catholicism that required its practitioners to settle doctrinal questions by going "across the mountain" to the Pope. Bourget linked his ultramontanism to French-Canadian nationalism, supporting the *Patriotes* who took to arms to demand responsible government in 1837–38. The connection of language, faith, and armed rebellion would remain vivid for Riel, while he dreamed of replacing the Pope with a Pope of the New World—none other than Bishop Bourget. Riel kept up a correspondence with the bishop even after leaving Montreal, and one of Bourget's fairly commonplace letters of encouragement to Riel became for the younger man a written guarantee that he was truly inspired by God as His prophet of the New World. Bourget's strong distaste for Durham's *Report*, written in response to the rebellions of 1837–38, became part of the intellectual underpinning of contemporary Quebec separatism, and the general tenor of that argument, with its emphasis on *pur laine* Quebeckers, is itself a kind of revitalization movement, albeit without the messianism. Although Bourget never accepted any of Riel's prophecies—and died on 8 June 1885, as Riel was awaiting his trial, without comment on Batoche— he clearly influenced Riel's ideas of what would be necessary for his New World prophecy and papacy. Since as Bishop of Montreal his residence was on Mont Saint Joseph and he organized a confraternity for the perpetual devotion to St. Joseph, he probably also influenced Riel in his devotion to the earthly father of Jesus, whom Riel would successfully petition to have installed as the patron saint of the Métis.[5]

The Ghost Dance also has a distinct and reasonably clear immediate intellectual heritage. James Mooney believed that Wovoka, the prophet/ messiah of the 1890 Ghost Dance, had been influenced by Smohallah and the Northwest Pacific Coast Shakers. He had definitely been influenced both by traditional Paiute ceremonies and by the 1870 Paiute Ghost Dance led by Wodziwob (an appellation that may be a title rather than a personal name). Wodziwob prophesied that if the Paiutes danced a variation of their traditional round dance, their beloved dead would return from the grave. This religion was fairly short-lived and Wodziwob seems to have given up on it, but it was taken up by California Indians who had recently suffered horrifying persecution and loss of life and, according to Russell Thornton, continues in the form of Bole Maru. By the late 1880s, Wovoka—or Jack Wilson, a younger Paiute man whose father had apparently been one of Wodziwob's associates—began to prophesy and, as figured by Mooney, on New Year's Day, 1889, during a total eclipse of the sun, he fell into a trance. After his return to consciousness, he reported a visit to a green land where the dead lived again and hunted and gathered plenteous game, nuts, and other traditional foods. Wovoka called for Paiutes to live peacefully and in harmony with their white neighbours, but he also called upon them to dance a version of the round dance that would allow them to visit their beloved dead in trances and that would eventually bring the green land of the spirits, with all the game and plants that he had seen, to replace the everyday world of white settlers and mines.[6]

If Bishop Bourget was Riel's spiritual teacher, Wovoka was the far more precise and proximate source of the Lakota Ghost Dance. His message spread rapidly to the south, east, north, and west, and interested Indians took the train to Mason Valley, Nevada, to meet this prophet or messiah and to bring his dance and message home to their own kin. While Riel certainly distorted Bourget's teachings far beyond anything the Montrealer would have recognized, the various messengers to Wovoka and home again were dealing with more syncretic traditions that allowed each group to compose its own Ghost Dance songs, develop distinctive forms of the round dance, and use individual symbols, but still have reference to Wovoka's teachings. Several Lakota men, including Short Bull, Kicking Bear, and Good Thunder, were among the delegates to Wovoka, and they returned as apostles of the Ghost Dance, instructing fellow Lakotas on most of the reservations in the songs, movements, and regalia of the dance. The Lakota form of the Ghost Dance utilized a centre pole reminiscent of the Sun Dance, which had been banned less than a decade earlier.[7] Because most Ghost Dance songs were either given by the spirits to the dancers while they were in the trance or were composed by the dancers to describe what they had seen in the spirit world, the Lakotas soon developed their own repertoire of music. A distinctive aspect of Lakota regalia was that the Ghost Dance shirts were thought to be bullet-proof. Although Indian agents, missionaries, and journalists often took this to mean that the Lakotas intended to attack whites in order to hasten the return of the old world of the buffalo and the beloved dead, the Lakotas' justified mistrust of the soldiers and even their own Indian police probably caused their interest in protection against guns.

Both the Riel uprisings and the various forms of the Ghost Dance, like other messianic movements worldwide, were responses to social, political, and economic forces, as well as to religious inspiration. Russell Thornton points out that the 1890 Ghost Dance was adopted most often by groups who were experiencing marked cultural change and unusually rapid population loss—as was the case among the Lakotas, who were suffering from epidemics of measles and other diseases and had not even had time to adapt to reservation life before their land base was once again halved by allotment and the sale of "surplus" lands. Like the Crees and Assiniboines during the starving winter of 1883–84 immediately preceding the return of Louis Riel to Canada, in 1889–90 the Lakotas found their promised rations slashed by a distant government that seemed willfully ignorant of both the treaties they had signed and of actual conditions on the Great Plains. Both the Lakotas and the Métis feared, with good reason, that their population would be swallowed up by immigrants from the East. In addition, the Saskatchewan Métis as well as the English-speaking settlers in Saskatchewan in 1884–85 were increasingly exasperated by Ottawa's failure to respond to their request for secure land titles, relief from crop failure, and the other issues they reasonably believed their government should address.[8]

In both Saskatchewan and the Dakotas, the connections among the demographic and socio-political factors, the religious revitalization movement, and armed conflict with government forces was not a simple case of cause and effect. Nor were there any clear lines of connection between the Métis and the Ghost Dancers. The Métis did not attack the police because of Riel's prophecies, nor did the army attack the Ghost Dancers out of pure religious intolerance. Louis Riel and Gabriel Dumont hoped that by commanding the police to retreat, they would get Ottawa to recognize their provisional government and to enter into negotiations about the land claims. Riel's vocation as a prophet became important only when the Exovedate, his provisional government at Batoche, recognized the Duck Lake victory as a sign that God favoured their cause. Despite US government sanctions against Native religious practices, the army did not attack any Ghost Dancers except the Lakotas. Not everyone who suffered extreme demographic and cultural deprivation joined a revivalistic movement. Although the Ghost Dance was adopted by Dakota groups in Saskatchewan and survived there until the 1960s, neither the Plains Crees and Assiniboines, who arguably suffered the worst of anyone during the 1883–84 winter, nor the various bands of the Blackfoot Confederacy ever adopted the Ghost Dance. Except for a relatively few individuals, neither did the Indians join Riel's call to arms, and most of those who did commit violent acts were inspired by the rumours of unrest to settle personal scores. As Stonechild and Waiser have shown, the Canadian Plains peoples maintained allegiance to their treaties, and even the non-treaty Dakotas avoided bloc support for the Métis. Since the Gros Ventres, near neighbours of the Blackfeet, were Ghost Dancers, like their Arapaho kinsmen whom Mooney studied, the Blackfoot would certainly have had a chance to hear about the Ghost Dance. The Crees might have learned about it from the Saskatchewan Dakotas, though Kehoe's chronology indicates that the Dakotas might have become converts so shortly before the Wounded Knee massacre that they would not have had time to pass it on before the tragedy abruptly stopped the quick adaptation of the ceremony. Sitting Bull and Riel may have known each other shortly before Sitting Bull's surrender and return to the United States from Canada.[9] Though the two exiled political leaders might have seemed to have had much in common, it is unlikely that Sitting Bull would have appreciated Riel's Catholicism or that Riel would have deigned to notice Sitting Bull's prophetic and religious traditions. Although Riel desperately wanted Indigenous allies, particularly at Batoche, he valued Indians as constituents of the Métis "race," not as separate cultures with their own traditions and aspirations. Hanged in 1885, Riel did not witness the Ghost Dance, but there is no reason to believe that he would have appreciated it or that he would have seen Wovoka's or other Ghost Dancers' visions as parallel to his own. As much as any of the missionaries, he hoped to see all the Indians become Catholics.

Even within each movement, there was not a clear relationship between secular and sacred quests. Logically, of course, there was little point in trying to protect "surplus" land if one expected to see a new green earth unrolled over all the land, not just the reservations. Unlike Euro–North American millenarians, however, who sold their possessions and awaited the end of the world, the Ghost Dancers and the Métis were pragmatic people who wanted their children to survive and prosper. They would pursue their goals by whatever means came to hand—sacred or secular. On the other hand, by 1885 and 1890, neither an insurrection nor calm, good-faith negotiations on land rights and treaty rights were of much use to Indigenous peoples in the Dakotas or Saskatchewan—or, for that matter, for Indigenous peoples in most of the world where room for colonizers was running out. What they needed was a miracle. Negotiation, armed resistance, and miracle would all play their parts.

The Saskatchewan Métis brought Riel back from Montana to lead a secular movement, to petition Ottawa to respect their rights and those of the white settlers. His original value to them was political, as it had been in 1869–70, when he had established a successful Provisional Government in Red River that negotiated Manitoba's entrance into Confederation as a province and secured land rights (though unfortunately not usable ones) for the Métis. Although Dumont and the others would have heard of Riel's mental breakdown and institutionalization in 1876 through his cousin Charles Nolin, if no one else, Riel had been largely out of contact even with his own immediate family for most of the period between 1870 and 1883. He had become a schoolteacher in Montana, taken out American citizenship, and entered into territorial politics. The Saskatchewan Métis were expecting a well-educated politician who was a pious and charismatic Catholic; they could not have been expecting a prophet of a New World Catholicism revitalized almost beyond recognition. Riel's work among the Métis of Montana was pragmatic and political, dealing with such topics as liquor sales and voting rights. It was this pragmatic leader whom the Métis summoned. According to J.M. Bumsted, the North West clergy opposed sending for Riel because "the good fathers were fearful of violence, but they also suspected Riel's prophetic tendencies, which were well-known among the western priests of the Church."[10]

For the Anglo-North American settlers, Riel's usefulness was also pragmatic, though it was perhaps conceived more cynically. "From the perspective of the European settlers, Louis Riel could serve either as a catalyst to shake up the dormant politicians of Ottawa, or as the sacrificial martyr/ leader of a failed rebellion that had made its point simply by existing. In either case, Riel was totally expendable."[11] Even while he was quietly working in Montana, however, Riel had been at work codifying his prophecies, à la *The Book of Mormon*, into a volume he called "the Massinahican, which in Cree means 'the book,' with particular reference to the Bible."[12] Riel would come back to its major precepts in his diary as he prepared for his death, but they seem to have played little part in his leadership in the North West until sometime after January 1885, when he received word from Ottawa that Macdonald's government was not in any hurry to act on the North West land question or Riel's own claims for recompense for his service to the Provisional Government of Red River or for his losses after he was forced to flee Red River.

Only in the spring of 1885 did Riel change his primary tactic from petitioning the government to forming a provisional government and calling for an armed rebellion in concert with any Indians he could persuade to join him. At this point, he also began to call publicly for a Catholic church that was separate from Rome—and from the missionary priests of the North West, if they did not accept his leadership—and for the creation of a new Métis federation that would welcome French and French-Canadian immigrants, settlers from all the Catholic countries of Europe, European Jews, and Scandinavians, all of whom would join in *métissage* with the Indigenous peoples of the North West. The most extreme elements of his messianic calling—renaming the days of the week, the sun and moon, the oceans, and so on—did not re-emerge until after his trial and death sentence.[13] Only after the first military engagement of the campaign, when the Métis under Gabriel Dumont had routed a company of North West Mounted Police and volunteers under the command of Superintendent Lief Crozier at Duck Lake on 26 March, did Riel assume the role of prophet. God, he believed, had sent a sign by delivering Crozier's men into Dumont's ambush. Riel prevented Dumont from following up his victory by annihilating Crozier's retreating column. With God on their side, he may have believed, the Métis did not need to send more souls to perdition. As Stanley asks, "Who would now challenge [Riel's] claim to be a prophet?" The Exovedate—the governing council plucked from "out of the flock" to be the provisional government of the North West Métis who followed Riel—resolved

> that the Canadian half-breed Exovedate acknowledges Louis David Riel as a prophet in the service of Jesus Christ and Son of God and only Redeemer of the world; a prophet at the feet of Mary Immaculate, under the visible and most consoling safeguard of St. Joseph, the beloved patron of the half-breeds—the patron of the universal Church; as a prophet, the humble imitator in many things of St. John the Baptist, the glorious patron of the French Canadians and the French Canadian half-breeds.[14]

But the priests saw him as mad.

Similarly, Sitting Bull, whom both Standing Rock agent James McLaughlin and the popular press erroneously portrayed as the main Ghost Dance leader among the Lakotas, was noted as a prophet because of, among other things, his accurate prediction of "white men [soldiers] on horse back descending to earth upon the Indian village"[15] before the Battle of the Little Bighorn. Yet, like Riel, he tried politics and diplomacy before prophecy or messianism. Like Riel, he was involved in a battle over land rights—the Métis to have their land claims acknowledged by the government, the Standing Rock Lakotas to avoid being allotted, allotment meaning that each family would have to select the equivalent of a homestead, with the "surplus" land being offered to non-Native newcomers. In 1888, the US government had sent Richard Henry Pratt, the founder of the Carlisle

Indian School, to gain the signatures of three-quarters of the adult men of the tribe, required by the treaty if the Standing Rock Lakotas were to sell any land. Sitting Bull successfully organized the people so that the requisite signatures could not be obtained. Sitting Bull was then part of a delegation that went to Washington at the behest of the government to negotiate another settlement by which the people would lose their land. Sitting Bull agreed to the new terms, trusting the Lakotas to once again withhold the necessary signatures. But in 1889, the Secretary of the Interior sent General George Crook to collect signatures. Agent McLaughlin and the Catholic missionaries pressed men to sign and threatened to cut off annuities and all future payments if they did not. McLaughlin eventually used the Indian police to line up men and move them by a table where each was required to register his X. The measure passed.[16] As with the Métis, patient diplomacy with the federal government had failed. As Riel had taken up his cross, Sitting Bull encouraged the Ghost Dance.

In Saskatchewan in 1885 and in the Dakotas in 1890, there was strong opposition to both the messianic movement and the taking up of arms from within and without the messianic community. The Catholic clergy firmly repudiated Riel, his messianism, and his call to arms. Other Métis settlements do not seem to have supported the uprising, and it is unclear how many of the people even of Batoche and St. Laurent fully supported the Exovedate. Although the white settlers had at first supported Riel as someone who could help them with Ottawa, both they and the English-speaking "half-breeds" or countryborn maintained their neutrality and repudiated any connection with Riel after the violence at Duck Lake. Like the farmers and merchants of northwest Nebraska, the white farmers and merchants of the North West panicked at what they convinced themselves was about to become a large-scale "Indian outbreak"—though some encouraged rumours of war, hoping to make a good profit from supplying the military who would be called in. As Stonechild and Waiser have shown, the "Indian leaders had their own agenda for addressing their grievances and were pinning their hopes on a large intertribal council to be held at Duck Lake that summer [1885]." Although individual hotheads favoured war, there was never widespread Native support for Riel. All the large-scale Indian hostilities in the North West were primarily functions of the panicked apprehensions of the Euro-Canadians. According to Stonechild and Waiser, Saskatchewan Indian leaders steadfastly refused aid to the Métis, for the most part simply by moving very slowly despite Riel's increasingly importunate cries for assistance. The "Siege" of Battleford was the Indian Agent's fearful refusal to meet with Poundmaker's people, despite their advising him of their intentions and following their usual pattern of approaching the town. The "Battle" of Cut Knife Hill was Colonel Otter's attack on a camp of families on their own reserve.[17] Even the Euro-Canadian praise of Crowfoot for keeping the Blackfoot Confederacy at peace and in Alberta was misplaced—the Blackfoot had nothing to gain by joining Riel and were far too astute to join a lost cause.

Similarly, among the Lakotas, according to Utley, support for the Ghost Dance varied from less than 10 percent to about 40 percent on the different reservations. Even on Pine Ridge, the reservation with the highest proportion of dancers, where the Lakotas called the inexperienced agent Young-Man-Afraid-of-Indians, where the actual Wounded Knee massacre would take place, fewer than half of the people were dancing, and observers like Santee physician Charles Eastman believed that their intentions were peaceful.[18] The Indian police generally opposed the Ghost Dance and definitely opposed all violence. According to Eastman, when Indian police attempted to arrest an Oglala man accused of cattle theft, Ghost Dancers surrounded police and prisoner, and threatened to burn the agency and take control. American Horse, a "progressive" leader, defused the situation by addressing the crowd:

> Stop! Think! What are you going to do? Kill these men of our own race? Then what? Kill all these helpless white men, women and children? And what then? What will these brave words, brave deeds lead to in the end? How long can you hold out? Your country is surrounded with a network of railroads; thousands of white soldiers will be here within three days. What ammunition have you? What provisions? What will become of your families? Think, think, my brothers! This is a child's madness.[19]

American Horse, like Poundmaker, Piapot, Big Bear, and other leaders who counselled patience, had an irrefutable point. A call to arms was simply doomed—at least without divine intervention. Even Riel, who pictured part of the divine intervention coming through the combined might of the Métis and all the Indians of the North West, was hoping force would lead to negotiations, not to victory. Thus, he held back Dumont's men at Duck Lake. The US Army did shoot down the Ghost Dancers at Wounded Knee, even if most of them were unarmed women and children. The desultory guerrilla campaign mounted by a few young Lakota men, including Black Elk, did not have the ammunition and supplies to last for more than a fortnight. While admirers of Gabriel Dumont, such as George Woodcock, have claimed that, were it not for Riel's messianic pacifism, the war chief of the Métis could have forced concessions from Ottawa by mounting a guerrilla campaign, that is doubtful. As Manfred Mossman writes, "Although guerrilla tactics often brought initial and impressive victories for the rebels [in messianic movements], they merely helped prolong the movements for a limited period of time and raised the overall number of casualties."[20] True, the Canadian government could not have afforded a protracted campaign, but nor could the malnourished Crees, even had they chosen to join Riel and Dumont, and the sedentary Métis villages would have been easy targets. In addition, the United States would never have allowed a successful guerrilla movement to operate anywhere near its borders. The Gatling gun and gunner that the United States ever so kindly lent to the Canadians to emplace on the steamer *Northcote* were hints of things to come. Arguably, a guerrilla campaign might have led to the US annexation of the North West, but it is unclear whether that would actually have helped the Métis. The United States would not have been likely to recognize Métis riverlot surveys, and the Métis and Indians who did flee to the States did not improve their fortunes.

If the Métis of the North West had stuck to petitions and messianism, they would probably have been left alone. They would also probably have lost their land. Although Thomas Flanagan argues that "it was a story of missed opportunities for reconciliation rather than rebellion provoked by unrelenting oppression," Gabriel Dumont might not have agreed. Certainly Riel's messages asking Indian communities to join him and his messages to the North West Mounted Police at Fort Carleton asking for surrender led most Canadian officials to fear an insurrection. Even if, as Don McLean has suggested, the 1885 "Rebellion" was deliberately fomented by Lawrence Clarke for the sake of the CPR and by Prince Albert merchants hoping to profit from a small Métis and Indian war, some threat of violence was necessary to set the troops and volunteers on their way west from Ontario. Duck Lake was very real violence, but according to Stanley, it was Lawrence Clarke and the Prince Albert volunteers who dared Crozier to return to Duck Lake, after a smaller party he had sent to secure stores had met the Métis and returned unharmed.[21] That the police chose to march again into Métis territory, where they were vanquished, confirmed the Exovedate's belief that Riel was a true prophet, just as the Little Bighorn battle had confirmed Sitting Bull's prophecy, but the police action was not the result of prophecy.

Future Prime Minister Wilfrid Laurier, speaking four months after Riel's execution, gave what may still be the most accurate apportionment of responsibility for the North West tragedy:

> Rebellion is always an evil, it is always an offence against the positive law of a nation; it is not always a moral crime.
>
> . . .
>
> What is hateful is not rebellion but the despotism which induces that rebellion; what is hateful are not rebels but the men who, having the enjoyment of power, do not discharge the duties of power; they are the men who, having the power to redress wrongs, refuse to listen to the petitions that are sent them; they are the men who, when they are asked for a loaf, give a stone.[22]

The same was probably true for the Lakotas. After all, though the United States had suppressed most Indigenous religious ceremonies and organizations and occasionally broke up Ghost Dances in other communities, it never attempted to arrest Wovoka, unlike Sitting Bull, nor did the army fire on large groups of Ghost Dancers except at Wounded Knee. Robert Utley's judgment is surprisingly similar to Laurier's:

> The dancers at Pine Ridge composed about forty per cent of the population, at Rosebud thirty per cent . . . These people were belligerent,

suspicious, and excited to the point of irrationality. They expected the white men to interfere with the dance . . . [and] it was only a question of time until another incident . . . ended in bloodshed. By the middle of November the lives of government employees at Pine Ridge, if not at Rosebud, were clearly in danger.

But the conditions that made troops necessary in November could almost certainly have been avoided if Congress had fulfilled its obligations to the Sioux earlier in the year.[23]

General Miles was even more blunt.

They signed away a valuable portion of their reservation, and it is now occupied by white people, for which they have received nothing. They understood that ample provision would be made for their support; instead, their supplies have been reduced and much of the time they have been living on half and two-thirds rations. Their crops, as well as the crops of white people, for two years have been almost a total failure. The disaffection is widespread, especially among the Sioux, while the Cheyennes have been on the verge of starvation and were forced to commit depredations to sustain life. These facts are beyond question, and the evidence is positive and sustained by thousands of witnesses.[24]

The greatest difference between Métis and Lakota messianism is in their doctrines and practices. Riel, as we have seen, developed his belief in his mission in the context of Bourget's ultramontane but specifically French-Canadian nationalism. Even after Riel had been expelled from Parliament and declared an outlaw, Bourget continued to see him, and in 1875, on Bastille Day, he wrote the thirty-year-old Riel a letter that became Riel's talisman, his sign of his divine mission. According to Stanley, "Riel never parted with this letter. He carried it with him every day, next to his heart, and he placed it at the head of his bed every night."[25] The text itself is relatively unremarkable, the lines that most moved Riel saying simply,

I have the deep conviction that you will receive in this world, and sooner than you think, the reward for all your mental sacrifices, a thousand times more crushing than the sacrifices of material and visible life. But God who has always led you and assisted you up to the present time, will not abandon you in the darkest hours of your life. For He has given you a mission which you must fulfil in all respects.[26]

What, exactly, was that God-given mission that Riel must "fulfil in all respects"? Presumably it was the "rejuvenat[ion] of French-Canadian culture, which [Riel's ultramontane supporters and friends] hoped would take on a new vitality in the young and idealized society of the Great West."[27] Bourget wished his protege well in establishing a western society that was Catholic, French-speaking, and obedient to its priests in all matters, including politics—a replication of the society Bourget had laboured, fairly successfully, to build in Quebec. But Riel, in exile in Washington, DC, his plans on hold until he could be allowed back into Canada, saw something more profound, especially after 6 and 8 December 1875, when he experienced, first in the US Capitol and then in St. Patrick's Church, something akin to a vision that produced great extremes of joy and sorrow. His host, Edmond Mallet, was becoming increasingly concerned about his sense of mission. "I would tell him that God's providence worked through natural means, except in very exceptional cases," Mallet wrote.[28] After these visitations, Riel's behaviour became increasingly unusual. He was passed through a succession of friends and family, none of whom could accommodate his strangeness, until he was finally admitted to first one and then another insane asylum in Quebec.

In some ways, Riel's behaviour was no stranger than that of various North American messianic prophets from Handsome Lake to Wovoka. He went into an altered state, returned, and began to prophesy. But most Indigenous societies, unlike Mallet, accepted visions, prophecy, and individual revelations from spirits, animals, or the dead as "natural means" of religious revelation and had holy men skilled in working with the visionary to interpret the vision and to make it accessible to all people or to all members of the community. Flanagan suggests that Riel would never have been considered insane had he had his visions in Saskatchewan, but that may discount too easily the power of the clergy and even Riel's own devotion to authority. The kinds of miracles enshrined in the various grottoes of Manitoba and the North West celebrate cures, visitations of the Virgin, and other "normal" miracles of the Church—not prophets. Wovoka's vision, coinciding with an eclipse and apparently enabling him to both predict and control the weather, was accepted among the Paiutes and among the many tribes who sent envoys to the Messiah. While some Christian traditions accept prophesying, speaking in tongues, and other visionary experiences as part of their religion, ultramontane Catholicism, with its complete dependence on the duly constituted hierarchy of the church, was probably the least favourable venue for Riel's belief that divine revelation could come directly to him. Some scoffers considered Wovoka a fraud and his successes with the weather only convenient coincidences or downright hoaxes (when, for instance, he caused ice to either fall from the sky or float down a river in the middle of July), but for the most part, they did not consider him insane.[29] While some of his enemies considered (and consider) Riel a fraud, many of his friends deemed him insane, the defence his lawyers used unsuccessfully at his trial. For Riel, there was a definite conflict between his personal religious experience and the Catholic ultramontane tradition within which he had to understand, interpret, and act upon his vision and his mission. The difficulty of balancing his sense of mission with his utter poverty and his banishment from Canada certainly left Riel emotionally vulnerable. Given his extremely pious nature and devotion to Bishop Bourget, it is not surprising that "insanity" would be the only way for him to balance the teachings of his faith and his powerful experiences of what William James, in the Swedenborgian tradition, called "vastation."

Wovoka seems not to have experienced any such conflict. He had a vision. It coincided with an eclipse of the sun, which increased its power in Paiute society. Although the Mason Valley Paiutes were reasonably well off in comparison to the Crees or Assiniboines or Lakotas or even Métis, their way of life was suffering from externally imposed change, including the cutting down of a major food source, the pinyon trees, for fuel and mine supports. Wovoka's vision of a green world must have been welcome to desert dwellers who were also suffering drought, and Hittman suggests that since he was a wood chopper (Wovoka translates to "cutter"), he may have been making amends for this destruction. The 1870 Paiute Ghost Dance followed epidemics of typhoid, measles, and other diseases, which made dancing to bring back the beloved dead extremely appealing. Although Wodziwob, the 1870 prophet, may have become dismayed and stopped dancing when no dead returned, the idea remained.[30] So when Wovoka told the people, once again, to modify the traditional Paiute round dance into a Ghost Dance to bring back their beloved dead, who had continued to die, it was not an outlandish idea—though it was out of the ordinary. Nor did the Ghost Dance seem out of the question to other tribes, particularly to people like the Lakotas, who were losing their families to malnutrition, overcrowding, and the attendant ills. Had French-speaking Catholics from across North America flocked to Riel to learn how they could participate in fulfilling his mission (as they do, for instance, in the annual pilgrimages to Lac Ste. Anne in Alberta), it is likely that he would have been able to bear the disjunction between experience and belief, and thus that he would not have been institutionalized.

Wovoka's message was also relatively simple. If Indians performed the Ghost Dance faithfully, a New Heavens and New Earth would in some way roll out. White people would disappear, and game and old-time Indians would reappear in a green world of plenty. No violence or threat would accompany this change—it would be surprisingly like *Looking Backward*, the 1888 bestseller by Edward Bellamy, in that change would appear almost organically. Peace, hard work, tolerance, and honesty—values common to most societies and religions, at least in principle—were the Ghost Dance virtues. Nor was hostility to the whites part of the doctrine. Wovoka had white friends and business partners. Though he was wary of whites who disparaged the Ghost Dance, especially, with reason, after the death of Sitting Bull and the Wounded Knee massacre, he had no objection to their studying or even joining the Ghost Dance. He just did not think—correctly—that many would. Anthropologist James Mooney, whose massive Bureau of Ethnology tome *The Ghost-Dance Religion and the Sioux Outbreak of 1890* is still the best text on the movement, had very little trouble persuading Ghost Dancers to let him observe and photograph their ceremonies. Sometimes he was even invited to participate. They translated Ghost Dance songs for him and taught him the tunes. Wovoka willingly granted him a long interview, after he introduced himself as a friend of the Arapaho Ghost Dancers, and gave him paint and other sacred objects to take back to his Arapaho friends. Even at Pine Ridge, where suspicion of whites ran high, Mooney was excluded from discussion not because he was white but because he had failed to believe. As he describes the conversation,

> On one occasion, while endeavoring to break the ice with one of the
> initiates of the dance, I told him how willingly the Arapaho had given
> me information and even invited me to join in the dance. "Then," said
> he, "don't you find that the religion of the Ghost dance is better than
> the religion of the churches?" I could not well say yes, and hesitated
> for a moment to frame an answer. He noticed it at once and said very
> deliberately, "Well, then, if you have not learned that you have not learned
> anything about it," and refused to continue the conversation.[31]

That the fearsome Sioux, who had wiped out Custer, had adopted a new dance and that part of its teachings included the belief that in response to devout dancing the Wanekiah (the Messiah) would send a whirlwind to blow away the newcomers and return both the buffalo and the Lakotas' beloved dead implied to most Euro-American observers that the Lakotas themselves might provide the "whirlwind" in the form of an insurrection. Certainly, not all Lakotas saw the Ghost Dance as purely spiritual. The young men who saw neither a future of achievements ahead of them nor a glorious personal past to cherish and remember were not averse to taking up arms. The main emphasis for the Lakotas, however, does seem to have been primarily spiritual. Lakota language accounts of the ceremony, collected by Catholic priest and Lakota linguist Eugene Buechel, all repeat the experience of dancing, falling into a trance, and seeing beloved kin, especially parents, and the Wanekiah. Ghost Dance songs, although individually composed or given to dancers in a trance, were taught to the entire group, and some that particularly conveyed the ideas of the group or had a catchy tune were frequently sung and taught to other groups. Even among the supposedly belligerent Lakotas, the songs are not vengeful or threatening and do not mention white people at all. Usually they record some action or saying of the beloved dead or a promise by the "father," the Messiah. One song that might seem to start ominously—"Give me my knife, . . ."—is sung by a ghost grandmother as she prepares to butcher a buffalo and dry strips of flesh to make *wasna*, pounded dried meat.[32] The bullet-proof nature of the Ghost Dance shirts that many of the Lakota men wore seems to be more defensive than offensive. The United States had outlawed the Sun Dance and other ceremonies, and the agents had repeatedly forbidden the Ghost Dance apostles to organize on the Lakota reservations. The dancers wanted to be prepared in case troops sent to break up the dance fired upon them. Unfortunately, the Ghost Dance shirts proved insufficient.

Riel's message is harder to discern, not least because he was trying to formulate it as his personal world was disintegrating. At the heart of his mission was the formation of a French, Catholic, Métis nation in the West—though he also hoped to include Jews and Scandinavians. Despite his desperate need for Indian allies and the many messages he sent out pleading with various bands to join him (plus the prayers that asked God to send him help from all the Indians of the North West and Montana),[33] Riel never seems to have realized that the Indians had their own agenda and had no reason to see themselves as any more connected to the Métis, who had been their rivals in hunting the very last herds of the buffalo, than to the white settlers. During his stay in Montana he had written "Memoir on the Indian Question," in which he wrote, apparently referring to himself,

> It is perhaps the one who, having enough white blood in his veins, honesty, experience, intelligence enough, would deserve and enjoy the confidence of a good majority of the american people: and who, at the same time, having some indians amongst his ancestors, would be allowed by public opinion, to say so and to have it known amongst the indians, as means of getting their confidence; and who, using his influence over them, would show them how to earn their living and would put them to work by all means.[34]

The "indians" had an abundance of Euro–North Americans only too eager to "show them how to earn their living" and to "put them to work by all means." They did not need Riel. A future of Catholic *métissage*, fused with different nations of European immigrants and settled into farming, was not what anyone was looking for, though the Crees and Blackfoot had insisted on having farming instruction and supplies written into their treaties, and the Dakotas, with fifty years of experience, were among the most successful farmers in Manitoba and the North West, given their lack of access to sufficient land. Riel believed that North American Indians were all descended from shipwrecked Jews (South American Indians were the descendants of the Egyptian masters on the same ship), and he seems to have had no knowledge of Indigenous religion in general, let alone the prophetic movements, such as the 1870 Ghost Dance, that presaged the Great Plains Ghost Dance of 1890. While the Lakotas would dance to bring about the return of the buffalo to the Great Plains, Riel prayed to his God to "keep wolves, bears, bison and other wild animals away from us."[35]

Unlike Wovoka, whose ability to direct the weather was acknowledged widely and who acquired disciples throughout the North American West, Riel, like the Lakota Ghost Dancers, found significant opposition from those near him, and his prayers, like the Ghost Dance shirts, failed at holding back the enemy. Bishop Bourget, Riel's intended Pope of the New World, died at eighty-six, before he could confirm—or more likely condemn—Riel's actions at Batoche. Riel attempted to recant his "heresies" after his trial, but as his sentence was appealed and affirmed and his hanging date set, postponed, and set again, he turned more and more to establishing a symbolic New Heaven and New Earth. The world he thought he had glimpsed briefly in the spring of 1885 might be beyond his physical grasp, but symbolic renaming and prayer could still, he hoped, bring it about. One of Riel's significant triumphs in Saskatchewan was in proposing St. Joseph and seeing him named as patron saint of the Métis, thus distinguishing them from French Canadians of European descent who claimed St. Jean Batiste. On 5 September 1884, Bishop Vital Grandin, visiting St. Laurent, named St. Joseph as the patron of a national Métis association and his feast day "a national holiday for people of Indian extraction." They could inaugurate the new association on 24 September, three months after the fête of St. Jean Batiste. The actual feast of St. Joseph the following year, 19 March, would end the novena Riel had proclaimed at the request of his cousin, Charles Nolin, and foreshadow the hostilities. Renaming, Mossman points out, is characteristic of utopian movements. As he awaited his final date with the hangman, the possibility of founding a Catholic mixed-blood confederation out of reach unless God intervened directly, Riel tried desperately to create a symbolic new future by renaming the heavens and the earth—the sun would become Jéan, with an accent, the Atlantic Ocean, Ṣaul-Paul, and the other heavenly bodies, continents, oceans, and so on would also bear new names.[36]

Riel's Catholic church of the New World may have been a heresy, but it was a recognizable heresy, produced by a particular kind of Catholic, though Flanagan has also proposed extreme Protestant roots in the United States. No one outside of Riel's community was likely to have envisioned a new papacy to be established first in Montreal and finally on the Red River at St. Vital. The belief that Indians were descendants of lost Jews was certainly not unique to Riel, but the idea that the Métis became both the new chosen people and inheritors of one-seventh of the land of the North West by virtue of their Indian blood was distinctive. (Riel would also claim that the Métis had rights to the land because God never created a people without providing a homeland for them.) The formulae of Riel's prayers were not only specifically Catholic but referred to various shrines and holy orders associated with his friends and family members. Even Riel's interest in European settlers seems to have derived specifically from his association with a Catholic colonization operation in Minnesota.[37]

Beyond the probable influence of Wodziwob and the earlier Paiute Ghost Dance and the traditional Paiute Round Dance, the influences on Wovoka's messianism are harder to trace. Mooney suggested that the Indian Shakers of the Pacific Northwest may have taught Wovoka some of his techniques. Various commentators have suggested that Mormon doctrine and ceremony may have influenced the Ghost Dance and that the Protestant pietism Jack Wilson learned from his white foster family may have informed the specifically Christian aspects of his teachings. As far as I know, novelist Leslie Silko was the first to tie the Ghost Dance to the apocryphal gnostic writings of first- and second-century Christians that were kept out of the canonized New Testament and only rediscovered in 1945, when an earthenware vessel buried for sixteen hundred years was excavated near Nag Hammadi, Egypt.[38] Obviously, Wovoka could have had no access to this esoteric material or even to other gnostic texts excavated in the 1890s. What I suggest, however, is that the similarities indicate that Christianity itself is a variant of many messianic religions that have sprung up in the intersection of a particular kind of cultural conflict and the rise of a prophet. The suppression of the gnostic tradition and its exclusion from

Christian sacred narrative also help explain why both the Ghost Dance and Riel's prophetic visions were seen as insane, heretical, or potentially murderous by nineteenth-century North American Christians.

Scholars seem to agree that Gnosticism predates and was an important part of early Christianity. Our understanding of Gnosticism has deepened since the discovery and translation of the Nag Hammadi texts; before that, it was primarily known through the writings of its opponents. Birger Pearson lists ten characteristics of Gnosticism, of which the first five are most significant. I quote these at length because they offer, I believe, striking parallels to the Ghost Dance. He observes

> first, that adherents of Gnosticism regard *gnosis* (rather than faith, observance of law, etc.) as requisite to salvation. . . . Gnosticism also has, second, a characteristic *theology* according to which there is a transcendent supreme God beyond the god or powers responsible for the world in which we live. Third, a negative, radically dualist stance vis-à-vis the cosmos involves a *cosmology*, according to which the cosmos itself, having been created by an inferior and ignorant power, is a dark prison in which human souls are held captive. Interwoven with its theology and its cosmology is, fourth, an *anthropology*, according to which the essential human being is constituted by his/her inner self, a divine spark that originated in the transcendent divine world and, by means of gnosis, can be released from the cosmic prison and can return to its heavenly origin. . . . The notion of release from the cosmic prison entails, fifth, an *eschatology*, which applies not only to the salvation of the individual but to the salvation of all the elect, and according to which the material cosmos itself will come to its fated end.[39]

The Ghost Dancers universally believed that only those who danced would survive the flood, earthquake, landslide, or other catastrophe that would bring back the green world of the beloved dead. The magpie and crow feathers that Wovoka gave or sent to his followers and the crow and eagle feathers that many Ghost Dancers affixed to their dance regalia were intended to lift up the wearer above the catastrophe and into the restored green world. Although it is impossible to generalize about Native

American sacred narrative, it is striking that many creation tales include emergences from successive worlds or the creation of *this* world from a pre-existing and continuous earlier world. Sometimes the existence of this world, perched on a turtle's back or hung by cords from the sky, is seen as precarious. If James Walker is correct, during the emergence of the Ghost Dance among the Lakotas, the frame of reference for many people who had been exposed to the constant pressure of agents, schools, and proselytizing Christians was changing from a complex system based on Iyan (Rock) and Skan (Sky) to one of a singular Wakan Tanka, who was more or less equivalent to Jehovah.[40] Iyan and Skan were not necessarily "transcendent and supreme," but they may have represented a theology of older gods related to the green world rather than the present one of religious persecution and material dispossession. Certainly the present of 1889–90 on the Lakota reservations—in the context of an engulfing Christian, individualistic society—was a "dark prison," and the human souls of the Lakotas (and other Ghost Dancers) yearned to break free.

The overpowering sense that the green world of the beloved dead was the real one and that the material world was a prison fated to pass away is absolutely central to the Ghost Dance and strongly parallels Gnosticism. The essential human being of the Ghost Dancers was the one who was freed in trances to visit the green world, and would be freed, through the dance, to return to the green world after the catastrophe. The Ghost Dance was to bring the material world of the white men to an end and to reunite the dancers with the pre-existing world of game and the beloved dead.

Not only were the beliefs of the Ghost Dancers similar to those of the Gnostics whose writings were discovered at Nag Hammadi, but so were some of their practices. The Round Dance of the Cross, in which the disciples hold hands and circle around Jesus, singing a hymn to the Father resembles the Ghost Dance and the many Lakota Ghost Dance songs that repeat the line "The Father says so, the Father says so."[41] According to Pagels, the Round Dance may have served as a sort of "second baptism" for its followers. Gnostic prophets, like the various tribal visitors to Wovoka, travelled around, exhorting others to fast and pray for visions and revelations. Gnostic believers prayed for "this world [to] pass away." Round dances and fasting for a vision are certainly common religious practices throughout the world (and perhaps other worlds, for that matter), but they are not usually seen as Christian. Osage scholar George Tinker suggests that Christians might look to Native American religions for a better understanding of early Christianity, before it was codified, Romanized, and, under Constantine, militarized.[42] My point is similar—instead of trying to find specifically Christian elements in the Ghost Dance, it might be useful to focus on universal (a word I usually eschew) human elements in both the Ghost Dance and Christianity—including those mostly purged from official Christianity.

This, of course, raises two issues—why do we find such striking parallels between first- and second-century Gnosticism and the Ghost Dance, and why was this dancing visionary tradition excised from canonical Christianity? Most contemporary people educated in Europe or North America conventionally see the Mediterranean world of the first and second centuries conveniently partitioned into Romans, Christians, Jews, and others, with the Christians being fed to the lions for the entertainment of the Romans. Pagels, however, suggests a far more fluid world that included a great variety of Christians, some of whom identified as Jews and some as Romans, and with many traditions of scripture and prophecy. It was definitely a dangerous world for followers of Jesus of all stripes, and the destruction of the Second Temple in Jerusalem may have made the material world feel as unsettled as the spiritual world. Gnosticism, according to Yamauchi, included non- or pre-Christian Jewish and Egyptian sources, among others, and so, of course, did Christianity in general. According to Pagels, the standardization of *a* Christianity began with Irenaeus, Bishop of Lyon, as a defensive movement to unify Christianity against its persecutors and to halt practices—such as women prophesying—that gave it a particularly bad name in the eyes of the Romans. The canonization continued under Athanasius and intensified when Constantine made Christianity the official religion of Rome.[43] Likewise, the Ghost Dance arose in a time of spiritual and material uncertainty, and was repressed both for material reasons and for the ways it differed from canonical Christianity—which was certainly the official religion forced upon American Indians, no matter what the First Amendment promised anyone else.

Captain H.L. Scott of the Seventh Cavalry, and others who interviewed the Ghost Dancers, recognized the similarities of the Ghost Dance to Christianity. Mooney includes a long section comparing it to everything from the "Biblical Period" to the "Adventists," including Islam.[44] Scott, investigating the Ghost Dance among the southern Arapaho, expected to find the leader a charlatan. Instead, he wrote, "he has given these people a better religion than they ever had before, taught them precepts which if faithfully carried out will bring them into better accord with their white neighbors, and has prepared the way for their final Christianization."[45] While one may cringe at the implication that to be Christian is the only way to be good, at least these observers could tell that feathers and round dances and songs were not always war dances and war paint and war whoops, and certainly not, as Standing Rock's staunchly Catholic agent McLaughlin described it, "absurd nonsense."[46] Would there have been a different reaction to the Ghost Dance by Christians raised on the Round Dance of the Cross? Perhaps.

Whereas the Ghost Dance resembled the gnostic traditions deliberately excised from the New Testament, Riel came directly out of the most rigid form of canonical Christianity in North America in the nineteenth century—ultramontane French-Canadian Catholicism. His sacred mission could not include prophecy, yet his experience did—a radical disjunctive that quite probably drove him "mad." If he has an analogue in early Christianity, it is with Irenaeus, who had to balance his own and his teacher's revelations with his belief that other prophecies, gospels, and interpretations were heresies that endangered true Christianity and its adherents.[47] Riel was right, from his point of view, in condemning the priests who opposed him in the North West—they did see his revivalism as a threat to their Christianity. Riel came at the wrong time for his religion. He probably would have been a superb follower for Irenaeus—but not for Bishop Bourget, for whom the questions Riel raised about the true church and chosen people had long been settled and clearly apportioned to Rome and the anointed priests. We miss the religious coherency of the Ghost Dance if we regard it as a hodgepodge dependent upon a learned Christianity for its symbols rather than as a basic human response to demoralizing change, drawing knowledge through many specific Native traditions. We also miss the coherency of Riel's theology by trying to see it, as he did, in terms of a canon that had no place for prophecy.

Partly because of their bloody immediate failures, we tend to see both the Northwest Resistance and the Lakota Ghost Dance as dramatic tragedies enacted as Anglo-Saxon Manifest Destiny rolled across the continent. The Ghost Dance shirts did not repel bullets. Sitting Bull was killed by the Indian police, some of whom perished with him. Approximately three hundred people, including women and children chased for miles by soldiers, were killed at Wounded Knee. The most often quoted lines from *Black Elk Speaks*—completely unlike, we should note, anything that Black Elk ever said—express this tragedy of unmitigated defeat: "And I can see that something else died there in the bloody mud, and was buried in the blizzard. A people's dream died there. It was a beautiful dream."[48] Similarly, Riel's dream is supposed to have died on the scaffold in 1885, after he was absolved from his sins by Father Alexis André, the missionary priest who had most strongly opposed his prophecy and his rebellion. Less often mentioned are the eight Indian men who were hanged a week later than Riel— the Crees Miserable Man, Bad Arrow, Round the Sky, Wandering Spirit, Iron Body, and Little Bear, and the Assiniboines Itka and Man Without Blood—and the leaders Poundmaker, Big Bear, and One Arrow, as well as the other men who died of their imprisonment at Stoney Mountain Penitentiary.[49]

Yet despite these appalling tragedies, they were not the end of the story. The big winners were clearly the national governments in both countries, and those who paid the highest price for the Ghost Dance and Northwest Resistance were the Lakotas, Crees, and Assiniboines. Prime Minister John A. Macdonald benefited so greatly from the Northwest Resistance that it is easy to be persuaded by the theories of McLean and Sprague that he deliberately fomented it.[50] It is no mere figure of speech to say that if Riel had not existed, Macdonald would have had to invent him, just as official Washington would have had to invent Custer. Certainly, Macdonald and the Indian Department and Interior Department bureaucracy laid the groundwork for anger among the Indians by refusing to honour the famine clauses of Treaty 6, especially during the starving winter of 1883–84; by cutting back on Indian rations generally; and by tolerating the misdeeds of individual agents, farm instructors, and civilians. Similarly, trouble with the Métis was fed by the slowness or downright refusal of response.[51] Macdonald was almost certainly right that a transcontinental railway was necessary to hold Canada together as a country—one can even make a case for the National Policy as a whole. But the cost—especially to the Indians, who were positively harmed by the railway—was high. The money saved on treaty obligations was poured into the Canadian Pacific Railway, and yet it was on the verge of bankruptcy by the time of Duck Lake. Only its use to speed Middleton's men west to quell an "insurrection" proved its indispensability and attracted the capital necessary to stave off bankruptcy and to finish the project. As Collingwood Schreiber, the government's engineer-in-chief, wrote to Charles Tupper in England, "The House and country are both in favour of the CPR and that should now be doubly the case when the fact is patent to the world [that] but for the rapid construction . . . Canada would have been involved in a frightful waste of blood and treasure quelling the rising in the North West."[52] Had the Canadian Pacific Railway gone bankrupt before its completion in the spring of 1885, Confederation itself might have failed. Saskatchewan settlement would have slowed down, and Canada as we know it would never have existed. Given the expansionist tendencies of the United States, however, it is unlikely that that would have provided any long-term gain for the Métis.

Another advantage to Macdonald from Riel's taking up of arms was the opportunity the "Rebellion" gave the Canadian government to include all the Indians of the North West as rebels and renegades, at least potentially. This provided Ottawa with an excuse to abrogate treaties, to continue to cut Indian Affairs budgets, and to fasten increasingly galling *Indian Act* restrictions—quite outside the treaties—on the Crees, especially.[53] The imprisonment and subsequent deaths of Poundmaker and Big Bear, important Cree leaders, completely doomed their plan of a confederacy and rights in keeping with the treaties. Even so, there was no people's dream trampled in the mud for the Crees, Assiniboines, and Dakotas. Their survival and persistence, particularly their ability to retain language and culture despite concerted efforts to assimilate individuals and break up families and communities, shows their remarkable resilience.

By the end of 1885, however, Macdonald must have been very satisfied. The CPR was both solvent and complete. The North West was at least superficially at peace. With British Columbia and the Prairies firmly tied by rail to Central Canada and no threat of an insurrection to tempt American troops to defend their own borders, the spectre of annexation that had haunted Macdonald since his young manhood in Kingston was finally dead. The government's abrogation of the famine clause of the treaties during the unbearable starving winter of 1883–84 was erased from the national consciousness—after all, the Indians who had starved were prospective murderers, and retrospectively deserved to die. Euro-Canadians had already developed an immunity to the spectre of starvation among Native peoples, reassuring themselves that Native lives were "brutish and short" and that it was "natural" for them to starve—an implicit part of the justification for the conquest and "civilization" of Indigenous peoples by imperial powers worldwide. Fur trade wife Letitia Hargraves had off-handedly remarked, nearly forty years earlier, that "no disaster has happened in the Northern Department this season . . . the Indians are starving in every direction and of course the dividends will be small."[54] Even in the twentieth century, Helen Anne English, the matron of a small Cree school run by her missionary husband, would calmly note in her diary after visiting a sick family, "Nothing very much the matter with any of them, just starving."[55]

If Macdonald won the jackpot and the Crees, Assiniboines, and Dakotas paid the price, what about the Métis? Only Riel was hanged. Most of the other Métis convicted of crimes against the government were imprisoned briefly and released. As Diane Payment has written, "Contrary to all studies on Batoche to date, which focus almost exclusively on the destructive impact of the 'North West Rebellion,' there was no final destruction nor dispersal of the local population, although commercial activities were interrupted and some were people [transposition *sic*] inevitably displaced in 1885."[56] The Métis continued to live successfully in the Batoche area for at least a generation after 1885, and their leaders played important roles in territorial politics. If we see Riel's messianism as a typical revitalization movement, it worked reasonably well for the Métis. St. Joseph and the ∞ flag are important symbols of Métis identity Canada-wide. The Métis are officially recognized as an Aboriginal people of Canada, along with First Nations and Inuit. In the United States, there are Indians with white grandpas and white folk with Indian grandmas, but there is no recognized and self-identified mixed-blood culture, in strong contradistinction to Canada.

Although the nature of the fur trade in Canada was different enough from that in the United States to account for part of this distinction, the nation-building ideals of Riel and the traditions and symbols resulting from the two movements identified with him are undoubtedly a strong source of Métis cultural persistence.

In the United States, one can discern a similar pattern of wins and losses—overall, the cavalry won and the Indians (at least the Lakotas) lost. But this is not entirely the case. Black Elk ended his "speaking" to John Neihardt not with the lugubrious plaint about the people's dream dying in the bloody mud of Wounded Knee but with a brief statement of the armed resistance he and the other young men mounted, and finally with the blunt and optimistic statement "Two years later I was married." Russell Thornton has argued that the Ghost Dance was "about" population decline. The 1890 Ghost Dance coincided with the nadir of North American Indigenous population. People wanted the dead to return because they were running out of the living. Although, as Alice Kehoe points out, the census figures upon which Thornton relies are hazy, Thornton argues plausibly that those groups who Ghost Danced enjoyed larger population increases than those who did not, and Kehoe herself has traced the value of the Ghost Dance for confirming cultural and group identity among the Dakotas in Canada. Despite the Wounded Knee massacre and the death of Sitting Bull, not to mention assimilation pressures, poverty, isolation, and a host of related social ills, the Lakotas have experienced population increase since 1890, and they even revived the Ghost Dance in conjunction with the 1973 occupation of Wounded Knee. For Ghost Dance tribes not involved in the massacre, the positive benefits were even higher. For the Pawnees, Kehoe states, the Ghost Dance allowed dancers to meet with medicine practition-ers who had died before passing on their songs and ceremonies. Thus, the Pawnees literally recovered their traditional liturgies and activities from the dead. Riel was also concerned with demographic revitalization, though it was to come from immigration and intermarriage rather than from the recovery of the existing population. The European settlers and métissage he dreamed of never happened, so to this extent, his movement did fail. Wovoka, meanwhile, grew to a comfortable old age, supporting himself mostly by selling Ghost Dance regalia to the faithful, usually by mail order.

He was recognized as a prophet by most of his Paiute community.[57] Both the Métis and the Ghost Dance societies received from their revitalization movements significant spiritual and demographic benefits with relatively minor material losses, except for the Lakotas—a result that is even more striking in comparison to the Crees, who had no revitalization movement and lost many of their best leaders.

If American Indians lost less in the suppression of the Ghost Dance than Canadian Indians lost in the suppression of the Northwest Resistance, the US government gained less than the Canadian government. If there was an individual winner, it was not the president, but Standing Rock agent James McLaughlin. McLaughlin had long been looking for an excuse to have Sitting Bull arrested and imprisoned, especially after Sitting Bull's fight against the sale of "surplus" lands. The Ghost Dance gave him an excuse, and in the event, the old chief was shot and killed by the Indian police. Despite the ensuing carnage and casualties among both the Indian police and Sitting Bull's followers, McLaughlin no longer had a rival for power on Standing Rock. Benjamin Harrison, an uninspiring president who, ironically, had gained a foothold in politics because he was the grandson of William Henry Harrison, "Old Tippecanoe," who had won the White House through his Indian-fighting prowess, had no particular need for an Indian war. Fourteen years before Wounded Knee, Custer's Last Stand had furnished Washington with an excuse to abrogate treaties and hunt down Indians. In many ways, the fact that Sitting Bull was back in the United States and on a reservation showed that Washington no longer considered him a person of interest. Custer's old regiment, the Seventh Cavalry, was the outfit that fired at Wounded Knee, gaining revenge for lost comrades (whom few of the 1890 soldiers had known), several Congressional Medals of Honor, and some rebukes for having killed so many women and children so far from the actual armed beginning of the action. The merchants and hay farmers of northwest Nebraska, like those of Prince Albert, Saskatchewan, benefited from the troops stationed in their vicinity, as was the case in virtually all Indian Wars in North America, but Gordon and Rushville and Prince Albert erupted in no major booms.

Both the 1885 Northwest Resistance and the 1890 Lakota Ghost Dance were messianic revitalization movements that arose from the combination of altered material circumstances and the availability of a prophet. Both attracted military intervention resulting in substantial loss of life. Paradoxically, the Northwest Resistance justified the completion of the CPR and solidified Canada as a nation. More darkly, it "justified" the suppression of the Crees, Assiniboines, and Dakotas, and even of the Blackfoot Confederacy, none of whose members had taken up arms. Riel's messianism, except for the naming of St. Joseph as patron saint of the Métis, vanished with him, but the two uprisings named for him did revitalize and cement Métis identity in Canada. And like the Holocaust for the Jews, Wounded Knee has become a symbol of "Never Again" for American Indians. Even for Ghost Dance participants like Black Elk, who eventually discarded the Ghost Dance for Catholicism and then subordinated Catholicism to return to his own Great Vision, the Ghost Dance served as a revitalizing force in circumstances that might otherwise have led to despair. Like Christianity, Riel's messianism and the Ghost Dance both gave courage to individuals to rebel against political and religious oppression and served as a powerful group identity for a living community.

Spiritual and Intellectual Resistance to Conquest, Part 3:   5
John Joseph Mathews' Wah'Kon-Tah *and John G. Neihardt's*
Black Elk Speaks

Although whitestream society in both Canada and the United States saw
Batoche and Wounded Knee as writing a firm "The End" to the story of
Indians in North America, Native nations were under no obligation to
share their point of view, and as we have seen, the material defeats were
in fact part of a renewal process. This chapter looks at the ways in which
two remarkable twentieth-century Indigenous American intellectuals,
John Joseph Mathews and Hehaka Sapa (Nicholas Black Elk, whom we
have already met as a participant in the Lakota Ghost Dance and a survivor of Wounded Knee), constructed accounts of Siouan religions that both
preserved beliefs for generations to come and introduced them, without
apology, to Amer-Europeans as land-based alternatives to Christianity and
other versions of whitestream religion and philosophy. *Black Elk Speaks*
is read as a kind of Lakota or even pan-Indian bible, is frequently taught
in high schools and universities, and has been reprinted many times, but
Wah'Kon-Tah, the book about Osage religion that was a Book-of-the-Month
Club selection in 1932, the same year *Black Elk Speaks* was published, is

now little read and is available in print only in an on-demand version from the University of Oklahoma Press. There are many reasons for the ongoing popularity of *Black Elk Speaks*, not the least of which is the extraordinary explication of Black Elk's Great Vision, which vivifies much of the symbolic language of nineteenth-century Lakota belief and ceremony. The dramatic tension between Neihardt's theory of the "fortunate fall" of the Lakotas to create America and Black Elk's pragmatic search for meaningful survival also gives great power to the text. *Wah'Kon-Tah* is a far more enigmatic book, and its author, John Joseph Mathews, is far more enigmatic than John G. Neihardt, but the book is just as fine in explicating a Siouan religion. While Neihardt recorded the story of the Lakota *wicasa wakan* Black Elk, Mathews novelized the journals of a Quaker agent to the Osages, Laban Miles. As unusual as was Neihardt's dedication to his epic of the West and to learning enough about the Lakota people to represent them as his Trojans against the American pioneers, he was not particularly unusual in his "Indian-struck" persona: many nineteenth- and twentieth-century Euro-American writers sought and tried to express an American Indian point of view, though not with such an important outcome.

Mathews was far more unusual—his nearest analogue is probably his fellow mixed-blood author, D'Arcy McNickle. Because Mathews and his work are so much less known than Black Elk's texts recorded with John G. Neihardt and later Joseph Epes Brown, I will write here in more detail of Mathews' teachings, and then summarize some of Black Elk's points for the purposes of comparison. An original Osage allottee, Mathews descended from an Osage great-grandmother and a Euro-American great-grandfather and their descendants, who were European-educated traders to the Osages. He grew up on the last Osage reservation, in Oklahoma, and spent most of the rest of his life in what had then become Osage County, Oklahoma. He lived among Osage ceremonies, went to school with Osage children, understood Osage, and spoke it at least passably. He visited many of the old Osages and wrote down their stories to safeguard them for the future, he would later found the Osage Tribal Museum and serve on the Tribal Council, but he did not write of himself as an Osage. A World War I flyer with a degree in geology from the University of Oklahoma and in humanities from Oxford (he had turned down a Rhodes scholarship because he could afford the freedom of paying his own way), he had lived and travelled in Europe and North Africa, and hobnobbed with the oil elite of Oklahoma, even writing a biography of one of them, but he did not account himself an Amer-European either. (He consistently used the term "Amer-European" to denote people of European descent who had failed to become naturalized to North America; he apparently found no use for a word that denoted any who *had* become naturalized.) In his memoir of ten years in nature, *Talking to the Moon*, he referred passionately to *my* blackjack country, but he never published a claim to *my* people of any sort. No tragic half-blood caught between two cultures, however, Mathews claimed, if not the right to speak *as*, certainly the right to speak *of* both Osages and Amer-Europeans. Thus, Laban Miles, a sympathetic but flawed Amer-European attempting, particularly in Mathews' accounts of him, to understand the Wah'Kon-Tah of the Osages, provides him with a particularly apposite mouthpiece.[1]

Osage scholar Robert Allen Warrior has linked John Joseph Mathews with Vine Deloria, Jr., as the mid-twentieth-century Native American intellectuals who have established a scholarly tradition to guide the programs, classes, and journals and other publications that have defined the field of Native American Studies since the "Red Power" movement and the beginning of the Native American literary renaissance in the late 1960s and early 1970s. In the context of this present study, however, there is a striking difference between Black Elk and Mathews, on one side, and Deloria and his Canadian contemporary Harold Cardinal, on the other. The Siouan intellectuals of the 1930s are both describing the sufficient, indeed exemplary, culture of the Siouan peoples. Deloria and Cardinal, writing after 1960, are examining the deficiencies in both governmental and academic whitestream attempts to understand, describe, and regulate Indigenous people and cultures. Even the titles of the volumes show the different foci. Neihardt describes Black Elk speaking, while Mathews writes of Wah'Kon-Tah and of talking to the moon. Similarly, Ella Deloria, Vine's aunt and an important intellectual in her own right, published an intermediate book, *Speaking of Indians*, in 1944, in which she, like Laban Miles, interprets the strengths of Siouan culture to whitestream readers, particularly those likely to read a book issuing from an explicitly Christian publisher. This slender, soft-spoken, and understated volume argues for the beauty and sufficiency of Lakota ways and the importance of their being adapted into whitestream society for the benefit of all who will inhabit North America after the cataclysms of World War II.

Vine Deloria and Cardinal, on the other hand, were writing as part of the rights explosion of the 1960s. Deloria's most famous titles—*God Is Red*; *Custer Died for Your Sins*; *We Talk, You Listen*; and the later *Red Earth, White Lies*—are brilliant, satiric, and as blatantly confrontational as the takeovers of Alcatraz, the Bureau of Indian Affairs building in Washington, and Wounded Knee, South Dakota, during the same time period. In addition, they are, as we shall see in chapter 15, bang-on critiques of colonial and federal Indian policy since the points of sustained contact between various whitestream and Indigenous North American peoples. Similarly, Harold Cardinal's *The Unjust Society* is a specific rebuke to Pierre Trudeau's proclamation of a "Just Society" at the same time as he proposed to do away with all the treaty and *Indian Act* rights that had been guaranteed in perpetuity to Canada's Native peoples. Both Vine Deloria and Harold Cardinal were prophets of Red Power, and their influence continues to be felt. Black Elk, Mathews, and to some extent Ella Deloria, on the other hand, are closer to Native Canadian intellectuals of the last twenty years: they acknowledge colonization and oppression but keep their focus on the exemplary nature of Indigenous North American philosophy and the need not only to accommodate it but to foreground it to create a satisfying twenty-first-century North America for Turtle Island and everyone here upon it, no matter how ancient or recent their occupancy.

For Mathews, all religion, literature, and anything that might be classed as culture was a species of "ornamentation" of the same sort as the species ornamentation expressed by flowers, by the dancing play of rabbits, or by the characteristic expression of something beyond mere survival that appears in every species. And he saw ornamentation as something that sprang from the nature of the place where the species lived. Two paragraphs from *Talking to the Moon* give the gist of his beliefs.

> I often think of the species *Homo sapiens* who was a part of the balance of my blackjacks. The Osage, while in perfect harmony, assumed that he had two natures; but, of course, he was almost as much under the

influence of his natural environment in his man-world of thought as he was in his animal-world of struggle and reproduction. His concept of God, springing from his ornamental expressions, was certainly colored by his natural environment and fear of the elements and his enemies. He built up in his imagination the Great Mysteries, and he walked, fought, hunted, and mated in the approval of them. When the Force urged him to expression, he turned his eyes to Grandfather the Sun; the colors he saw under his closed eyelids he put into beadwork, quillwork, and painting, as inspirations from one of the greatest manifestations of the Great Mysteries, the Sun, father of Father Fire, impregnator of Mother Earth.

He thought of his tribe as symbolical of the universe, and he divided himself and his universe into two parts, man and animal, spiritual and material, sky and earth, which he called Chesho for the Sky People and Hunka for the Earth People, because he felt this duality. With his Chesho thoughts, his ornamental expressions, however, he was colored by the processes of the earth in general and by his own struggle in particular.[2]

Mathews first introduces his main character, Laban Miles, to the reader as a young man of Quaker faith, committed to what he sees as William Penn's beliefs in fair dealing with Native Americans. He quotes Miles's non-fictional reaction to the offhand Indian-hating remarks of a temporary roommate at the University of Iowa, an incident that stayed with him and eventually prompted his entry into the Indian service, leading to his posting among the Osages. What Miles embarks upon is no mission to "civilize" the Indians but rather, at least in Mathews' telling, a patient education in the ways of the Osage Wah'Kon-Tah. The entire book is informed by Miles's inarticulable desire to understand the values of the Osage people.

For Mathews, the land itself was the teacher. Osage society depended deeply on the clan structure divided between Earth and Sky, as described above. The relationship of specific communities to landforms in Missouri was, Mathews believed, replicated as far as possible in Oklahoma.[3] Mathews came home to the blackjack and post oak country he loved—not, in his account, to the people or even to his own family. Whereas *Black Elk Speaks* defines a sacred landscape—Harney Peak in the Black Hills is the centre of the universe—Mathews defines a beloved landscape that acquires its meaning from its dialogue with the people. Chapter 1 of *Wah'Kon-Tah* begins with a description: "The impression was one of space; whispering space. . . . When a line of blackjacks became the meeting place of sky and prairie, their rounded tops became black and cut definitely into the blue in such a way as to suggest adventure beyond."[4] The entire short chapter describes the land, and, very briefly, in passing, a few of the people. Only in the third chapter does Mathews take up the people, and while the descriptions of the land had flowed easily, Mathews focuses on Miles's inability to describe the people.

> He could never write about them as he wished to write. In the first place he could not express what he had begun to feel, and in the second place his understanding and friendship with men like Big Chief, Hard Robe and Governor Jo was something that one couldn't write about. How could he make people [Miles, or perhaps Mathews, implicitly defines his audience at this point as Amer-European] understand a man like Gray Bird, for example? (33)

This note of uncertainty does not appear in the passages Mathews quotes from the actual diary, but it may well be something that Miles expressed to Mathews in the long talks they had together in the year after Mathews had returned to the blackjacks before Miles died. It runs throughout the book, even as Miles seems to be moving to a better understanding of the Osages and their Wah'Kon-Tah religion.

Although place is necessary for developing a sense of meaning, it is certainly not sufficient. As a geologist, Mathews had a great deal of respect for the oilmen who could read the land to find underground pockets of oil. A major focus in his biography of the oilman and later governor of Oklahoma, E.W. Marland, is on Marland's growing understanding of the processes that had created coal and oil, and thus of the surface structures that would alert drillers to the presence of oil. Mathews did not approve of the wasteful exploitation and abandonment that characterized the oil boom in Oklahoma, but he did not associate it with the geologists, whose respect for the land was real, if quite different from that of the Osages. Instead, he shows Marland as being ruined by the Morgan interests, the banks that managed to oust Marland from his own company and to change its focus from geology and respect for the workers and the community to market manipulation and respect for nothing but profit for the bankers themselves.[5] Most of the actual oilfield workers, in Mathews' descriptions of them, are interested only in fleecing the land and fleecing the people. Even the Osages themselves are demoralized by the many forces working against them. They no longer gain solace from the land, and the ancient traditions of the Mourning Dance and the Making of a Medicine Man are not only outlawed but no longer seem to fulfill the needs of the Osages. The peyote religion as directed by Moonhead, one of the precursors of the Native American Church, seems to have functioned, in Mathews' eyes, as a kind of halfway measure that maintained Native ideas behind an ostensibly Christian facade. For other people, alcoholism, assimilation, or, during the days when oil made the Osages the richest people on earth, conspicuous consumption provided people with, if not meaning, at least something to do. All of these alternatives, however, ignore the land or commodify it. They are essentially Amer-European, Christian, and capitalistic responses. At the same time, there is nothing essentialist about Mathews' beliefs. He chose a Latin rather than an Osage motto to carve upon his fireplace mantle, and it expresses a world view that is consonant with the blackjack prairies: "Venari Lavari Ludere Ridere, Occast Vivere" (To hunt, to bathe, to play, to laugh, that is to live).[6]

As Robert Allen Warrior points out, Mathews' philosophy was distinctive. All living creatures, he believed, went through cycles from juvenescence to senescence that included a period of flourishing (he called it "virility," an indication of his lack of appreciation for both real women and "feminine" principles) that varied from species to species and included the flourishing growth of a young post oak, the mating dance of a prairie chicken, and the spiritual and intellectual flourishings of humans. Although he shared in Miles's distaste for the deaths involved in the Mourning Dance and the payments for knowledge in the Making of a Medicine Man ceremony, he preferred the Osage spiritual ornamentation to that of Amer-European Christianity.

Both *Talking to the Moon* and *The Osages* offer more penetrating interpretations of Mathews' own philosophy and of Osage ceremony and belief than does *Wah'Kon-Tah*, but his first book invites Amer-European readers to understand through Miles's slow initiation. Mathews quotes from Miles's journal, and then interpolates those things that Miles was unable to express in writing, though it is never entirely clear whether these are thoughts that Miles actually expressed to Mathews, Mathews' interpretations of Miles, or Mathews' attribution of his own thoughts to Miles. Although Mathews' description of Miles's thoughts does not completely avoid the aura of the "noble savage," Mathews does avoid the clichés and arrives at the images experientially.

> [Miles] was afraid of being sentimental, but he knew he was beginning to understand these people who were certainly not European, but possibly Asiatic in their origin. Their customs, their conception of God, their quiet dignity and courtesy and sincerity as compared with the aggressiveness and hypocrisy of his own race, made the understanding of them difficult. It seemed to him that they did not assume virtues as did the white man, or attempt to control the destinies of others. They were individualists in that respect, though they lived by the harsh rules of the herd. . . .
>
> It seemed to the Major that the two races would never meet, and that there would be no one with sympathy and understanding sufficient to interpret the Indian. He knew what he himself had begun to feel, and he knew what the better class trader [which would have included the Mathews family] felt about them; a sort of respect and admiration that was almost inscrutable. (40–41)

The major's respect and understanding grow the longer he stays in Osage country, and for him, the country and the people and their religion belong together and reinforce each other.

> He loved the blackjacks and the prairie because they were the home of a people whom he loved and respected. He often thought that the wild prairie with its temperamental changes of weather was a perfect home for the children of Wah'Kon-Tah, the Great Mysteries which was the sun, the wind, the lightning; that which lived in all things which had life. The just, cruel, vengeful god visualized by these people.

To the Major, Wah'Kon-Tah was more than the god of a so called primitive people. In his strict consciousness, he had seen, in his contacts with the children of Wah'Kon-Tah, how many of the credos of his own belief of Brotherly Love had become mere form, and without meaning. In his contact with primitive virtues he had realized this. This realization had broadened him and given him tolerance. He was never the monitor, nor did he like to be didactic, but he often thought he would like to hold the worshippers of Wah'Kon-Tah up as an example to some of the people who worshipped as he worshipped. . . .

. But he could not lose himself in a few years. He was European and understanding of the people came slowly. . . . There had been too many generations of the stern teaching of Right and Wrong, for the Amer-European iron in his soul to have dissolved so quickly. But he was no simpering sentimentalist, and therein lay the value of his sympathy and understanding. (62–63)

These passages are, I believe, crucial to understanding Mathews' presentation of Osage religion. Wah'Kon-Tah is, as he says, the Great Mysteries, the same basic words and concept expressed in the Lakota Wakan Tanka of Black Elk and Neihardt, a concept perhaps inexpressible in European languages where "great" has so many hierarchical connotations that it does not represent either "large in size" or "diffuse through the universe" very effectively. Nor is Wah'Kon-Tah truly imaginable without the land and the people. Mathews' presentation of Miles's very slow process of understanding—which includes years of living on the land, learning the Osage language, and developing longstanding instrumental friendships with Osage men—contrasts with Neihardt's mysticism and his apparently quick and uncomplicated acceptance of his role as Flaming Rainbow, Black Elk's amanuensis and spiritual "son."

At the same time, Mathews is not an essentialist. Although he postulates an Asian rather than European origin for the Osages (currently a rather hot topic I won't discuss here), his point has to do with cultural heritage, the manicheanism of Mediterranean religions. Mathews sprinkles all his work on the Osages with the patronizing word "primitive" and later "neolithic," but both of these seem to be in contrast not to their usual opposites, such as "progressive" or "civilized," but to "decadence" and "corruption." Neihardt met Black Elk in the context of the poet's research for the "Messiah Dance" book of his *Cycle of the West*. As Julian Rice has shown, despite Neihardt's dissociation with organized Christianity, a Christian metafiction of sacrifice joins with the more obvious Homeric tradition so that the Lakotas figure as both Christ crucified (and resurrected in America) and as the Trojans (defeated in Troy but rising again in Aeneas to found Rome). David Young has made a compelling case that Neihardt's account of the death of Crazy Horse is based on the death of Hector in the *Iliad*.[7]

It is exactly this kind of analogizing that Mathews portrays Miles as trying to escape. Unlike Neihardt, Miles never assumes any entitlement to the stories, and Mathews himself, during his work on the Osage Tribal Council long after the publication of *Wah'Kon-Tah*, accepted the fact that his mixed-blood heritage and the European education that made him valuable to the council was also highly problematic to the very old full-blood men that he (and Miles) most admired. Mathews gave his fictional alter ego in his novel *Sundown* a full-blood mother, and the young Chal Windzer sometimes longs for the world view of his full-blood uncles at the same time that he fits himself to live in the corrupt Amer-European world that he, like Mathews, sees as flourishing. When Chal proudly shoots and presents to his mother an English sparrow, he has, for the moment, killed this imported species that expands onto the blackjack prairie, where neither the English sparrows nor the Amer-Europeans have natural predators and where each is imposing upon rather than adapting to the land.

When Miles finally has the understanding to appreciate it, he is invited to a Making of a Medicine Man ceremony. While at first he had worried about ever describing a man like Gray Bird, he now describes his host confidently. And so he is ready to become a witness. Because Mathews is careful not to give a direct description of the ceremony, it is impossible to know exactly what it is that Miles attends. I am assuming that it may be what Francis La Flesche carefully researched and described, and that has now been published by Garrick Bailey, the Songs of the Wa-xo'-be.[8] The one aspect that Mathews describes in detail, the artificially elongated and stuffed dove from which he derives his title for the ceremony, does not appear in La Flesche's account. Since La Flesche's songs came from the Buffalo Clan, the Dove probably belonged to another clan proper to Gray Bird.

Black Elk Speaks describes and explicates Black Elk's Great Vision, describes the Horse Dance that Black Elk caused to be performed to transmit the power of the vision to the Lakota people, and explicates some of the significance of the Sun Dance, especially when the same symbols appear in it as in the Great Vision—the importance of the four directions, for instance, or of the four virgins. Mathews, on the other hand, describes the carefully hidden but nonetheless penetrating looks that some of the celebrants direct at Miles and Miles's courteous understanding that they question his right to attend. He has already raised this question with his host: "It will be all right if I go to this place? I do not know this because I am white man" (117). Gray Bird reassures him it is all right: "You are my friend" (117). What Mathews describes, probably following Miles, but confirmed by his own experiences and conversations, is only the preliminary part of the ceremony, the testimony of various men about the virtuous exploits of the postulant, recited before the artificially elongated and stuffed dove, a symbol that "faced the sun . . . with an air of aloofness and gravity" (127). Although Mathews mentions "the songs [the postulant] must learn," he does not quote them or even hint at what they might say. After the stories and the songs, Gray Bird tells Miles that he himself has been through the ceremony. He fears that Miles finds it "not good" (130). Gray Bird seems to doubt the usefulness of having to pay witnesses to attest to his good deeds and confesses that although he has learned all the songs, his head is still not clear. Yet this is far from a condemnation of the knowledge, for Gray Bird then launches into a discussion about the nature of generosity, thus undercutting any hesitancy Miles may have about the good of the ceremony. The chapter ends with Gray Bird joking with his wives.

Until Miles's actual journal—supposing a copy of it still survives— reappears, it is impossible to tell what is Mathews and what is Miles in this chapter, except that both seem to respect the privacy of the ceremony. Miles is welcome, but as a friend of Gray Bird, not as an ethnologist who might describe the ceremony to outsiders. La Flesche, arriving among the Osages at a time when the ceremonies were no longer being performed, was able, over a number of years, to establish his trustworthiness to the old men who knew the songs and to record them for the sake of future Osages as well as anthropologists.[9] The songs that La Flesche quotes are achingly beautiful, and they relate directly back to the land and to the animals and birds that inhabit it and pass on their virtues to the Osages. Thus, the teaching of the land, which Mathews describes and employs, becomes a way to direct Miles—and the reader—toward Wah'Kon-Tah without giving away the songs that the real postulants are required to buy at the price of many ponies. In *Talking to the Moon*, Mathews offers his own synthesis of meaning within the framework of the Osage "year" of moons, all based exactly on what game and plants do in that particular moon in that particular place.

Miles seems to have begun his official writing *for* an Amer-European audience but *to* have shifted, as much as possible, to attempting to express an Osage point of view. In his correspondence as Indian agent, he is supposed to be writing *for* the Osages and *to* Washington, which forces him to make clear his translation of Osage ways and of how they are different from those of the encroaching Amer-Europeans.

> He once wrote boldly in his notes that the government was only the mirror of the people, of people who thought of nothing else except the mad exploitation of the natural resources, which included the senseless destruction of the forests and the game; things which the Indian had considered as gifts from Wah'Kon-Tah, and as such were revered. (77)

Miles believes that if he can just express what he has come to understand about the Osages, "the swarming Europeans who thought of gold and land, and razed forests" (78) would also come to understand and to practice at least some of the Osage virtues. But he also fears sentimentality and hates the "'trash' that had been written about . . . the 'poor Indian'" (79). At the same time, the Major hated to be intolerant toward the Amer-Europeans: "He really believed in the land of opportunity" (92). Like Neihardt, Miles sympathized with both sides of the story—as did Mathews, though perhaps to a lesser degree. Miles talks to Lame Doctor, a traditional man who, in response to a vision that came to him after witnessing his father's murder at the hands of a white gang, chooses to give up his childhood vow to kill ten white men.

They sat for a long time and talked of the Great Spirit and his children, and of the white man and the white man's God. They came to the conclusion that they were the same; that there was one God for all people, but that the Indian saw him one way because he was an Indian, and the white man saw him another way because he was a white man. (166)

Writing in 1931 and 1932, Mathews could not avoid the sense of the "Vanishing American," the dream trampled in the bloody mud, that Neihardt expressed so strongly in his oft-quoted coda to Black Elk's own account. Yet Mathews, even as he feared loss and "vanishing," also expected the people and the people's understanding of Wah'Kon-Tah to live on. He was instrumental in founding and supporting the Osage Tribal Museum, which, if it was an attempt to hold onto the past, was also an expectation of a future, since the museum was explicitly established to be run by the Osage people.[10] *Wah'Kon-Tah* ends with the morning chant of the old man, Eagle That Dreams, to Wah Tze Go, the Grandfather Sun; the chant ends as the lower edge of the rising sun clears the horizon—"and the early morning world seemed to be listening, except for the coughing of the oil pumps carried from the oil fields on the heavy air" (342). Despite the ominous oil pumps, the sun returns—as do the Osages. Since the primary audience of the book was—and seems to have been intended to be—those Amer-Europeans whom Miles himself had fretted at not being able to reach so that they would truly know and learn from the Osages, it would not be appropriate for him to transcribe private ceremonies, but the morning chant to the sun was for everyone to hear. The sun still streams over the blackjack hills every morning. And the Osages still greet it, eighty years after the death of Laban Miles and fifty years after Mathews' own death.

Nonetheless, the mood of the book is valedictory, and Mathews increasingly focusses on the ceremonies of death—on funerals, the Mourning Dance, and the Ghost Dance. The only ceremony for which Mathews, in the voice of Miles, gives a full description is of the funeral of Miles's beloved friend Big Chief. Unlike Neihardt's description of Black Elk's Great Vision with its intricate symbolism, or James Walker's elaboration of Lakota creation stories, or La Flesche's need to justify Osages to anthropologists, Mathews did not feel he (or Miles) had to show anything particularly complex to make Osages comprehensible and admirable to Amer-Europeans. Miles needed only to observe and to report. Intercut between his memories of his friend and his description of his friend's funeral are Miles's observations of the birds around them: "On the dead top of a sycamore two red headed woodpeckers quarreled with each other about store houses for winter. . . . A flock of crows had found a barred owl in the gloom of the tall trees of the bottom, as he dreamed away the day, and were cursing terribly; darting at him or sitting above him, calling him thief and murderer" (234–35).

Garrick Bailey explains that he was only able to understand Osage cultural continuity when he began to take part in the peyote meetings and the I'n-lon-schka dances, because his Osage friends kept insisting that the ceremony or the dance "shows you" or "teaches you."[11] Robert Allen Warrior points out that although Mathews was not a ceremonialist, he was, as a writer, doing what he did best and thus was never inauthentic in writing about what he did not practice.[12] I would suggest that for Mathews, in addition, what he did know was the same place and creatures that informed the ceremonies themselves, and that in addition to his courtesy in not describing what his readers could not earn, through actual friendship, the privilege of seeing, his descriptions of the land and wildlife provided the context for understanding without the sentimentality or just plain trashiness that both he and Miles despised in books about Indians. Thus, the birds, in their setting of river and trees, are the appropriate context for understanding how Big Chief is dressed in death:

> He had his necklace of bear claws and at his throat was the shell gorget
> made from the fresh water mussel and representing the sun at noon; the
> symbol of the god of day. Over his shirt was his bone breastplate with
> wampum on each side. His face had been painted with red; a symbol
> of the dawn, symbol of the god of day; the Grandfather. On this were
> alternating lines of red and black on each side of his face representing
> the tribe and clan and family and the symbol which designated him as
> peace Chief, or chief of the Chesho division of the tribe, the division which
> represented the sky. (235)

Big Chief's death is, at his request, not followed by a Mourning Dance, a practice that involved killing an enemy to accompany the beloved dead on his last journey and one that Mathews represents the Osages themselves decrying as no longer appropriate.

Like the Lakotas, the Osages practiced the Ghost Dance in 1890, but Mathews presents it as less important to the Osages than to the surrounding whites who liked to scare themselves with dark tales of Indian "savagery." Black Elk also comes to regret the Ghost Dance as a distraction from his own Great Vision, but it continued to play an important role for Neihardt. For Miles, the Ghost Dance seems to have little meaning, since none of his particular friends are associated with it; for Mathews, it had little power, since it was not directly associated with the blackjack prairie. In *Wah'Kon-Tah*, the Ghost Dance is only a faintly ironic image of vanishing, even though the basic premise was certainly attractive to the Osages: "it seemed good to have buffalo back on plains and deer in blackjacks back in their great numbers; it seemed good to them if all white men were to leave Reservation" (317). But they could not help doubting the vision and worrying that any fighting of whites might simply give the whites an excuse to seize what the Indians had left. Some people, though, went ahead and erected a dance lodge, and "each day there was dancing and sometimes the drums were heard far into the night" (318). The white men

> talked about it as a child might talk of ghosts; gaining a certain thrill out of imagined possibilities. Some of the United States marshalls . . . assumed to know about the trouble with Sitting Bull and the ghost dancers on the Sioux reservation, and they led their listeners to believe that there was the same trouble on the Reservations of the Territories. . . .
>
> But after a short time the camps at the head of Sycamore Creek were deserted and the dried leaves on the branches, which formed the roofs of the open structures, rasped softly, and the wind sang little songs in the framework of the lodges.
>
> Due to the sanity of the older men of the tribe, there had been doubt and a lack of fervency in the ceremonials, and the Osages were lost to the Messiah from the land of the west. The head waters of Sycamore

Creek saw the last feeble gesture of the Great Osages; it was a ghost dance; the white man had named it well. (318–19)

With his valediction on the Ghost Dance, Mathews introduces Major Miles's retirement from the Osage agency and the Great Frenzy of the oil years, but, as we have seen, the valedictory was premature. The Osages have outlived the oilmen, outlived Miles, outlived Mathews himself. When Miles died, he was buried as a white man, in a white man's coffin with no paint on his face to identify himself to Osage friends he expected to meet. Our last image of religion in the text is of the peyote man "praying fervently to a god who was a composite of Wah'Kon-Tah, the stern god of his fathers; Christ, the god of the white man who had proved so powerful; and the peaceful, dreamy god of Peyote, the god of resignation" (386).

None of these three ways, though, really seems to express the Wah'Kon-Tah that Mathews admired. Rather, understanding must come out of living on the land, with its creatures, not out of the secondary observances of stories, ceremonies, and symbols that the Osages had constructed out of symbiotic relationships. Although not completely supported by the Osages, The Nature Conservancy has obtained Amer-European title to a sizable and surpassingly beautiful portion of Mathews' blackjack hills, with Mathews' own stone cabin in view, and has populated it with buffalo, leaving room for the deer and coyotes, and for the various birds, including Matthews' beloved red-tailed hawks. The Osage Tribal Museum provides classes in Osage language and arts. The people continue. Garrick Bailey attends peyote ceremonies and the I'n-lon-schka dances to learn enough to edit Francis La Flesche's observations, and the Songs of the Wa-xo'-be are once again available to the Osages. Just as *Black Elk Speaks* serves as a cultural "bible" for the Lakotas despite Neihardt's "bloody mud" comments, the blackjack hills and their birds and animals continue to be the cultural "bible" for the Osage people (and, I believe, for Amer-Europeans who, like Miles, are striving to replace the "iron" of European imposition and accept the responsibility and vulnerability of living gently on the land), revealing and mirroring Wah'Kon-Tah, despite Mathews' depressed comments on "ghosts."

Intellectual Justification for Conquest: *6*
Comparative Historiography of the Canadian and US Wests

During the last quarter century, the New Western Historians and a growing turn to regional studies have made the history of the American West a
particularly vital part of the profession. Meanwhile, extraordinary strides
in Canadian western women's and, particularly, Aboriginal history have
revised and revitalized the history of the Canadian West. During the same
period, first the Free Trade Agreement and later NAFTA have focussed both
Canadian and American attention on continental issues and the differences and similarities between the two enormous land masses that make
up the bulk of North America. Comparative histories and monographs of
the Canadian and American Wests are now beginning to flourish, including two volumes of essays edited by Carol Higham and Robert Thacker, *One
West, Two Myths* (2004 and 2007), Beth Ladow's *Medicine Line* (2001),
Sheila McManus's *The Line Which Separates* (2005), and Andrew Graybill's
Policing the Great Plains (2007).[1]

On the one hand, there is no difference between the Canadian and American Wests. There is one unbroken geographical entity (though it may be called Prairies or Great Plains) that changes gradually from north to south and from east to west, and that includes a vast range of microclimates and microgeographies. The "United States of America" and the "Dominion of Canada" have divided this region between them for less than two centuries, but the impact of their citizens upon it has been great and largely similar. In both countries, the slaughter of the great bison herds led to land treaties with Aboriginal peoples. Domestic cattle replaced the bison, and railroads brought thousands of commercial agricultural settlers who ploughed the land and planted cereal crops. The newcomers used the federal government and the courts to separate even more land from Aboriginal peoples—for farmland, mineral development, urban growth, and hydroelectric and irrigation dams. Both north and south of the forty-ninth parallel, almost all of this region is now commercially cropped grasslands, producing grain or meat. In both countries, this agriculture is one of boom and bust, with fewer and fewer people on the land and more and more relocating outside the region or to cities that, except for those in Saskatchewan and Texas, are only on the fringe of the region. The extraction of energy resources, especially petroleum, also continues to transform the land.

On the other hand, the two Wests are so different in the context of their current political identities and intellectual histories that almost no comparison is possible. To find a true parallel, we would have to discover that Sitting Bull was George Washington's primary antagonist, or that Americans still hotly debated whether Mexican general Santa Ana should be called a Founding Father or a vicious renegade. Canada needed its West to bring about Confederation; the eastern United States claimed its West as Manifest Destiny. Canada's West is separated from its eastern population centres by a thousand miles of rugged Canadian Shield, while the United States deployed a continuous frontier of Amer-European settlement—despite historian Walter Webb's contention that it toppled over for a moment at the hundredth meridian. Euro–North American traders traversed Canada for centuries via the empires of the St. Lawrence and Hudson Bay, working in a symbiotic relationship with Aboriginal trappers and fur preparers. Euro-American traders in the Mississippi basin gave way to American trappers, mountain men, who wiped out the beaver as far as they could reach and supplanted the Aboriginal trappers with whom they were not unusually at war. The United States, it seemed, waged war against all in its path—the land, the animals, and the Aboriginal, Hispanic, and mixed-blood peoples. Canada prided itself on its avoidance of US-style violence and waited for disease and starvation to reduce its Indigenous westerners to acquiescence in treaties and dispossession.

These, at least, are the broad strokes that most contemporary western historians in either country would agree to. The historiography of the two Wests is also different. Neither Frederick Jackson Turner nor Harold A. Innis was, by any means, the first historian of his respective West, but each in some way encapsulated the work of those who had gone before him and laid out the major theoretical approaches that future historians, explicitly or implicitly, would follow—even those American historians who insisted that they were not following Turner and those Canadian historians who did not realize they were following Innis. J.M.S Careless's 1954 article "Frontierism, Metropolitanism, and Canadian History" lays out two lines of interpretation, the Turnerian frontier thesis as opposed to the metropolitan theory that Innis included, almost as an afterthought, in *The Fur Trade in Canada* and that Donald Creighton expanded upon in *The Commercial Empire of the St. Lawrence* and other texts. While Turner, in 1893, was reacting to the "germ" theory of Herbert Baxter Adams and others, that American society was merely the development of European "germs" in American space, Innis was reacting to the ideas of Turner and his followers that American democracy was born out of the forest, the individualism of the frontier. Innis, however, stressed not "germs" but the continuing effect of the metropolis— whether it be London, Montreal, or other—on the economic, and hence social and political, nature of Canada. Rather than the remote medieval antecedents of democracy in Saxon forests, Innis looked at European fur markets and innovations in the manufacture of European goods for the fur trade to show how even this most far flung of markets developed in relationship to Europe and Montreal. "The importance of metropolitan centres in which luxury goods were in most demand was crucial to the development of colonial North America" because they manufactured trade goods and provided relatively high prices for raw materials such as fish and fur.

This relationship between metropolis and hinterland would continue to determine Canadian western development. The United States featured an interlocking network of large and small metropolises, stretching from New York and Albany west and north to Buffalo and Cincinnati, from New Orleans and St. Louis west and north to Des Moines and Council Bluffs, from Chicago, from Minneapolis, from San Francisco . . . and on and on. Even a major centre, as William Cronon has so magnificently shown in his study of Chicago, *Nature's Metropolis*, is in the middle of a web. Canada's fur trade and frontier development, however, was much more linear, given the nature of the country's geography. Two great pathways arced west, the Kingdom of the St. Lawrence and the string of forts that eventually sprouted westward from Hudson Bay. Instead of fur trade forts that grew into cities surrounded by agrarian settlements, the slender Canadian network of fur posts and missionary churches and schools remained isolated from eastern Canada and its Euro–North American population centres and lifeways.[2]

As Turner would tell his audience at the World's Columbian Exhibition in Chicago, "The existence of an area of free land, its continuous recession, and the advance of American settlement westward, explain American development."[3] Turner's insight, which seemed to encapsulate perfectly what other European-descended Americans were thinking and saying in 1893, was that there was no longer a discrete line of settlement in the United States. The network had become so far flung that it overlay the entire continental span of the nation. If there was still public land (and more was available for homesteading after the disappearance of the frontier line than was taken up before), the days of a line of settlement and its "continuous recession" were gone. Turner, of course, did not consider the effect of this moving line on those who were either scraped along before it or marooned as it rolled past them. Innis, especially in his work on the fur trade, was continually aware of the Indian role in the fur trade fiefdoms he described and in the resulting nation. The formative value of the US frontier became obvious only in hindsight, when Turner claimed it. Innis, as an historian, obviously studied the past, but its pastness was more provisional. The Canadian fur trade was smaller, proportionately, to Canada's economy in the 1920s when Innis was researching and writing his book than it had been before Confederation in 1867, but it still flourished throughout much of the northern Shield country that, Innis claimed, had defined Canada as a nation. In fact, the fur trade continued until it came upon metropolitan limitations. Just as the collapse of the fashion of beaver felt hats in Europe had removed the market for the staple of the classic fur trade Innis and Turner (in his dissertation) had described, animal rights groups in the 1980s and 1990s in Europe and to a lesser extent in North America have reduced the fur market almost to nil (though, except in Alaska, it had largely declined to hobby status in the US before the frontier line disappeared). The role of the metropolis has reverted from the fur trade markets to Ottawa, which funnels payments through the Nunavut Tunngavik corporation to subsidize hunters and trappers willing to remain on the land. Innis is best known for his work on staples—fur, fish, and the others that characterize the Canadian economy. Not primarily a historian of the West, his work on the fur trade nonetheless laid out a basis for understanding the nation that other historians have developed (like Creighton) or debated (like W.J. Eccles).[4] Turner is known for his Frontier Thesis—it might be a question on *Who Wants to Be Millionaire?*—but also for his discussion of the United States as a set of regions. Both Innis and Turner were American-trained economic historians who became, almost by accident, the writers who determined the formulas for their respective Wests.

One of the important distinctions between Innis and Turner for understanding the historiography of the two Wests is that Innis is very particular in talking about *place*—he was careful to travel to as many as possible of the places he discussed—while Turner is most concerned with talking about *process*—place is virtually irrelevant, something that the New Western Historians have noted. In some ways, all of Turner's frontiers were simply his own Wisconsin pine lands dressed up in another environment. One of the most oft-cited passages from Turner's celebrated "Significance of the Frontier" essay reads: "Stand at Cumberland Gap and watch the procession of civilization, marching single file—the buffalo following the trail to the salt springs, the Indian, the fur-trader and hunter, the cattle-raiser, the pioneer farmer—and the frontier has passed by. Stand at South Pass in the Rockies a century later and see the same procession with wider intervals between."[5]

Turner's process is also hierarchical. Buffalo are the first and lowest element in the procession, closely followed by "the Indian," the only figure in the whole procession who is defined by essence and not by profession. Even the buffalo is doing something (looking for salt), and "the Indian" as a real person was also as likely as not to be a "fur-trader and hunter." The final element, the apotheosis for Turner, is the "pioneer farmer"—imagined, of course, as white and male, unlike the unmentioned Indian women who had tended crops, at least in the vicinity of the Cumberland Gap, literally from time immemorial. Innis's scene is by no means as orderly and hierarchical. (And nor is that of contemporary fur trade historians.) Euro–North American fur traders feuded endlessly against one another. Different Aboriginal groups struggled with each other, changed roles, and played Euro–North American companies and nationalities against each other. This process, however, was not telocentric in Innis's telling—it was not "going somewhere," not evolving into some "higher" form of land use, such as farming. It was sufficient in itself. While in the United States, even the fur trade worked toward the displacement of Native people—with Anglos themselves taking over trapping, as is evident from Turner's conflation of "fur-trader" and "hunter"—in Canada, during any phase of the fur trade, as Innis said, it was never the case that the only good Indian was a dead Indian. Native people in the Canadian fur trade, as in the eastern United States, fulfilled vital economic roles as trappers, fur preparers, canoe builders and paddlers, hunters and provisioners. Even Plains peoples performed a vital role in the Canadian fur trade—as buffalo hunters and pemmican makers, feeding the fur brigades as they stretched out to the mountains and the great northern rivers.

For Turner, "the wilderness" and "the Indian" were versions of one another with no positive values except to strip Europeanness from the settler and set "him" on the way to becoming an American.

> The wilderness masters the colonist. . . . It puts him in the log cabin of the Cherokee and Iroquois and runs an Indian palisade around him. Before long he has gone to planting Indian corn and plowing with a sharp stick; he shouts the war cry and takes the scalp in orthodox Indian fashion. In short, at the frontier the environment is at first too strong for the man.[6]

In addition to various inaccuracies about Indians, what is arresting about this oft-quoted statement is that the Indians have vanished, leaving their role to the frontiersman, who Turner believed would soon leave Indian deficiencies behind. Upon the former European, now reduced to a blank slate, a new and improved American would be drawn. While Innis, no less than Turner, subscribed to the idea that European contact doomed the Aboriginal way of life, his account recognized the indispensability of Native people to the fur trade and pointed out the continuing Aboriginal influence in Canada. "We have not yet realized that the Indian and his culture were fundamental to the growth of Canadian institutions," he wrote.[7]

In fact, Innis, by 1930, clearly recognized three founding peoples of Canada. "The Northwest Company was the forerunner of confederation and it was built on the work of the French *voyageur*, the contributions of the Indians, especially the canoe, Indian corn, and pemmican, and the organizing ability of Anglo-American merchants"—labour, land knowledge, and management. As Doug Owram points out, western Canadians of European descent have before and after Innis shown decided preferences for either the Nor'westers or the Selkirk settlers as the founders of western Canada, and I will discuss the various arguments below.[8] The big difference, for Americans, is that the United States never had a choice. Daniel Boone and Kit Carson are both, as Henry Nash Smith pointed out, Sons of Leatherstocking. The mainstream of American culture, academic or popular, has not really come up with alternatives to these Sons of Leatherstocking as founders of the West. Mixed-bloods like James Bordeaux are certainly available as colourful characters, Tatanka Iyotaka and Ta Sunka Witko (Sitting Bull and Crazy Horse) are anti-heroes, and Santa Ana is the bad guy, but there is no question about who the hero of the settlement saga is.

Both Doug Owram's *Promise of Eden* and Henry Nash Smith's *Virgin Land* are histories of the ideas of the respective Wests, and they provide useful contexts for examining the ideas of Turner and Innis in the contexts of the intellectual history of each country.[9] In seeing the North West Company as the precursor of Confederation, Innis, an Ontarian himself and teaching at the University of Toronto, was definitely following in the footsteps of the Ontario expansionists of the 1840s to 1860s, who had determined the particular manner in which the great territories to the northwest of the Canadas in the 1840s would be enfolded into the Confederation of the 1860s. According to Owram, until the 1850s, the North West appeared to

Canada West to be a fur trade hinterland, connected, after the 1821 merger of the North West Company and the Hudson's Bay Company, not to Canada but to England through the Bay, thereby seeming both distant and arctic. By the 1850s, however, as the Canadas ran out of agricultural land and it became clear that their internal improvements would never win them much of a share of the commerce of the American Midwest, Ontario expansionists began to look to the West not only as a solution to the problems of land and commerce but also as a kind of ballast to the struggles between East and West, Catholic and Protestant, French and English in the Canadas and the Maritimes. In the United States, most Euro-Americans had always seen the West as their destiny (something that called forth the 1763 proclamation and separated Canada from America in what would turn out to be a decisive way). The only question had been how far West. The annexation of Texas, the Mexican-American War, and the *Kansas-Nebraska Act* were all ways in which the United States tried to use its West to balance out pro- and anti-slavery tensions in the East—as the Canadas used their West to balance French and English tensions. The Ontario expansionists rediscovered the North West and claimed it for the United Canadas, as the heirs of the North West Company. This claim, of course, required "liberating" Rupert's Land from the Hudson's Bay Company, whom the expansionists portrayed as despotic, much as the American expansionists portrayed the Mexicans. The Selkirk settlers were the object of the Ontario expansionists' concerns in much the same way that the Austin settlers were the concern of the American expansionists. Ontario expansionists also accused the Hudson's Bay Company of blocking the Protestantization—and hence the "civilization"—of the Aboriginal peoples of the North West. Because of their monolingualism, unconscious racism, conscious anti-Catholicism, and barriers of distance, the expansionists did not consult Métis or Indian residents of Red River but simply assumed that those who, for whatever reason, opposed the Hudson's Bay Company spoke for all. In the same way, proponents of the Lone Star Republic never asked the opinion of the actual Mexicans in Texas, and still less the Indians. As for Kansas, as Paul Gates noted, it was all Indian Territory when the *Kansas-Nebraska Act* was passed. There was no squatter sovereignty or right to consultation provided for the Indians, many of whom were hustled out of Kansas before they had even been paid for the eastern lands they had surrendered to the government when they were moved *to* Kansas. For the Anglo expansionists in both Canada and the United States, any title other than their own was so deficient as to be incomprehensible. Both American and Ontarian expansionists were perfectly aware that the claims of the Hudson's Bay Company had counted for very little (and the claims of the Indians to nothing at all) when it came to Oregon country. American settlers, intending farmers, had held the land for the United States up to the forty-ninth parallel. The Ontarians were right in assuming that unless white Canadian farmers settled in the Red River valley, the hungry expansionists of St. Paul would gobble up that country, too.

Americans expected Mexican resistance to US expansion into Texas and the Southwest, and most were content to deal with it through warfare. Ontarian expansionists, on the other hand, genuinely did not expect Métis resistance to Canadian control of Red River and were puzzled and offended when the Métis resisted the immediate (and illegal) annexation of their land without their consent or even notification. The expansionists did not know what to make of Louis Riel's Provisional Government. They theorized that the Métis must be pawns of the Americans, the Catholic church, or Quebec foes of Confederation. But future historians, especially those who were either in blood or in sentiment the descendants of the Ontario settlers of Manitoba, dealt with the confusion in another way. If the Selkirk settlers, the Scots crofters whom Lord Selkirk had transported to and granted land in Red River starting in 1811, rather than the Nor'westers, were posited as the true founders of Manitoba, then the English-speaking Protestant Ontarians could claim to be their natural heirs. If the Nor'westers were the true founders of the West, then the Métis were, if not their only heirs, certainly their senior heirs. Anglo Texans had never hesitated in claiming to be the heirs of the Austins, of Sam Houston and Sam Maverick and Jim Bowie and, of course, of Davey Crockett. No other choice was even visible.

Writing in the first decade of the twentieth century, George Bryce, who had moved to Red River in 1870 and had become a booster, proclaimed *The Romantic Settlement of Lord Selkirk's Colonists*, who, he claimed, were the true first settlers of the area, the ones who had toiled and suffered, had survived hardships that eclipsed those of the Acadians, and had made the West Canadian—and Canada possible. Since most of the settlers' tribulations had come at the hands of the North West Company and their Métis allies, Bryce, then, had to complete the rearrangement of good guys and bad guys. The Selkirk settlers were no longer the dupes and pawns of the Hudson's Bay Company, as the expansionists had cast them, and the North West Company was not, as Innis would have it, the pre-figuration of Confederation. Instead the North West Company became the half-civilized predators on the noble agricultural settlers of Red River. John Thompson uses the dramatic C.W. Jefferys painting "The Massacre at Seven Oaks" to demonstrate how the Anglo-Canadian tradition had developed into a sort of "Remember the Alamo" rendition of the conflict between the Selkirk settlers and the Métis in 1816. Interestingly enough, Governor Robert Semple, who commanded the settlers and was killed at Seven Oaks, was an American by birth. According to Hartwell Bowsfield's entry on Semple in the *Dictionary of Canadian Biography*, Semple at first misjudged Métis intent—and then sent for cannon, a little too late. According to Bryce, more men of European descent were killed at Seven Oaks than in the whole preceding two centuries of fur trade rivalry in Canada.[10]

It has never occurred to Anglo Americans, either in the academy or in popular culture, to make Santa Ana a hero of regional resistance to federal domination nor to cast him as an Indigenous leader valiantly resisting imperial domination. Although many American academics, such as Richard Slotkin, have criticized the American obsession with "regeneration through violence," and although the New Western Historians and most of their immediate academic predecessors have shown that America's frontier epic was neither as predestined nor as admirable as Turner had portrayed it, Americans have never had the choice of heroes and founding fathers that Canadians have had.[11] While popular writers like Bryce and academics like W.L. Morton have chosen to be the sons of the Selkirk settlers rather than of the Nor'westers, in Innis the possibility exists to claim a very different descent. Despite the flaws in Canada's claims to be the mosaic, not the melting pot—to be the first multicultural society—this is not merely a self-serving rhetoric dreamed up sometime around the Trudeau years, but a potential that has been in the idea of western Canada since long before Confederation and is inescapably part of Canada's historiography.

While there have certainly been large, synthetic histories of the Wests since Innis and Turner—most importantly Walter Prescott Webb's *The Great Plains* (1931), Ray Allen Billington's many times revised *America's Frontier Heritage* (1966), and Arthur S. Morton's *A History of the Canadian West to 1870-71* (1939) as well as various accounts, such as George Stanley's, centred on Riel—the historiography of the mid-twentieth-century Wests belongs mainly to monographs and articles.[12] The rise of the New Western Historians in the 1980s, however, has required a new framework for understanding western history and hence the publication of several ambitious overview histories. I would like to conclude this chapter by looking at four of the most influential, two Canadian and two American, in the context of the Innis-Turner dichotomy I have sketched above and in terms of their incorporation, or lack of incorporation, of monographic texts that have substantially stretched these interpretations.

Both of the American texts, Patricia Limerick's *Legacy of Conquest* (1987) and Richard White's *"It's Your Misfortune and None of My Own"* (1991), cover the entire trans-Missouri US West, which means that the Great Plains region tends to drop out of consideration for long periods and there is no attempt to look at Canada, either comparatively or as part of a larger region. Both Canadian texts, Gerald Friesen's *Canadian Prairies* (1984) and John Herd Thompson's *Forging the Prairie West* (1998), on the other hand, are focussed on the Prairies and do, particularly in *Forging the Prairie West*, offer comparisons to the US West.[13] Both of the American books present themselves with metaphoric titles that suggest, as the texts reveal, what Limerick calls the "injured innocence" that the westerner addresses to the East and particularly to the federal government. The Canadian texts use more straightforward titles that simply announce their subject. *Canadian Prairies* and *Your Misfortune* are texts designed to be used in western-history classes. *Forging the Prairie West* is part of a series of regional Canadian histories published by Oxford University Press for the general reader but also for use as university texts. *Legacy of Conquest* is more personal and idiosyncratic, less designed on the coverage model. It is more concerned with establishing a vantage point on the West that, in contrast to Turner, is present-centred, concentrates on the West as place rather than process, and eschews the triumphalism inherent not only in

Turner but in the whole Manifest Destiny, Indian-fighter popular culture of the American West.

Instead of discussing a linear process moving from east to west, Limerick instead provides a drama with many players coming from many different directions or originating in the West: "Everyone became an actor in everyone else's play; understanding any part of the play now requires us to take account of the whole. It is perfectly possible to watch a play and keep track of, even identify with, several characters at once, even when these characters are in direct conflict with each other and within themselves."[14] Limerick's directions do not include the North, however, because she never looks at Canada, nor does the book become quite the multivocal text that she promises, as her own voice remains quite determinant. Although White does look at the urban twentieth-century Plains West, as in his discussion of World War II and the subsequent aircraft industry in Wichita,[15] he, like Limerick, tends to discuss the Great Plains primarily as the nineteenth century of Indians and homesteaders and the 1930s' disaster of the Dust Bowl. Both authors address the bulk of their coverage of the twentieth-century West to the Southwest and the Pacific Coast.

Since the Prairies region has been occupied by human societies for millennia, contemporary historians, working in English and using written sources, face a substantial problem in dealing with all but the most recent four centuries—what to say about all those preceding centuries? White announces at the beginning of *Your Misfortune* that his definition of region relies on political geography and delineates the West only in terms of its Euro-American occupation. Although this is in many ways a sensible decision that keeps the book from either becoming impossibly long or shrinking Indigenous occupation to a relatively few pages that trivializes Indigenous longevity and impact on the land, it also seems to presage the loss of focus on Native peoples after the end of the Indian Wars. Both Limerick and White discuss John Collier and the "Indian New Deal" as well as AIM and the Red Power movement, but both largely skip over the continued and insidious dispossession of Indians during the twentieth century. Given the controversy over Angie Debo's *And Still the Waters Run* (1940) and its enduring fame, as well as the publication of Michael Lawson's tellingly titled *Dammed Indians* (1982), it is unfortunate that this chapter in history rates so much less ink than the Indian Wars. Custer and his less vivid peers, though advancing what they saw as a well-deserved Anglo-Saxon empire in America by means that were cruel and vicious, at least displayed a physical courage and panache that were completely lacking in the lawyers and "guardians" who made an industry out of cheating mixed-blood orphans out of their land and even dispossessing whole communities of successful small farmers in Oklahoma in the twentieth century. Bureaucrats who consistently flooded reservation land along the Missouri and other western and northern rivers with unneeded dam projects and relocated successful villages from riverine forests to unprotected grasslands devoid of game are similarly bland and faceless. Sincere child welfare advocates who, through various programs, ensured that more than half of Native American children would not be raised in their own families during the 1950s and 1960s also deserve more scrutiny, especially as their policies are still continuing, with Aboriginal children severely overrepresented in foster care and other types of state guardianship in both countries. To show the nineteenth-century military defeats of the Cherokees, Lakotas, and others without showing the economic and social destruction of their descendants who had *successfully* made the transition to reservation life trivializes both the suffering and the resentment that fueled Red Power in the latter part of the twentieth century. Serious discussion of Native American sovereignty and other contemporary issues in Native American society and culture is underground in the United States, but it is becoming increasingly audible and effective in Canada, largely enabled by Supreme Court decisions that affirm Indigenous rights. Fiction writers have probably done a better job than synthetic historians with these tales, as witness Linda Hogan's *Mean Spirit* and especially Thomas King's border-crossing *Green Grass, Running Water*, with its reference to Debo's and Lawson's issues.[16] Similarly, Beatrice Culleton (Moisionier) with *In Search of April Raintree* (1983) and filmmaker Alanis Obomsawin with *Richard Cardinal: Cry from a Diary of a Métis Child* (1986) showed some of the tragedies of foster care before the historians did.

Friesen's *Canadian Prairies* and White's *Your Misfortune* are probably the most comparable of the four books, as both are textbooks for university classes. Friesen begins with Indigenous societies on the Canadian

Prairies, a somewhat more manageable topic than Indigenous population on the less-glaciated Plains, since much of the area emerged from glaciation only four thousand or so years ago. Nonetheless, his use of oral sources is limited, and archaeological and ethnohistorical sources are not particularly clear or definitive, and all are focussed most intensely on the last four hundred years. (The explosion of monographs since Friesen was writing, more than twenty years ago, will make things much easier for future historians.) While both Limerick and White explicitly repudiate Turner, Friesen mentions Innis, as he does many previous historians, only in passing, as "the economic historian who first perceived the pattern in the Canadian staple trade."[17] Friesen does, however, use Innis's framework for his extensive discussion of the fur trade era. While White and Limerick shift their twentieth-century interest away from the Plains, Friesen, tasked with the Prairies, simply limits his coverage of the more recent past. Like Limerick and White, he does not discuss the post-treaty dispossession of Native peoples, but unlike the two Americans, he had no sources such as Debo's from the 1930s and was writing before the important monographs on the subject, such as Sarah Carter's *Lost Harvests* (1990), were available. Canadian Indian policy gained immeasurably—from the point of view of the federal government and non-Native intending settlers—from lagging behind American Indian policy in the nineteenth century. In the twentieth century, Canadian Indian historiography lagged far behind that of the United States, in part because of the protective myth of Canadian benevolence but even more because of the lack of a European-educated Native or mixed-blood intellectual corps that still identified as Indigenous, such as the La Flesche family, John Joseph Mathews, D'Arcy McNickle, Gertrude Bonnin (Zitkala Sa), and others who wrote and published in the United States. Pauline Johnson stood alone until Olive Dickason began publishing Aboriginal history in the 1970s. The field now supports many talented Native scholars, including Taiaiake Alfred, James (Sakej) Henderson, John Borrows, Leroy Little Bear, Antoine Lussier, and Blair Stonechild, some of whom are discussed below. The lack of an Angie Debo in Canada, however, is perhaps less inexplicable than her existence in the United States.

Of the four books, *Forging the Prairie West* is perhaps the most successful. For one thing, it is the latest and has access to the most monographic materials. For another, it is the shortest, and although Thompson does not oversimplify, he has room only for broad strokes, leaving less room for quibbles. Thompson gives the most emphasis to the twentieth-century Prairies, and although his US/Canada comparisons are relatively few, he uses them effectively to frame Canadian controversies, such as those over Macdonald's National Policy, in ways that make strategies possible for using comparative data to generate new answers to old questions. Like the other authors, he does not look at the systematic white-collar aspects of fleecing Native peoples during the twentieth century.

History is inescapably presentist, if only because no one would write, or would bother to read, a study that had absolutely nothing to do with the lives they are living and the thoughts they are thinking. History is also presentist because it is cumulative. Even purely archival research rests not only on what has been kept into the present but on the kinds of questions raised by past historians. Similarly, we can only read past historians from our vantage point in our own particular moment. Historiography, however, is a kind of cross-focussing device that allows us to look at the kinds of ideas that have framed our understanding of the past and to refine the questions we will ask in the future. The Wests of Friesen and Thompson are as different from the Wests of Limerick and White as are the Wests of Innis and Turner. But all the cross-border comparisons remind us that we write not about the past but about our ideas of the past. Ideas, as Henry Nash Smith showed us more than half a century ago, determine the actions of historical players, but they also determine the actions of historians. Cross-border comparisons force us to examine our whole frame of reference, and when we do that, we may decide that the play we have been watching is even more complex and amazing than we had thought.

Homesteading as Capital Formation on the Great Plains 7

Dedicated to Paul Wallace Gates,
America's greatest public lands historian

The frauds against Native peoples—such as the withholding of rations, the unilateral abrogation of treaties, the use of "Rebellion" to justify outright treaty violations, and myriad personal get-rich-quick schemes for the benefit of unscrupulous individual Indian agents and other officials—were certainly not the only frauds of the late nineteenth century. Frauds against Indians were only one part of an ethic, particularly in the United States, that involved turning the public domain to private gain. Mark Twain's strand of *The Gilded Age*, the novel that gave the era its name, details the life and death of a particularly egregious scam—based in Kansas in real life, though moved to Tennessee in fiction. "An Act to Found and Incorporate the Knobs Industrial University," supposedly for the education of freedmen, is simply an elaborate scam to get the US government to spend millions of dollars buying worthless land for the enrichment of the family that owns it—and the congressmen and senators who have been bribed into supporting the bill. Ironically, in the context of this chapter, it involved taking land back into the public domain, but the net result is the same—public domain is to be transmuted to private capital. Twain based his story on the successful plan by Kansas Senator Pomeroy to convert Ottawa Indian land into private gain through the founding of a university that, predictably, did not educate many Ottawas.[1] But many of the truly useful public works of the mid- and late nineteenth century also succeeded through creative financing—that is to say, fraud. The extralegal Credit Mobilier provided the flexibility necessary to attract capital to the economically premature enterprise of the Union Pacific Railroad, the first of the transcontinental iron horses that enabled homesteaders. It, too, was pilloried by a literary wit, John W. DeForrest, who straightfacedly parodied the Union Pacific as the Great Sub-Fluvial, a tunnel to be built *under* the Mississippi River from Minneapolis to the Gulf. The Pacific Scandal that temporarily removed John A. Macdonald from the office of prime minister was an even more premature attempt to fund what would become the Canadian Pacific Railway.[2]

Given that the original *Homestead Act* was passed in 1862, not long before the Gilded Age, it is surprising that it has never been treated in the same irreverent way by either historians or novelists. Paul Wallace Gates, our most influential public land historian, concluded a 1962 talk, at the centennial of the first of the Homestead Acts, by saying that "their noble purpose and the great part they played in enabling nearly a million and a half people to acquire farm land, much of which developed into farm homes, far outweigh the misuse to which they were put." I will argue something somewhat different: the Homestead Acts were only indifferently successful as instruments for creating family farms—probably less efficient than the various sales regimes that they partially replaced—but they were a howling success at moving the public domain into the private sector and turning "free land" into capital for the rapid development of the West. That success, like the building of almost all the transcontinental railways, depended on various frauds, some more egregious than others, and a kind of "wisdom of crowds" that involved tweaking the rules so that an intending farmer could, by selling successive relinquishments and pre-emptions or even proved-up claims, finally amass enough capital to establish a successful farm. Those with less aptitude or taste for farming could similarly prepare themselves to set up a store in a nearby town or more distant city, secure an education, or do anything else capital allowed. As Senator Robert Stanfield of Oregon asked ruefully in 1926, the tag end of the homestead era in the lower forty-eight, "It is a matter of historical record, is it not, that it has taken about three migrations everywhere in the western movement to bring about permanent settlement?"[3]

But it did not take until 1926 for people to recognize that the *Homestead Act* had not turned "free land" into farms overnight. Opponents and even some proponents of American homestead laws had recognized as early as the 1840s that "free land" would be a great boon for speculators. Canada had no real choice but to authorize the *Dominion Lands Act* in 1872 to match the American offer and even to raise it by requiring only three rather than five years on a homestead before the intending settler could obtain it in fee simple, which tended to limit debate on the drawbacks of the act. Discussion about the *Dawes General Allotment Act* of 1887, a bizarrely *reverse* Homestead Act that divided American Indian reservations into individual or family plots and confiscated any remaining non-allotted lands for the benefit of non-Native people, was clearly debated and passed in the context of turning common lands into capital and individual Indians into aspiring capitalists.[4]

Our understanding of the various Homestead, Dominion Lands, and Allotment Acts shifts if we change our focus from the formation of *homes* to the formation of *capital*. In a famous 1938 article, "The Homestead Act in an Incongruous Land System," Paul Gates discussed the ways in which the *Homestead Act* contradicted the more confirmed uses of the public land as a good to be sold for the benefit of the US treasury or to be granted—as to soldiers and railroads—in payment for services. An even more incongruous fit, however, is that of "free" land into an almost maniacal nineteenth-century American obsession with private property and capitalism. While the millions of homesteaders who staked claims—and the half of them who actually proved up—were undeniably real, and while many of them undeniably wanted to turn "the little old sod shanty on the claim" into a farm home that would endure for generations, we cannot really understand or evaluate the workings of the various Homestead Acts without understanding them as mechanisms for capital formation, nestling securely within the ideology of the Gilded Age.

The Homestead National Monument outside Beatrice, Nebraska, commemorates the site of what is generally accepted as the first homestead filing. Daniel Freeman, a Union soldier on Christmas leave, persuaded the general land office in Brownville to open just after midnight on New Year's Day, 1863, so he could file on a homestead for his family and still get back to his unit before his leave expired. Or so the official story goes. Dan Jaffe, a poet who became fascinated by his famous, but unrelated, namesake finds that Daniel was not listed as a member of the unit he claimed. More likely, he was a fraud, a queue-jumper who managed to file before the land office opened so that he could oust the squatter who had already taken up residence on the quarter section in question and even built a small log shack on the creek bank.[5] That the Homestead Monument apparently celebrates a fraud is actually quite appropriate. Like the 1887 *Dawes Allotment Act*— which broke up reservation holdings and distributed them to individual families, in the process opening up "surplus" land to non-Indian settlers— the various versions of Homestead Acts in both the United States and Canada had contradictory goals. The first, the explicit, goal was to situate farm families on all the land that might be suitable for farming and to encourage them to develop a huge agricultural empire on the Great Plains, giving rise to towns and cities and filling in the centre of the continent until both nations were settled and prosperous from sea to sea.

For some families, the Homestead Acts and the *Dawes Act* were successful. Some homestead families have even persisted to the present day on the land that great-great-grandpa or great-great-grandma homesteaded. They are the people who are celebrated in newspapers and at county fairs or town reunions as "century families" or "heritage farms." That they are celebrated shows how rare they were and are. Even fewer Native families farm or ranch on original Dawes-era allotments. The land still held by the descendants of original allottees is now usually leased out to someone else, and individual title is obscured in "undivided heirship lands," administered by the Bureau of Indian Affairs and providing little accountability or return to the ostensible owners. (A settlement painstakingly worked out over fourteen years between Native landowners in a class-action suit against the Department of the Interior has finally passed the Senate, allowing for the issuing of substantial, but nonetheless token, cheques to allottees in the summer of 2011.) Many homesteaders did not prove up, and most allottees lost their allotments. Of the homesteaders who did prove up, most were not there ten years later. Of the ones who succeeded, almost none could get by with just the original quarter section, or even with an adjacent quarter thrown in through pre-emption (a process by which the homesteader could buy a quarter section). Many of the most successful homesteaders were large families, often European immigrants, who settled sons and sons-in-law and, in the United States, daughters, on adjacent claims. Or, in Canada, they settled in large block settlements that provided them the flexibility to develop village sites, water, timber, cropland, and pasturage. Others succeeded by buying up the homesteads of neighbours who proved up and moved out or by buying railroad lands or sections reserved to provide for schools or other socially valuable institutions. The large family settlements contradicted the idea that it was the individual, or at least the nuclear family, that was the ideal, market-based unit for the land. The families that succeeded by buying out their neighbours required that their neighbours fail. As Paul Gates points out, Iowa was largely turned into farms without the use of the *Homestead Act*, and farm-making proceeded more quickly in Iowa, Illinois, and Indiana immediately after the Civil War, in the first years of the *Homestead Act*, than it did on the Great Plains. Thus, it is quite probable that many of the same kind of people who succeeded in creating farms in the cornbelt could also have succeeded in the same way—through buying land—on the Great Plains. The "safety valve" theory of homestead proponents—and opponents—that free land would pull the oppressed labourers from the eastern cities to happy farm homes on the Great Plains, raising the wages of those who were left, did not work. Oppressed labourers could not afford to travel west, let alone to amass the capital needed for a house and farm equipment.[6] But homesteads—and especially the extralegal flexibility of the land system, the opportunity for even a small settler like Daniel Freeman to speculate—definitely speeded up the privatization of the national domain.

Nor is this a view from hindsight. Beneath all the regional differences about settlement of the West, the debates about land for settlers that occupied Congress from the 1840s through the early 1860s openly discussed the likelihood that free homesteads would be a boon for speculators. "To offer each [landless man] a quarter-section of Public Lands as a free gift with liberty to sell the fee simple to anyone, would be simply enabling the speculator to obtain at second hand for a few dollars what now costs him hundreds, and thus to monopolize Counties instead of Townships," wrote a commentator in the *New York Weekly Tribune* in 1845.[7] That is exactly what happened, either by dummy entrymen working to monopolize access to water for large cattle spreads or timber for lumber companies or minerals for mining companies, or by small settlers just trying to get enough money to start over and finally own a farm. Thomas Flanagan, describing how speculation in Métis land scrip in Manitoba worked, wrote, "The Métis might be compared to producers of land, the middlemen to wholesalers, the land companies to retailers, and the settlers to consumers, in order to understand the chain of relationships and the profit margins at each stage."[8] Greed and self-interest were clearly part of the system created by the Homestead Acts and foreseen by both supporters and detractors. In fact, a certain self-interested shrewdness, if not outright greed, was necessary for even the most idealistic intending farmers to succeed.

The truly acquisitive nature of the land laws, however, is most clear in the debates about the Allotment Acts that would reduce Indian-held lands from about 175 million acres at the passage of the *Homestead Act* in 1862, to 138 million acres at the passage of the *General Allotment Act* of 1887, to 52 million acres in 1934 when allotment was officially rescinded as policy. In 1880, the Committee on Indian Affairs of the House of Representatives reported favourably on an allotment bill. Republican Russell Errett of Pennsylvania joined with two other members (Charles Hooker and T.M. Gunter) of the nine-member committee to submit a minority report.

> The main purpose of this bill is not to help the Indian, or solve the Indian problem, or provide a method of getting out of our Indian troubles so much as it is to provide a method for getting at the valuable Indian lands and opening them up to white settlement. . . . The provisions for the apparent benefit of the Indian are but the pretext to get at his lands and occupy them. With that accomplished, we have securely paved the way for the extermination of the Indian races upon this part of the continent. If this were done in the name of Greed, it would be bad enough; but to do it in the name of Humanity, and under the cloak of an ardent desire to

promote the Indian's welfare by making him like ourselves, whether he will or not, is infinitely worse.

Errett went on to say that Indian progress toward "civilization" had hitherto been made under the "tribal system": "Gradually, under that system they are working out their own deliverance, which will come in their own good time if we but leave them alone and perform our part of the many contracts we have made with them."[9]

Errett's arguments, like the earlier arguments that free homesteads would lead to land speculation, got nowhere, but they clearly show that even during the 1880s, many people knew that the allotments had nothing to do with Indians. Nor did they have anything to do with homesteaders—the "surplus" land was to be sold, not homesteaded. Despite a glaze of humanitarianism that would have done Mark Twain's real-life Indian-swindling Senator Pomeroy proud, the *General Allotment Act*, or *Dawes Act*, of 1887 and its predecessors and successors were about making land available for purchase, about privatizing land in favour not of Indians or homeless white farmers but of people with money—people whom almost anyone would call land speculators. According to Flanagan, Métis scrip worked the same way. But then, most non-Native western landseekers were speculators. Attempts to stem fraud in the homestead system were largely unsuccessful, mostly because most westerners, no matter how upstanding, condoned the various lies and extralegal activities that were necessary for the Homestead Acts to work to commodify land and turn it into private property as quickly as possible. Whatever people said the Homestead Acts and the allotment acts were for, opponents, at least, made it very clear that they would actually function to get public land and Indian and Métis land into private hands. The various land acts actually did function very efficiently to privatize and capitalize the land. Let us look at how that happened.

If the most important goal of the *Homestead Act* was the unstated one of commodifying the land as quickly as possible to get rid of the commons and facilitate the accumulation of capital, then the success of individual families on the land was irrelevant. While most of the successful farms in the eastern, reasonably well-watered sections of the Great Plains could have been settled, as Iowa was, over a longer period of time using loans, that would have concentrated capital in the hands of those who already had it rather than bringing more capital into the country by converting land directly into capital. If the ideal homesteader was actually the entrepreneur, the intending capitalist, then Daniel Freeman is a most appropriate homesteader to celebrate, though it is unlikely that many were quite as crafty as he. The tall tales of claim shacks that met minimum-size requirements—but in inches rather than feet—or were merely hauled in for an inspector before being hauled off to "prove" the next claim, or of timber culture claims that were "planted" with twigs rather than saplings, however, make one wonder. Freeman is also an appropriate man to commemorate because he dealt in townsite speculation as well as homesteading, at one time claiming the lot on which the Gage County courthouse was to be built.[10] Townsites were far more lucrative operations than farmland for turning the land into capital, but the fiction of the farming community—and usually a railroad terminal—was necessary to add value to the putative townsite.

For many entrepreneurs, the purpose of homesteading was to establish a claim that could then be turned to capital. One could sell a relinquishment—another extralegal frill added to the Homestead Acts by inventive pioneers—by which a title holder "relinquished" his or her claim to another, who immediately filed on the land and took over the improvements. One could pay cash to commute a homestead (obtain title before the requisite five years) or to buy a pre-emption, or one could actually prove up and then sell the patented claim or mortgage the land. As Paul Gates writes, even more important than the large-scale speculator

> was the small man with no capital for the arduous task of farm making who nevertheless took up a piece of land to which he expected to acquire a preemption right. Frontier custom assured that his claim of one hundred to two hundred acres was his to do with as he wished. With patience and little labor he might improve slightly, sell, and then move to another tract and do the same thing.[11]

Legally, each individual was entitled to only one pre-emption, but custom held that serial pre-emptions were perfectly fine, and there was really no way to track an individual, especially if he used different versions of his name or even different names. Relinquishments were even more ext-ralegal and just as important. Again, Paul Gates: "Undoubtedly, the business of selling relinquishments was carried beyond all justification, but it should be emphasized that it permitted persons who lacked the means with which to begin farming to acquire some cash, farm machinery, and stock and after two or three false starts and sale of relinquishments to succeed finally in establishing ownership of a going farm." Or the seller might start a store, move to the city, or do any number of things that required ready money. Such ground-level tinkerings with the law are but another example of the informal knowledge acquired by doing that allows any large and overly generalized scheme to work in the real world. According to Benjamin Hibbard, the possibilities of speculation in land really did draw to (or keep in) the Dakotas many who had no intention of becoming farmers but who were glad to acquire some cash. He quoted a Senate report on twentieth-century homesteading in North Dakota: "Thus a veritable multitude of farmers' sons and daughters and servant girls as well as ne'er-do-wells have sought land in the Dakotas." The gender equity in this quotation is also intriguing, for it shows that homesteading was one of the few places in early twentieth-century United States where women (as long as they were unmarried) could accumulate capital on the same footing as men. They lacked this opportunity in Canada.[12]

For capital to "accumulate" on the Great Plains, it had to come from somewhere else, and come it did. Since mortgage rates were high on the frontier, sometimes—through creative financing and in excess of usury laws—banks, mortgage companies, and other lenders in the East and in Europe were eager to make as many loans as they could. The land would always be there, they reasoned, and the world would always need food. Great Plains land would not, however, always produce food. It behooved the astute pioneer entrepreneur to mortgage a homestead for more than the land was worth even on speculative markets, take the money, and leave the land to the hapless mortgage holder. As Paul Gates writes in the *History of Public Land Law Development*:

> High interest obtainable in the West attracted farm loan companies which might be either western or eastern but in either case drew their funds

from the East. On occasion there was a plethora of such funds and agents competing with each other in placing their funds, paying little attention to the actual improvements on the quarter section or the reliability of the settler.

Again, to quote Paul Gates, "Many western settlers had larceny in their hearts when it came to dealing with the government, and it did not stretch their consciences unduly to take advantage of the insurance companies or other absentee sources of capital." And again, "Many westerners had no compunction about taking the loan and skipping, leaving the abandoned tract to the mortgage company." The newly capitalized ex-farmer could, like the others, either start a new career or head to another frontier and start the process over, this time with enough money to develop an actual farm—and maybe buy out some neighbours. High interest rates and loans that were much too big in proportion to the productive capacity of the land certainly led to high rates of mortgage default and foreclosure, and in many cases, these marked the human tragedies of families who had given their all to build farms in places or in times when their know-how and technology were insufficient for the soil or climatic conditions. David Jones amply demonstrates this in his *Empire of Dust*, set in southeastern Alberta during the 1920s. In other cases, the tragedy, such as it was, befell the bank, the mortgage company, and the proverbial "widows and orphans" who had invested in farm mortgages.[13]

While most historians have focussed on the public face of the Homestead Acts, the extent to which they succeeded or failed in terms of filling the land with yeoman farmers, few scholars have looked at all carefully at the ways in which homesteads succeeded as instruments of capital accumulation in an environment that could support neither densely settled farms nor other forms of economic development commensurate with the enormous amounts of money that flowed into the Plains. Paul Gates, however, was explicitly aware of how the Homestead Acts worked. He conceded that "a very considerable portion of the misuse of the public land laws resulted, it appears, from the credit needs of actual settlers." Yet curiously enough, he ends his centennial homestead speech condemning the useful fraud and alluding to the ostensible purpose of the acts:

The old evils of careless drafting of land legislation, weak and inefficient administrations (inadequately staffed), and the anxiety of interests to take advantage of loopholes in the laws, all brought the Homestead Acts into contempt and censure. But their noble purpose and the great part they played in enabling nearly a million and a half people to acquire farm land, much of which developed into farm homes, far outweigh the misuse to which they were put.[14]

I would argue, rather, that their success was *in* their "misuse," that their real utility was in the quick and efficient privatization of land and its conversion into capital. Whether or not privatization was a good idea is another question, one to which I would probably answer no. By "success" I mean that the laws succeeded in doing what I believe they were intended to do—turn land to capital and convert the commons into the privately owned. The success or failure of individual families of white homesteaders or Indian allottees or Métis scrip recipients was irrelevant to the underlying goal of the acts, though not to the families themselves. The successes, as Paul Gates himself shows, however, have allowed historians to take the Homestead Acts at face value rather than as the greatest middle-class entitlement ever promulgated by a federal government, especially in the United States, whose ostensible rules were stricter than those of Canada, but whose loopholes gave the intending capitalist more freedom and levelled the playing field between (unmarried) women and men. As for those who really did fail or for the genuine hardships that some successes endured, they only burnished the legend of "free land" and farms.

The "success" of allotment in privatizing and whitening reservation lands bespeaks a far greater tragedy, one that continues to blight people's lives today with the extreme poverty of reservations and the small percentage of land that actually remains in Native hands. Far more than homesteading proper, allotment reveals the urge for extreme privatization and the bureaucratic urge to have owners to tax, the final act in the millennium-long "loss of the commons," as Irene Spry points out, that forwarded nineteenth-century capitalism at the expense of the land and the people who most loved it.[15] Although Canada for the most part avoided allotment and the loss of lands through sale and tax foreclosure, the smaller initial size of the reserves forced growing families off reserves, while appropriations for highways, parks, bombing sites, and the like, plus the actual confiscation of lands during World War I, led to a similar loss of lands and to continuing reserve poverty.

Let us see what the development of the homesteading West looks like through spectacles focussed on privatization and capital formation. Capital flow and homesteading coincided to an extraordinary degree—in the United States from the end of the Civil War in 1865 and the completion of the Union Pacific Railroad in 1869 until 1914, with dips around the crashes of the early 1870s and of 1893, and from 1896 to 1914 in Canada. Canadian scholars—not wed to the idea of Manifest Destiny and, in the West, far more suspicious of the federal government at the time of the homestead boom—have been a little bit less triumphalist than their colleagues in the United States. In their *History of the Canadian Economy*, Kenneth Norrie and Douglas Owram point out that dry farming and Marquis wheat made the Prairies suitable for agricultural development and that wheat was the staple of not only the Prairies but the nation as a whole. According to the Rowell-Sirois report, the settlement of the West after 1896 brought prosperity and a sense of accomplishment to Canada. All of this is received opinion, but then Norrie and Owram point out that the expansion of the economy after 1896 was not fueled by wheat exports, which didn't really rise until 1905 and after, but by investment. Conventional economic development theory focusses on export as the engine for generating economic expansion, but there is no absolute reason this must be. The wheat boom and the "winning of the West" contributed to Canadian growth, but it also overshadowed other major long-term developments in hydro, mining, and manufacturing. Wheat, however, accounted for half the rise in the Canadian standard of living between the censuses of 1901 and 1911.[16]

But in some ways, wheat was a proxy for the value of the land. Until the 1990s, wheat was still king in Saskatchewan and Alberta, the Dakotas, and the Plains of Montana, and a major crop in Kansas, Nebraska, Colorado, Oklahoma, and Texas as well. But since 1919, wheat has been a surplus crop, an export for which markets must be painstakingly won and held. While American farmers assail the Canadian Wheat Board as wielding an unfair monopoly power, Canadian farmers protest by trucking their grain illegally to elevators in North Dakota or Montana to dramatize their claim that they are being unfairly prevented from selling their wheat for higher prices south of the line. A minority Conservative government attempts to dismantle the Wheat Board. The bounteous harvests of the golden years between 1896 and 1914 turned out not to be sustainable when dry weather returned, and not to be marketable in any case. Pierre Trudeau's famous and ill-timed rhetorical question to Prairie farmers, "Why should I sell your wheat?," was legitimate in substance if not in tone. Why should either federal government prop up a farming system that did not fulfill any real market, was increasingly driving rural population from the land, and was ecologically unsustainable? As Jackie Skelton said on a CBC Radio *Ideas* program rebroadcast on 27 September 2005, the people who homesteaded western Canada saw their role as feeding the Mother Country and a hungry world, but they are not needed for that purpose any more. The hungry world is, in the current round of World Trade talks, demanding that US, Canadian, and EU governments stop supporting their farmers so that developing world farmers can compete in world markets—even markets on the Great Plains. As James Scott and Raj Patel have shown, Great Plains grains, over the last century and a half, have played a somewhat sorry role in impoverishing farmers and whole societies in the Global South, presaging unfortunate consequences for the Plains as well.[17]

The rapid development of biofuels from corn, soybeans, and canola at least temporarily changed the equation on farm profitability and stabilization, though the oil price plunge of late 2008 and early 2009 may be changing it back. Some of the newly built ethanol plants are now bankrupt and shut down, but most continue to function and ethanol remains a popular formulation at gas stations, particularly in grain-producing states. Although the amount of petroleum used to fuel farm machinery and to fertilize biofuel crops, and the amount of water used to raise them and process them into ethanol, makes the ecological advantage of ethanol problematic, government minimum requirements of ethanol in fuel mixes and outright subsidies have given ethanol an increasing market share. Because the profit margins on food are larger the greater the processing employed—wheat is cheaper than flour, which is cheaper than breakfast cereals, and so forth—crops intended directly for food are the ones with the lowest profit margin and thus those most likely to be replaced by biofuel plantings. Wheat and corn is no longer dumped on poor countries in foreign aid, but the small farmers in South and Central America and in Africa who used to grow food crops have been replaced by larger plantations specializing in crops like coffee, cocoa, rubber, and narcotics for export. They cannot quickly be replaced. Corn used to produce high fructose corn syrup to sweeten soda pop and to "finish" cattle in feedlots to produce fatty "marbled" beef, however, continues to compete with corn for ethanol plants. Farmers astutely plant fewer acres to crops that humans eat—macaroni requires far less processing than soft drinks—and more to corn for biofuels or highly processed foods. This does solve the problem of world grain surpluses, but it also degrades the promise of the "breadbasket of the world" without helping to rebuild staples farming in the Global South.

We seem to have entered a new and differently destructive grain boom, complete with its own bust, without answering any of the questions raised by the earlier booms or busts on the Great Plains. The homestead mission was part of the ideology of the Garden of the World, but as that promise now seems out of reach, we need a new reason to maintain the monocrop commercial agriculture that was the basic premise of the *Homestead Act*. If our farms are no longer needed to feed the hungry world, we need to redirect the sense of heroism at the root of the Great Plains self-image. Our biofuels, then, can protect us from dependence on the oil policies of the Middle East. In fact, biofuels concentrate the Plains economies even more in the energy sector than did prior extractive energy missions. If feeding the world with our wheat—or even our canola oil—is not the answer, what then *is* the purpose of rural culture? What can the rural Plains and their people offer to the world? What can they learn from their Native neighbours? The early golden years of wheat, when settlers poured in to turn Saskatchewan into the third largest province by population turned out to be an anomaly. The land itself was the commodity that brought investment to the Plains, and when the land was gone, put into private hands more or less firmly at the end of the feasible homesteading era, the investment dried up and both people and capital began their steady hemorrhage away from the Great Plains region.

Like travellers on the overland trails, many homesteaders recognized that they were part of a great folk movement and self-consciously recorded their experiences in diaries and memoirs. Community histories, mostly from the 1950s, 60s, and 70s—especially in Canada, where they were commissioned wholesale as part of the centennial celebrations—also offer plentiful reminiscences of homesteaders, and they have been mined by historians, particularly John Bennett and Seena Kohl, and Paul Voisey. In addition, several writers either spent childhoods on homesteads or homesteaded as young adults and wrote effectively about their experiences. The community histories mostly celebrate the folk who stayed on the land, so there are discrepancies in emphasis among the different sources. Once a *New York Times* reporter called me, looking for a regional image to go with the dead mule of southern literature. I suggested "leaving the homestead," a motif running through both memoirs and fiction and grounding what we are to mean by asking how the Homestead Acts, in aggregate, succeeded or failed.

Hamlin Garland, born in Wisconsin and raised on an Iowa farm, came to Brown County, South Dakota, with his parents as a teenager. He later settled on his own quarter section pre-emption claim long enough to be able to sell (he says "mortgage," but that appears to be a euphemism) what was apparently a relinquishment, since the land had apparently still not been surveyed, for $200 and to move to Boston, where he educated himself to be an American man of letters, apparently his goal even as a schoolboy in Iowa. Garland's first claim to fame came as the author of short stories about the grimness of farm life. He focussed especially on the plight of women, the evils of landlordism and land speculation, and the guilt felt by the boy who leaves the farm, makes good, and returns to see the poverty and drudgery still suffered by the people he left behind. There is a certain irony in the fact that Garland, that devout follower of Henry George and the scourge of the land speculators, got his own professional start by an extralegal bit of speculation, holding down a pre-emption claim during a Dakota winter for the sake of selling a relinquishment. On the other hand, Garland was making a rational choice. His stories make it abundantly clear that, despite his love for the beauty of the land, particularly in the Wisconsin Coulee country, he had no aptitude for being a farmer but rather a burning desire to be a man of letters. Selling his claim gave him a tidy nest egg that he could not easily have obtained in any other fashion. Garland was unusual only in that he used his nest egg to go to Boston and to read through the public library rather than attending the University of Wisconsin or Iowa State or one of the various normal schools in the Midwest. A self-made (and intensely self-conscious) man of letters is always a rarity. Selling either land or a relinquishment to fund post-secondary education for oneself or one's children was probably a more common pattern, but the most common pattern was turning homestead land to money in order to fund some kind of business—as Garland's father did, running a store in town.[18]

In *Vulcan*, Paul Voisey paints a detailed and vivid picture of the amazing fluidity of the homestead community in and around Vulcan in southern Alberta. While American homesteaders were required to hold their land for five years and live on it nine months out of every twelve, Canadians were required only to live on their claims for five months out of twelve for three years, rendering their situation even more fluid than that of the Americans. Most of the pioneers who came to the Vulcan area in the decade before 1914 were already highly mobile, both geographically and professionally. Many had moved through several frontiers, several farms, and several off-farm trades, even if they were only in their early thirties by the time they arrived in Vulcan. Not surprisingly, they continued their mobility after landing in southern Alberta. As Voisey notes, "Generally, fewer than half the farmers remained in the township for as long as five years before 1920." And farm owners were the most stable class—farm tenants moved more rapidly than farmers, businessmen more rapidly than tenants, institutional managers such as elevator agents, bankers, and clergymen more rapidly still, and labourers were the most mobile of all. Homesteaders gambled their time against Ottawa for 160 acres of land. In addition, almost all the men in the Vulcan area gambled on card games and ball games. And particularly they gambled in land. As Voisey points out, their behaviour was fully as ironic as Garland's—at the same time that they were trying to take the gamble out of agriculture itself with dry farming and other "scientific" farming methods, and the gamble out of marketing with the wheat pools, they were gambling in everything else. "They did not oppose gambling so much as large competitors; mostly they only opposed losing."[19]

Voisey's frank appraisal of speculation in homesteads and also in railway lands is still fairly unusual among historians, who see the home-steaders who pulled up stakes as both failures and victims. For instance, Paula Nelson, whose study of West River Dakota settlers was published two years before *Vulcan* (1986) and covers a period only slightly earlier, almost seems to be fighting her own data to stick with the victim conclusion. Like Voisey's, Nelson's book is an expanded dissertation developed from a won-derful array of both printed and manuscript sources. While Nelson tends to pay better attention to women than Voisey does—especially since she was writing about the United States, where single women could homestead in their own right and thus had a lot more to say about homesteading from the point of view of actual proprietorship—she is scornful of "absentee" as opposed to "real" homesteaders, and particularly disparages women who started cultural institutions and then moved away. She seems to believe that these women unfairly raised and then dashed the community's hopes and suggests that they were congenital quitters who lacked the right stuff to become permanent pillars of the community. At the same time, Nelson sees most of her settlers as victims or failures, for the land was not as conducive to farming as the Vulcan area has proven to be. A homestead of 160 acres was not likely to prosper. Where Voisey looks at how mobility and highly "creative" application of homestead laws allowed frontier communities to raise capital and to establish mercantile and professional establishments, Nelson sees only broken dreams. She does not examine why the women who were cultural leaders would have been likely to have found it in the best economic interests of themselves and their families to engage in pri-mary economic development and then to leave.[20]

The Homestead Acts did create some 160-acre farms, but they also enabled some homesteaders to make something for themselves out of the land: those who were deliberately speculating in either farm land or town-site lots as well as those who took land either as a lark or in good faith, only to decide that a farmer's life was not the life for them. Those who sold out often became the small business owners, and the teachers and clergymen who came to a newly settled area were often inspired to stay because they could homestead on the side. If the first purpose of the Homestead Acts was the creation and development of private landholdings—and not family farms—from the public domain, then they, like the *Dawes Act*, represented stunning success, whether or not one agrees with the desirability of such goals. This aspect, however, almost never appears in celebrations or other discussions of the principles of homesteading.

While some intending farmers undoubtedly gambled their all, worked hard, and came out with nothing—victims of government hype and an incongruous land system—many others simply redesigned the system to fit their personal needs and the development of the community. Garland's creative cashing in of his homestead prospects, for instance, gave the "middle border" a short story writer who excelled at depicting frontier conditions and was a stalwart fighter for the rights of those who did want to stay on the land. Homesteaders selling out also provided elasticity to the land market, breaking out of the 160-acre straitjacket more effectively than pre-emptions could—since they depended upon unclaimed land—and less expensively than purchase of railroad, school, or other privately or publicly owned land. Even there, as Voisey again points out, small settlers speculated on railway land, hoping its value would go up enough before their final payments were due that they could sell it for a profit. Thus, instead of making money for the government or individual proprietors, as eastern lands had done, and instead of giving the railways the increased value of the land created by the social development Henry George describes, the various Homestead Acts and even the railway land grants distributed land to farmers and others who would use it in a reasonably efficient manner and included in the capitalization of the land many other families and individuals who would never be farmers. Ironically, as Voisey again points out, "the real victims [of land speculation] were those efficient expanding farmers who failed to sell out before 1920 when the falling grain prices and environmental difficulties that foreshadowed the catastrophe of the 1930s sparked a long-term fall in land values."[21] The winners, however, were probably their children, for school lands sold last and highest and thus provided a reasonable endowment for public schools.

Yet if reasonably successful small-scale speculators made up the bulk of Great Plains homesteaders and even purchasers of railroad land grants, some settlers really were the victims of hype by governments and speculators, particularly during the nineteenth-century panics and the Dust Bowl and Depression. Other families truly put roots down and prospered. Voisey suggests that it was not ethnicity or experience that formed the "stickers," but individual taste and, particularly, timing. Most Vulcan-area settlers had the support of family or at least friends from "home"—however such peripatetic people might define the term. But Europeans in block settlements probably had less chance for speculation and more of a cultural ethos that valued community.

Who were the folk that fitted into the "Little House on the Prairie" mode? Before answering that question, it is well to note that it is a trick one, rather like the question about who the first homesteader was. The Little House on the Prairie was not a homestead. The Ingalls family was squatting illegally on the Osage Diminished Reserve in southern Kansas. And it seems probable that they left the Little House rather than pay the government the $1.25 per acre that the land office would try to collect to pay the Osages the obligations that the United States had incurred in inducing them to move to Oklahoma. Our images are tricky. Another family that homesteaded in Kansas, immortalized by son John Ise in the memoir *Sod and Stubble*, is a more realistic choice. Here, we see a family suffering real tragedies, such as the death of their first child and numerous faithful and beloved farm animals, but building up the farm bit by bit through homestead, pre-emption, and purchase until it was profitable enough to send most of the surviving Ise children to college.

Mother Rosie Ise is in many ways the central character of *Sod and Stubble*, and with good reason. Not only was she personally the indomitable core of the family, but it was woman's work in subsistence agriculture and in raising a family that enabled the homestead frontier—and had from the beginning. It was also often women and children who maintained residence on the homestead during the winter while the man worked elsewhere to raise cash for stock, equipment, seed, and other necessities. This domestic work was particularly important on the earlier homestead frontiers in the United States, where child labour, such as Garland remembered from his own childhood, substituted for hired labour before the advent of highly mechanized farming. In fact, as Deborah Fink shows, the elasticity of child labour has been essential to the "family farm," and even in the twenty-first century, it is not uncommon for children under the age of twelve to drive tractors, grain trucks, and other machines at peak labour times such as harvest, haying, and seeding. Women birthed, raised, fed, and clothed all these labourers, so subsistence and reproductive work fused completely. Women's vegetable gardens, chickens, and milk cows fed the farm family and the hired man or hired girl, while butter-and-eggs money customarily funded a good many basic purchases, such as coffee, tea, sugar, and clothes.[22]

Wheat, usually the first cash crop on most North American frontiers because it produced a higher value per bushel than most other crops, required that the tough prairie sod be broken and perhaps sown to sod corn for a year to break down the tough roots and rhizomes of the prairie grasses. As Voisey points out, most homesteaders even in the twentieth century could not break a whole quarter section for several years after proving up—unless they were willing to gamble on hiring someone to do custom breaking and paying it off with a bumper crop. Like web-based marketing, wheat farming was real—but wheat farming did not live up to the investment in Great Plains lands any more than web-based marketing lived up to the early dot.com bubble.

Women's work was the "day job" that let the homesteader survive. Investment was what sent money into the area and made possible the land speculation that, rather than actual wheat sales, accounted for most of the profit of homesteaders. Even in the twenty-first century, it is often women's off-farm work that keeps family farms viable—especially in the United States, where health insurance is frequently a more important consideration than salary. Sinclair Ross wrote of a little prairie town that knew only two years—the year it rained all June and next year. The Next Year Country of Alberta and Saskatchewan ironically represents the eternal and unfulfilled hopefulness of the quintessential prairie wheat farmer. The stories that we still tell ourselves of the enormous prosperity that wheat brought to the Great Plains—and especially to Saskatchewan and Alberta between 1896 and 1918—have left us believing that this could happen again. If not in wheat, then in canola, in beef, in hog-confinement operations. This is, we insist, the breadbasket of the world (or at least the non-imported petroleum source), and federal governments are duty-bound to sell our wheat and our corn, even if it forces peasant farmers off the land in Africa and South America and creates for their countries a food dependency on North

America. Even if biofuels eventually capture enough acres to absorb the grain that used to be exported as surplus and cause riots as food prices soar beyond the reach of Third World consumers.

Like Voisey's farmers, people of the rural Prairie Provinces and the US "red states" are gamblers, supporting politicians who vow to protect private property and the rich—because most people would like to become rich themselves. The idea of self-sacrifice to achieve modest comfort for people, animals, and the land, is more of a working-class—even, as in the case of Tommy Douglas, a British working-class—ethic, not the proper frame of mind for a horizon-to-horizon universe where expansiveness is the obvious lesson of earth and sky, as well as popular history. At the same time, as a matter of policy, neither federal nor state and provincial governments value a settled rural life. Oil, gas, and coal continue to be more valuable than farming in parts of the Great Plains, and as Roger Epp has shown, to create costs and risks for rural residents as they enrich city residents.[23] Manufacturing on the edge of the Plains in centres like Winnipeg and Kansas City and in places like Wichita that developed with federal money during World War II still lags behind Sarnia, Quebec, the West Coast, and the US Southwest. The rural Great Plains has certainly not turned out to be what the intending homesteaders and eager townsite promoters of a century ago forecast, and it is not really working as an economic society. That does not, however, necessarily mean that the Plains are uninhabitable by all but eco-tourists and their guides and hosts. It is useful to think seriously about the way in which land became money during the homesteading period in order to think about how it might become land again and what that would mean to all of us who live here. Because many of us have chosen the Plains as our home.

Although the *Homestead, Dominion Lands,* and *Allotment* acts applied to land other than the Great Plains—and although about half of the Great Plains land that moved from federal to private hands was actually sold by the feds, the states or provinces, or the railroads—the image that the Great Plains has of itself and that is largely accepted, both by historians and popularly, is of a free land, homestead frontier. But the economic basis of this region is not, no matter how much we would like to tell ourselves that it is, the product of idealism and sweat equity. The Plains was settled in the way that it was because of a belief that private property was the most important aspect of landholding, especially when that worked against the principles of mobility and diversity of land uses as means of dealing with the extremes of a Great Plains climate. Native people looked for allotments that combined riverine forest with space for both grazing and horticulture, and land locators supported themselves by pointing intending homesteaders to the "best" land. Metes and bounds surveys of the East and Europe and the long riverlots of the Métis on the Prairies both encouraged an appreciation of differences in the land, but the square survey implied that land was interchangeable units and encouraged, even required, people to stake out and to stay upon land that usually could not provide the flexibility to support families from season to season or in years of drought or flood. The square survey originates in Europe and was imposed there to create order from a maze of interacting, usually informal, usufruct rights that made it hard for rulers to tell who "owned" property and thus whom to tax.[24] In North America, it dates from before the Constitution and was imposed first on eastern lands whose lack of interchangeability was shown by such features as hills and bogs. While early settlers lamented the monotonous darkness of the woods, they did not have the visual cues of the grasslands, where one could actually *see* "miles and miles of miles and miles." The challenge of simply "taking" a claim rather than buying or even squatting on land also encouraged the concept of interchangeability and commodification of land on the Great Plains more strongly than anywhere else. Daniel Freeman's homestead is an anomaly, in that it was chosen for its combination of creekbed and upland, as pointed out by the previous squatter. One aspect of hinterland status is that it is defined by lack—and lack is by definition uniform. All of the land open for homesteading "lacked" cities, "lacked" farms, and was therefore indistinguishable, especially on the Plains, where variation is perceptible either very close up or very far away, but not on the middle ground of, say, 160 acres. Garland describes picking out his land simply by going beyond the tripod claim stakes of earlier hopefuls and, since the surveyors had not yet come, measuring out a half-mile by half-mile claim by counting the revolutions of the wagon wheel. Satisfied, "we turned and looked back upon a score of the glittering guidons of progress . . . I turned to the west where nothing was to be seen save the mysterious plain and a long low line of still more mysterious hills, [and] I thrilled with joy at all I had won."[25] The tripods of fresh new lumber represent progress, while the unclaimed land is indistinguishable.

The very terms of the Homestead and Allotment Acts stemmed from a belief in deficiency—the applicant was required to "improve" the land. Though on later ranching homesteads, "improvement" with fences and buildings was sufficient, for most of the Great Plains, "improvement" meant ploughing up native grasses and replacing them with what James Malin has called domestic grasses: wheat, corn, and so on. None of these were native to the Great Plains and, except for corn, none were native to this hemisphere or had been grown on the Plains before the advent of non-Native settlers. While the whole rationale for the settlement of the Americas by whitestream immigrants was that the land was in need of "improvement," the *Timber Culture Act* of 1873 implied the deficiency of grasslands in particular by requiring the planting of trees to hold a timber culture claim—even while homesteaders, often dummy entrymen for lumber companies, were cutting down every tree in sight in Minnesota and the southeast. Although writers as different as John Ise and Willa Cather, and even Hamlin Garland, mourned the loss of the tallgrass prairie, they tended to express their sadness as nostalgic sentimentality, akin to the feeling of a mother mourning the loss of her little child in the happy bloom of a healthy mature young adult. They did not express a practical respect for how the land consistently worked for millennia to manufacture food from soil and water despite the extremes of climate. People like Cather's Ivar or W.O. Mitchell's Saint Sammy or John Joseph Mathews himself, concerned with living *with* and observing the land rather than "improving" it, were seen as eccentrics or just plain nutcases, however attractive to their authors. Ivar and St. Sammy lost their land to more practical neighbours, while Mathews' blackjacks preserve was dependent on his considerable personal wealth, derived from oil. Mountains were sublime and eastern landscapes picturesque, but the Great Plains was merely a blank slate, which is why the tallgrass prairie is the most degraded ecosystem in North America, with only approximately 1 percent still alive in any fashion. Although land surveyors did attempt to distinguish classes of soil and to point out land they believed to be unsuitable for farmsteads, and locators like Old Jules flourished by trying to find appropriate homesteads for intending farmers, much "location" had more to do with the placement of townsites, railroads, and potential speculative advantage than with close understanding of the land itself. This interchangeability of land was totally appropriate for a commodity that was most valuable in terms of its interchangeability with instruments of capital. Like money. The emphasis on interchangeable terrain in need of improvement to become more like the humid east, however, made it impossible for intending settlers to study and learn from the intact ecosystem. Given the amount of speculation that took place on earlier frontiers, it was not just the concept of "free land" that hindered settlers (and investors) from focussing on things like microclimates and grazing patterns, but especially in free market terms, gamblers are frequently less careful than buyers. The deficiency model and the free market reinforced each other during and after the homestead period in ways that make it almost impossible to imagine or find an understanding of the Great Plains as a totally satisfactory heartland, as it had been for Native peoples especially since the return of the horse. The Homestead Acts were incredibly successful at changing land, privatizing it, and devoting it to the production of surpluses of periodically unmarketable products. In the process, they totally changed vegetation and dwelling patterns, and dispossessed the people who were using the land sustainably. This seems to me an odd definition of success. The future of the Great Plains, perhaps, depends not on learning to perfect the square survey and the humid-area farm, as reformers in the 1930s believed, but on learning to look beyond the deficiency model, the square survey, and private property to see what the land does well (*not* what the land is "good for") and to learn to mimic it.

Many scholars have studied the "Women's West" since 1973, when I completed my dissertation entitled "Women in Frontier Literature" (Cornell University). Sarah Carter is particularly astute in her many books and articles on the subject, including *Capturing Women* (1997) and, more recently, *The Importance of Being Monogamous* (2008). Here Carter argues that both the United States and Canada invested much national identity in a distinctive "civilized" view of marriage that was particularly contested and defined by western settlement and was in contradistinction to both Indigenous marriage patterns and Mormon polygamy. "Claiming to have superior marriage laws that supposedly permitted women freedom and power was (and continues to be) a common boast of imperial powers," writes Carter.[1] "Civilized" white marriages thus boosted the righteousness of the Manifest Destiny of both the United States and Canada to occupy their respective Wests from "civilization" to sea. Marriage in the custom of the country was clearly deficient.

Contrasting "civilized" to "savage" also served to make so-called civilized women accept their dependent and subordinate status with the rationalization that things could be worse—even though flexible Indigenous systems of marriage and divorce, land and personal property holding, and social hierarchy often netted Indigenous women more autonomy and power than whitestream women. The various homestead provisions, especially in Canada, not only resulted in the commodification of land but in its property values being assigned to men who commanded domestic dependent labour in the persons of their wives and children. This sharp gender division (along with the existing race divisions) more than halved the potential propertied class, thus creating a privileged elite that led to the kind of social stability necessary to creating a material- and market-driven form of economic development. As Carter notes, "The policy of making it nearly impossible for women to homestead in Canada was not an oversight of policymakers; it was deliberate and in contrast to the United States, where single women were permitted to homestead, and did so in the thousands." Canada's more tenuous National Policy required, at least in the eyes of the policy makers, more social restriction if it were to develop successful continental nationhood.[2]

Before the separation of the Plains into two national hinterlands, the roles of women varied based on the economic pursuits of their people. The introduction of the horse led to economic dependence on the buffalo, which meant both competition with other mounted buffalo hunters and the leisure and mobility for pursuing war as an avocation. Both the hunt and the wars tended to reduce women's prestige in relation to men's. Horticultural women had different roles from those who lived primarily from the buffalo herds and from gathering plant foods, and women in the fur trade lived quite differently from the other two groups. But once the forty-ninth parallel was established as an actual boundary, it represented a much larger disjunction for women than for men.

Euro/Afro/North American men could homestead freely on either side of the border, whatever their nationality, notwithstanding racism against African North Americans. Asian and Indigenous North American people for the most part were denied any homesteads at all. South of the border, Euro/Afro/North American single women could homestead.

North of the border, they could not. South of the border, married women had dower rights to the homestead. North of the border, they did not. Voting rights for women—at least Euro–North American women—came to the continent first in Wyoming Territory in 1869. For Euro-Canadian women, they came first in the three Prairie Provinces in 1916. Canadian governments accepted Indian marriages but not Indian divorces, leading to populations of women who were not married or who were appropriately remarried under their own understandings, but who lacked official rights to such necessities as land and rations. In Canada, under the *Indian Act* of 1876 and its various revisions until 1982, Indigenous women who married non-Indian-status men lost their legal Indian status forever, no matter what happened to the marriage, while non-Indigenous women who married status Indians gained legal Indian status. The United States accepted Native marriages and divorces—but demanded that men be married to only one woman at a time, creating more anomalies.[3] In the United States, women kept their ethnic status despite marriage. (And sometimes, as during the Osage Reign of Terror in Oklahoma, they were killed by their non-Native husbands for their land rights.)[4] Asian women were systematically barred from both Wests, though Canada's *Chinese Exclusion Act* did not become absolute until 1923, fifty years after that of the United States. In the States, the Indian service hired Indigenous and mixed-blood women as field matrons and teachers as early as the 1870s and accelerated the policy after 1934, allowing Indigenous women, especially those of what became a mixed-blood elite, to assume important bureaucratic roles in Indigenous communities. In Canada, the Indian service stayed almost completely white, while the *Indian Act* prohibited women from voting in band elections or filling band offices, thus effectively removing women from almost all bureaucratic and governance roles in Indigenous communities.[5]

On the other hand, in many ways the border made no difference for women. On both sides, Indigenous women coped with enormous change in their lives as the Great Plains moved from heartland to hinterland. The various epidemics that swept through the Plains disproportionately killed children. Smallpox also lowered fertility drastically. Women who retained their fertility needed to give birth more often for the population to rebound and thus needed to nurse and nurture more children. The advent of the horse made it easier to move camp but harder to find firewood after firing the prairies to create better pasture for the horses became more common. The pemmican trade and especially the buffalo robe trade increased women's commercial rather than subsistence work and decreased their leisure. The hide trade among the Blackfoot/Blackfeet lowered women's age at marriage and increased the likelihood that women would be in plural marriages, as one hunter could kill more buffalo than one woman could process.[6]

The advent of whiskey increased spouse abuse, and the advent of soldiers, traders, and mounted police posts increased or introduced prostitution. The starving years of the early treaty era made it increasingly difficult for Indigenous women to feed their families. Even though the advent of "civilization" changed men's roles more than women's, the move from tipis and earth lodges to log cabins and the introduction of stoves changed women's everyday chores. The rise in warfare among Indigenous groups in response to overcrowding and overhunting rendered most Plains societies more patriarchal than they had been. The balance between men and women was skewed by the need for warriors. The resulting higher death rates for men left more women than men and fostered plural marriages— until those were outlawed by whites. Missionaries and schools imposed another form of patriarchalism, further undercutting women's power: the forced enrollment of children in residential and boarding schools severely undercut the roles of mothers and grandmothers as caretakers and teachers. The advent of Euro–North American brides, especially in Hudson's Bay Company country during the governorship of George Simpson, damaged both the social and economic position of mixed-blood women in the fur trade. The fact that Nancy Ward was the last of the historically important "Beloved Women" among the Cherokees indicates that the forced relocation of the southeastern US peoples to the Plains undercut the sacred relationship between women and the land, and probably the public significance of women in the society.[7]

The early days of the Euro–North American settlement frontier on the Plains were marked by a predominantly masculine society. Young men came out by themselves to stake claims, then returned east to pick up a wife or sweetheart. This created a profitable niche for prostitution, a relatively equal-opportunity profession for women, without strict regard for race or ethnicity. In the early years, when women were working for themselves or for madams, they seem to have been relatively successful, often moving out of the trade to set themselves up as madams, marrying and blending into the general population, or perhaps setting themselves up in "respectable" trades. Like the mixed-blood wives of the fur trade, prostitutes and former prostitutes suffered a drop in social status when white wives began arriving from the East. Although this was usually stated in moral terms, its roots were also economic. As long as marriage remained the most secure career choice for most women, wives or intending wives could not afford to see their valuable sexual wares undercut by freelancers.[8]

Once the main bulk of Euro/Afro/North American settlement came onto the Great Plains, the majority of female people were either married or children themselves. Although Carroll Smith-Rosenberg and others have examined female friendships in the nineteenth century in some detail, very little research has been done on the woman homesteaders who often "batched" together. Given what we know of such women as Grace Hebard and Willa Cather in the small university communities of the Great Plains, we can certainly assume that female homesteaders found as much comfort and satisfaction in small same-sex communities as the more studied cowboys and settlement house workers did.[9] Teaching at all levels was also in some cases a haven for women's same-sex relationships, though it is probable that colleges and universities were safer than small local schools, where a teacher's life was the property of the community. My great aunt, Norah Power, does not seem to have had a companion during her short tenure as the first classics professor at Mount Royal College in Calgary, though after she left the Plains and eventually moved to Louisiana she did meet her life partner—possibly because the American South remained oblivious to the "discovery" of lesbianism longer than most urban areas of North America. Nor, of course, did marriage remove the possibility of same-sex relationships for women. An obliging husband would sleep in the barn when his wife's best friend came to visit and to share a bed—for a month or more at a time. Visiting among friends, sisters, and cousins not only relieved the isolation of farmsteads and brightened the social life of small towns and cities, but also allowed same-sex relationships to be sustained despite distance and marriage.

Women's work on homesteads was exhausting. Women typically rose before dawn to cook breakfast. They were also responsible for a midday dinner, sometimes carried to the fields, and an evening supper. Depending on the size of the family or whether custom threshers, neighbours trading work, itinerant harvesters, or others were expected, women baked enormous quantities of bread and pies every week. Women customarily milked one or more cows, separated the cream, and made butter. Frequently they tended hens and almost always, large kitchen gardens. Although spinning and weaving had moved out of the household by the mid-nineteenth century and men's clothes were usually ready-made, women sewed clothes for themselves and their children, either by hand, or later, with pedal sewing machines. Most of the time, women were either pregnant or nursing and were tending several small children.[10]

Both women and men expected women to be the keepers of the culture, responsible for establishing churches, schools, and other social institutions. While culture-keeping was certainly important for Indigenous women, too, they had less institutional infrastructure to re-establish, even when they underwent forced relocation, and they shared the cultural duties more evenly with the men. Although most school boards were made up of men, women's earliest experiences with suffrage often came in elections for school boards, school inspectors, and so on, and the first women elected to public positions tended to be school inspectors. The fairly widespread belief in maternal feminism in even otherwise conservative areas often held that "the mothers of the race" ought to have a say in areas like the education of children. Similarly, although only men served as ministers and priests and members of church boards, Catholic sisters were crucial in providing schools and health care, particularly in the Canadian North West or in areas where there were large Métis, Irish, German, Polish, or Czech Catholic communities, as well as in the Hispanic communities in the southwestern Plains of the United States. And Protestant Ladies' Aid societies frequently provided much of the funding for building, maintaining, and particularly furnishing churches.[11] Middle-class women in small towns and cities on the Plains customarily employed some hired help for routine cooking, cleaning, and child care, freeing themselves for work with the Ladies' Aid, the school, women's institutes and clubs, and other cultural and service obligations. Because cities offered single women more choices than the country—particularly in Canada, where they could not homestead—single women moved to urban areas to work in offices, as journalists, as seamstresses and milliners, as shop clerks, and in a range of other occupations, often supported by clubs. The Canadian Women's Press Club in Winnipeg, for instance, fostered the careers of agricultural journalist E. Cora Hind, Nellie McClung, and the Beynon sisters, Lillian and Francis.

Mennonite and Hutterite communities on the Great Plains, especially in Canada, where block settlements were allowed, provided a communal experience for women but remained within the paradigm of a single patriarchal head and a dependent wife in each household. The much more radical and egalitarian Doukhobors were sometimes dispossessed of their land when they would not follow Canadian models. Lacking sufficient draft animals, Doukhobor women, working as a team, sometimes pulled ploughs to break ground. Although this was a sensible arrangement, since a group of women could pull a plough through tough sod, but the greater weight and upper body strength of the average man made it easier for him to hold the ploughshare down, a photograph of the practice was widely used to "prove" that Doukhobor women were abused and treated as cattle, again demonstrating the evils of anything but an Anglo-Canadian "civilized" marriage.[12]

Obviously, women's roles changed with time as well as space—American women were more likely to homestead in their own right after 1900 for instance—but the differences I have noted between the two countries also seem to have induced women to seek to change laws. Although Wyoming was the first full-suffrage polity in North America, the organized women's movement in the United States began in and always stayed in the East. The 1848 Seneca Falls meeting launched Susan B. Anthony and Elizabeth Cady Stanton as the leading women's rights advocates. Most organizations on a national level stayed in the area from Ohio east. This does not mean there were no outliers of the movement: Clara Colby, a particularly strong and devoted feminist, published an important suffragist paper from Hebron, Nebraska.[13]

While Ontario had its suffrage leaders, the movement was not nearly as strong there or in the Maritimes as it was in the West, and Quebec's ultramontane Catholicism meant that it lagged behind the rest of North America by at least twenty years. On the Great Plains, two somewhat contradictory rhetorics were advanced for women's rights. One was the narrative of the self-reliant "rugged individual." Although the West in both countries was a creation of the federal government and its railroad building and land distribution policies, and although co-operation and community were essential in a region of vast distances and extreme and unpredictable weather, the free market aesthetic and the need of governments to tie persons to plots of land for taxation and other bureaucratic conveniences firmly argued for the importance and independence of the individual. Women internalized this as well as men, especially if they were holding down claims in their own right, managing a homestead while a husband worked away, or simply riding astride or guiding a team alone under the big sky. On the other hand, since women were almost universally acknowledged as the guardians of civilization in a "wild" landscape, it was evident that they—or at least their Euro–North American strand—deserved a fair bit of public power to do their duty.[14]

Canada's most successful suffragists came from the West. The Famous Five, who successfully pursued their court case to have women declared "persons" in the meaning of the *British North America Act* (specifically for eligibility for appointment to the Senate), were all westerners with long experience in women's issues. Their careers show some of the possibilities and some of the pitfalls for women in the intellectual milieu that produced them. Of the five, Nellie McClung was the most outspoken. Ontario born, she was raised on a homestead in Manitoba, which her parents had taken to provide a better chance for their children. Like many middle-class girls, Nellie became a schoolteacher when she was still in her teens, and throughout her life, she continued to use the tactics that worked with her students—faith in a Protestant meliorism and lively stories tinged heavily with self-deprecating humour. One of the strange legacies of male Plains writers from Hamlin Garland, to Frederick Philip Grove, to Wallace Stegner, to Rudy Wiebe and on is the erasure of the humour that lubricated life on the Great Plains—particularly women's humour. Luckily, Nellie McClung makes that impossible. Her novels and short stories combine sentimentality with an infectious undercutting of sentimentality and cant, and her suffrage essays unerringly puncture the rhetoric of male leaders like Manitoba Premier Rodman Roblin, whose insistence that "the hand that rocks the cradle rules the world" was supposed to obviate the need for woman suffrage. Nellie was overtly political: she organized and starred in a parody "Women's Parliament" in Manitoba in 1914 that helped unseat Roblin and win woman suffrage two years later, and in 1921, she won a seat in the Alberta legislature. The wife of a pharmacist turned insurance salesman and the mother of five sons, Nellie McClung was in many ways a typical urban prairie matron in her involvement with the Social Gospel, the Women's Christian Temperance Union, and social betterment in general.[15]

Emily Murphy, probably the most famous of the five after McClung, was also a wife, a mother, and under the pen name Janey Canuck, a best-selling author. Murphy made her influence felt as the first woman magistrate in the British Empire and wrote an influential, if stereotyping, book on illicit drugs. The arguments against her right, as a woman, to hold any position in the judicial system and her ambition to become the first Canadian woman senator eventually led to the launching of the Persons Case. Henrietta Muir Edwards, the oldest of the five (eighty years old by 1929), worked for women and children all her life. Her main intellectual capital was her extraordinary grasp of what little family law existed, and her main instrument was the National Council of Women. She was particularly concerned with dower rights and other laws concerning women and property. Because her husband had long been a physician on reserve communities, she had a deep understanding of the issues of Aboriginal women, though she does not seem to have worked particularly to advance them. Louise McKinney, an American immigrant and a temperance leader, was elected to the Alberta legislature in 1917. For her, women's rights seem to have been primarily a stepping stone to prohibition. Irene Parlby, a well-connected Anglo-Indian immigrant, became the leader of the United Farm Women of Alberta and was elected to the Alberta legislature, where she served as Minister without Portfolio, but effectively supported women's issues.[16]

It is tempting to read these women only in a purely celebratory way—foremothers of whom we may be unashamedly proud. Unfortunately, the meliorist rhetoric that they and many others used was coercively assimilationist and easily congruent with a eugenics movement that rapidly became ugly; however, Patricia Roome argues persuasively that there were distinctions among the five, and Nellie McClung and Henrietta Muir Edwards were less likely than the others to see non-Anglo-Saxon people as lesser, though they certainly saw them as different but able to change.[17] The Social Gospel movement, the United Farmers of Alberta, the CCF, and the whole intellectual context in which the feminists moved welcomed a benevolent social Darwinism in which society would peacefully evolve into a co-operative commonwealth, the kind of utopia envisioned by American Edward Bellamy in *Looking Backward*, a tremendously influential book. Birth control, in some form or other, was definitely part of the feminist movement, just as eugenics was part of Margaret Sanger's plan. Although eugenics was sometimes directed against visible minorities, its main goal was to breed "mental defectives" out of the gene pool. Indeed, Emily Murphy and Alice Jamieson as magistrates and Nellie McClung in her varied reform activities were correct in noting the pain and suffering that fear of coerced pregnancies and the pregnancies themselves could cause developmentally delayed girls and their families. The belief that science could cure anything—from surveying a railway across the Shield and over the Rockies and Selkirks, to scientific farming and dams and irrigation ditches for drylands agriculture, to the plant genetics that developed prairie-perfect Marquis wheat out of Eastern European strains—was an integral part of the intellectual baggage of Great Plains settlement and of colonization in general. That science could cure social problems seemed to be a given. But people are not plants, and eugenics was as false a science as the theory that rain follows the plough. Alberta's infamous eugenics law resulted in the forcible sterilization of young women for no other reason than that someone else thought of them as "defective." It had its seed in the same intellectual currents that produced the early feminists, but it long outlived them. Although the law was repealed in 1972, it was not until Leilani Muir won her case against the province in 1996 that any reparations or apologies were offered for forcible sterilization.[18]

On the American Great Plains, the conjunction of feminism and agrarian discontent peaked in the 1890s, before North Americans in general were willing to accept woman suffrage. In Canada, however, feminism and agrarian discontent peaked together, around the time of the First World

War, so suffrage, agrarian discontent, and provincial and regional third-party strength all coincided. After the Civil War and the Civil War amendments enfranchising blacks but not women—and, indeed, in the Fourteenth Amendment introducing the word "male" to the US Constitution for the first time—the coalition of feminists and abolitionists that had worked extremely well together before and during the Civil War was shattered. The coffin was nailed shut in Kansas in 1867, when George Francis Train, an articulate and determined feminist with a strong white-supremacist tinge, sponsored a speaking tour featuring himself and Susan B. Anthony advocating for suffrage for women but not for blacks.[19] Suffrage thus became a conservative, rather than a radical cause. Wyoming suffrage was likewise conservative. In a territory overwhelmingly populated by young unmarried men seeking their fortunes as miners or adventurers, only one set of women was to be entitled to vote—married white women and their adult daughters. Native women were, as Indigenous people, disfranchised. The Chinese who had come to Wyoming for the building of the Union Pacific Railroad were overwhelmingly male and were disfranchised, men and women alike, as "Orientals." Prostitutes, since they frequently moved from town to town and had no fixed address, were also disfranchised, and public opinion forced even the most stable of madams to stay home from the polls. Married men as a class were wealthier and far more stable than unmarried men. Because their wives for the most part accepted Victorian ideals of social control, woman suffrage in Wyoming was an essentially conservative movement, as shown by the lionizing of Esther Morris, the first woman Justice of the Peace in the world.[20]

Agrarian feminism was also conservative in its acceptance of maternal feminism, but far less conservative than the earlier models. For one thing, as we have seen, Great Plains farming was economically dependent on women's work of both subsistence and reproduction. Butter, eggs, and gardens kept farm families alive, and women ran the farm and homestead when men worked off the farm to make money. Daughters who worked as schoolteachers or as hired girls off the farm were also likely to send some of their pay back to their families, perhaps more likely than boys who were off working. While neither the law, the more patriarchal farmers, or even economic historians have fully understood and recognized farm women's roles, many of the agrarian reformers, trying hard to understand exactly what farms needed to succeed within North American society and a world economy, recognized the need for women in agriculture and the need for those women to represent themselves. As early as the 1840s and 1850s, the *Ohio Cultivator* published women's columns that spoke vividly for women's rights and honed the talents of two important abolitionist feminists, Hanna Maria Tracy Cutler and Frances Dana Gage, who is now best remembered as the amanuensis for Sojourner Truth's "Ain't I a Woman" speech. The Winnipeg *Grain Grower's Guide* followed in the tradition, publishing the columns of Francis Marion Beynon, in many ways the most original thinker among the prairie feminists; unfortunately, she moved to New York and disappeared as a writer after World War I.[21]

The Grange was an early supporter of women's rights, and under its auspices, Iowa struggled unsuccessfully for years to pass woman suffrage. Although Carrie Chapman Catt did not herself come from an agricultural background, she did graduate from an agricultural college that is now Iowa State University, and her rise to leadership of the National American Woman Suffrage Association to lead it to eventual triumph was partially enabled by the agrarian feminist tradition of her adopted state.[22] (Like the Alberta feminists, she was also a eugenicist.) Populism was an even more important forum for another woman of the Plains, Mary Elizabeth Lease—though her colourful and oft-cited (but perhaps apochryphal) admonition to farmers to raise less corn and more hell sets her a bit outside the domain of maternal feminists. She was a powerful speaker for the Populists, sharing the platform with Hamlin Garland in Omaha in 1892, the last convention of the Populists before they fused with the Democrats and nominated the socially conservative (but pro–woman suffrage) William Jennings Bryan. Lease did not need to argue for suffrage—her leadership position attested to the importance and power of women. Annie Diggs, more conventional and less colourful than Lease, worked more in the maternal feminist mode and advocated woman suffrage to clean up the "dirty pool" of politics, which she envisioned as something like a cesspool rather than a backroom game of cues and balls.[23]

An anomalous Great Plains feminist was Clara Bewick Colby, who for many years published the *Woman's Journal*, first from Hebron and then from Beatrice, Nebraska, a relatively small agricultural and industrial centre on the Little Blue River, and ironically the same town where Daniel Freeman had homesteaded. Colby's husband, an alcoholic and probably abusive Army officer, brought his wife to the West and then went off to duty, leaving her to her own devices, probably to her great relief. One of his commands involved the massacre at Wounded Knee, where he kidnapped an unharmed Lakota baby found among the wounded and slain, and brought her home to his wife, more, it would seem, as a souvenir or pet than as a child. (One is eerily reminded of the girl Laura Ingalls's imperial whining for a "papoose" in *Little House on the Prairie.*) The ordeal of Lost Bird, as the little girl from Wounded Knee was called, undoubtedly complicated Colby's life and her ideas on society and women's roles, but she had folded up her journal by this time and thus did not publish on this era of her life.[24] Lucy Stone and Henry Blackwell, whose journal eventually absorbed Colby's, were easterners in the mainstream of the American feminist movement. Colby, by contrast, was marginalized in space, by her marriage, and by her increasing identification with Lost Bird, who herself was marginalized as an Indigenous girl and woman, by the massacre of her family, and finally by her abduction from the remainder of her people. She was perhaps sexually assaulted by white relatives as a teenager, another effect of marginalization that she shared with many Indigenous girls and young women who were taken from their own families and culture to be "civilized."[25] Given Lost Bird's anguish, the disfranchisement of middle-class white women must have seemed trivial—except as it increased Clara Colby's impotence to free herself or to help her daughter.

The relatively sunny meliorism that Canadian feminists like Nellie McClung and Emily Murphy espoused may well have been an antidote to the despair that Clara Colby could not ignore. Certainly, McClung would suffer later when her son Jack committed suicide, an after-effect, she came to believe, of the horrors he had been forced to witness and participate in overseas during World War I.[26] The barriers of class and ethnicity undoubtedly kept Murphy and McClung from looking too closely into the darkness, despite their experiences. Henrietta Muir Edwards, the oldest and most silent of Alberta's Famous Five, may have had a different sense of the tragedy of Indigenous women of the Plains that was playing itself out during her lifetime. Her husband was the government doctor on several reserves and was frequently removed from his posts when he complained that the government was starving the Assiniboines and Blackfoot to whom it had made treaty promises, letting them die from the diseases of hunger and poor housing that no doctor could cure. Henrietta Edwards—who visited with many Indigenous families and who, with her husband, commissioned art and artifacts from Indigenous friends, introducing a meagre bit of cash into reserve economies—must have known, as Clara Colby did, of the despair and displacement, and may have tried, at least on a personal level, to assuage it. Yet if this deepened and complicated her feminism, she seems, unfortunately, to have kept it to herself.[27]

If we go back to the mainstream of Great Plains women's movements, we see them continuing to focus on Euro–North American women. Perhaps because Canadian women faced more legal restrictions and fewer economic opportunities than their sisters across the line, Canadian Prairie feminists were more visible than American Plains feminists during the first three decades of the twentieth century. The exceptionally able journalists Lillian Beynon Thomas, Francis Marion Beynon, and E. Cora Hind, along with Nellie McClung, formed a powerful activist nucleus in Winnipeg before World War I. Their enormously popular "Women's Parliament" of 1914, starring Nellie McClung as a parodic version of Premier Rodman Roblin, benignly denying the vote to men, both popularized the cause and helped weaken Roblin's government. The efficient backstage management of the Beynon sisters guaranteed that after Roblin had been defeated, the Liberal government under T.C. Norris provided women with full provincial suffrage in 1916.[28]

Saskatchewan feminists were not as showy as those in Manitoba and Alberta, but they persuaded Premier Walter Scott to enact woman suffrage if they could demonstrate widespread support for it among women. The resultant petition drive, with thousands of signatures collected from mostly rural women all over the province, more than fulfilled Scott's requirement, making Saskatchewan the only polity to enact woman suffrage in direct response to women's own petitions. Alberta, like the other two Prairie Provinces, enacted woman suffrage in 1916, but its major feminist claims to fame came both earlier and later. Emily Murphy in Edmonton and

Alice Jamieson in Calgary were the first female magistrates in the British Empire. Irene Parlby, elected in the UFA sweep of 1921, became the second (by only months!) female cabinet minister in the British Empire, and she and Nellie McClung, elected by the Liberals, were the first two women to serve together as MLAS.[29]

Alberta's Famous Five, however, are remembered best for the Person's Case, in which the English Privy Council in 1929 reversed English common law and declared that Canadian women—and indeed all women in the British Empire—were "persons" in the meaning of the *British North America Act* that had founded Canada in 1867. It is hard to discuss the impact of what seems to be such a self-evident ruling, but it overthrew centuries of common law (plus a specific 1876 ruling) and would be part of the basis for recognizing women's individual claim to status as citizens and, under Bill C-31, as status Indians under the *Indian Act*. Unlike the arguments of maternal feminism that largely won woman suffrage, the Persons Case was argued and won under the aegis of equal rights for women as individual human beings—as, quite specifically, persons. Although it is very unlikely that the Alberta women knew anything about it, they were recapitulating another important Great Plains civil rights case, that of *Standing Bear v. Crook* in Omaha in 1879. The Ponca leader, arguing his right to return to his original home from relocation in Oklahoma, was denied habeas corpus, and he sued to have that common law staple recognized for Indians by the United States. He won. Indians became "persons" under the meaning of the US Constitution, though the practical aspects of the win were denied to most other American Indians and Standing Bear's "personhood" depended upon his explicit assimilation and renunciation of his Indian status.[30] In this sense, his victory resembled the forced "enfranchisement" of the *Indian Act*. Finally, in 1955, in *Brown v. Board of Education*, a third set of Great Plains residents, in this case African American, carried what was essentially a third "persons" case to the US Supreme Court, which struck down the 1896 "separate but equal" doctrine and proclaimed the equal rights of all Americans, including those of African descent.

It is not surprising that Standing Bear's case was brought on the Great Plains. By the late nineteenth century, most people whom the federal government officially recognized as American Indians were from either the

Southwest or the Great Plains, or had been relocated to the Plains. Nor is it surprising that the Persons Case came from the Prairies. Women's political power had been institutionalized in the West before anywhere else. Even though Agnes MacPhail, the first Canadian woman MP, was elected from Ontario, she represented the United Farmers of Ontario and later the CCF. "Rural" and "western" are not unusually stand-ins for each other. As Walter Stewart points out, the CCF received a large bulk of its votes, if not its seats, from Ontario. The strength of women in agriculture translated into political power.[31] While *Brown v. Board of Education* seems at first more anomalous, the Great Plains has always represented—and sometimes delivered—a greater equality to North Americans of African descent than have other parts of the continent. The Exodusters to Kansas, blacks in Indian and Oklahoma territories, blacks on the railways, and black homestead settlements from Nicodemus, Kansas, to Amber Valley, Alberta, all provided outlets from strict segregation even if they did not deliver equality. Despite the claims of North Carolina, the lunch counter sit-ins that marked the beginning of the 1960s civil rights movement started in Kansas and Oklahoma.[32] Living with unfulfilled promises is more conducive to revolution than is living with constant and unwavering denial and suppression.

Although the American Great Plains was not as significant in the suffrage fight as were the Prairie Provinces, there were many local woman leaders who, for whatever reason, attained a place and a voice in the Great Plains. For instance, Jeannette Rankin of Montana, both a suffragist and a pacifist, was the only member of Congress to oppose US entry into both world wars. She supported the militant suffragettes, but she herself campaigned behind the scenes with her fellow members of the House of Representatives.[33] As we have seen, women's subsistence activities carried Great Plains homestead agriculture until the family was able either to sell out or to acquire enough land for a successful commercial operation, and women's willingness and ability to have children supplied the labour for the farms. Yet in both Canada and the United States, property rights rested in the husband, not the wife, until the 1970s. An Alberta divorce case (*Murdoch v. Murdoch*) reached the Supreme Court in 1973 and resulted in the ruling that despite all her work on the family ranch during twenty-five years of marriage, Iris Murdoch was entitled to nothing when she had to leave her husband. The ensuing outcry led to changing the law in all provinces, with Alberta's *Matrimonial Property Act* specifically declaring, in gender-neutral terms, that "the contribution, whether financial or in some other form, made by a spouse directly or indirectly to . . . a business, farm," or any other enterprise, as well as contributions as a homemaker or parent had to be taken into consideration when splitting the enterprise upon divorce. Meanwhile, in the United States, a group called Women Involved in Farm Economics (WIFE) finally succeeded in changing American inheritance laws so that a surviving husband or wife inherited a farm on exactly the same terms. Previously, the law had provided that when a wife died, her husband automatically inherited the whole farm with no inheritance taxes, but when the husband died, the wife was liable for all estate taxes and frequently ended up losing or selling the farm to pay the taxes, especially as the capital-intensive days following World War II meant that farms were often worth millions of dollars.[34]

Historians have paid a great deal of attention to Prairie women suffragists—with good reason. Nellie McClung's wit and verve alone make her remarkable in feminist annals, while Francis Marion Beynon's sentimental but incisive *Aleta Dey* and her columns in the *Grain Growers Guide* are both rhetorically and intellectually complex. Jeannette Rankin, University of Wyoming professor Grace Hebard, and Clara Colby all represent original strains of argument that do not show elsewhere in the United States. Prairie writers from Nellie McClung and Emily Murphy themselves, to Willa Cather and Meridel LeSueur (who could not be more different from each other), to Margaret Laurence and Jane Smiley and Sharon Butala both show and create important aspects of the intellectual history of the Great Plains and have attracted considerable study. Deb Fink has shown how the ideal of the family farm has required great sacrifices from women. Sylvia Van Kirk and others have demonstrated how completely the fur trade depended on women's work. The whole tradition of agriculture on the Great Plains has relied on women. The Hidatsa and Mandan women, with their shoulder blade hoes, were the horticulturalists of their day, and homesteading women kept the gardens going before land speculation or wheat could pay off. Even today's "farming the mailbox" has primarily and invisibly depended on the off-farm work of women as well as their efforts in the work of both field and management. The only economic exception is that of ranching (though one could point out that the cows do a good deal of reproductive work), and that, too, began to depend on the labour elasticity of women and children when fences and careful herd management replaced free grass. When Euro/Afro/North American women began to settle on the Great Plains in the nineteenth century, they replaced the "deficiency" of Native and mixed-blood women, but they were still deficient compared to men, with an inferior capacity to hold property and thus a diminished personhood. In general, recognition of women's rights has certainly improved, though Native women on the Great Plains, as elsewhere in North America, still suffer from gendered racism that negatively affects everything from life expectancy to professional advancement, despite the successes of many individual women.

Historians have tended to miss the central economic role of women on the Great Plains using a rhetoric that viewed Indigenous people as hunters—not also gatherers and horticulturalists—and homesteading as a failed program of making family farms rather than a successful program of commodifying the land and parcelling it out to private owners. They have also, in general, treated 1885 and 1890 as if they represented the defeat of the Indians and the end of the Indian way of life instead of the switchover point from public and military ways of dispossessing Native people to private and bureaucratic ways of dispossessing Native people. By 1934, when the United States repealed the *Dawes Act* allotting land in severalty, Native people had lost all but 47 million of the 138 million acres of land guaranteed to them by treaties at the end of the Indian Wars. They would lose another 3 million acres, including some of the most valuable remaining timber lands, to "Termination" in the 1950s and 1960s.[1]

On the Canadian Plains, Indigenous people would see—and barely survive—a considerable effort to disallow and undercut virtually all subsistence activities, both traditional and innovative, that they attempted. Like Native Americans, Canadian Natives would continue to hemorrhage land to nearby municipalities, for highways and reservoirs, for bomb testing, and for energy development. Only in the early 1990s would they begin to get it back, especially in Saskatchewan and Manitoba, through the Treaty Land Entitlement process. The original Prairie treaties had allowed for reserves with a requirement of a certain number of acres, usually 120, per person. Miscounts, late adhesions, and subsequent births resulted in reserves that were substantially too small for the people—without even considering land that had been taken from the reserves after the treaties. Treaty Land Entitlement agreements allowed the First Nations entities to recalculate the land owed to them and to receive either Crown lands or special funds to buy lands on a willing seller basis. Lands that had been ceded back to non-Native governments or individuals after the original establishment of the reserve could also be counted as part of the calculation of land owing to the First Nation.[2]

In both countries during the late nineteenth and most of the twentieth century, Indigenous people faced monomaniacal government onslaughts on their culture and religion, their families, and their economies. The idea that *only* through Christianity and private property could Indians join the North American market society was demonstrably untrue, but both countries insisted upon enforcing it, even to the extent of barring Indigenous Canadian farmers from using the equipment that was absolutely required to harvest crops during the short growing season and liquidating Indian horse and cattle herds in both countries during World War I.[3] Both governments banned essential religious ceremonies—particularly, on the Great Plains, the Sun Dance, or Thirst Dance, and the giveaways. Most demoralizing of all was the deformation of the idea of "education" to become an excuse for removing half or more of all Indigenous children from their families of origin and their nations from the 1870s through the 1950s or even beyond.[4] Yet when dealing with this period of several generations of displacement and devastation, historians mostly ignore Native people in their histories of the Great Plains, and, except for the Dirty Thirties, the Plains drop almost completely out of national histories of this period, especially in the United States.

Until recently, although Canadian historians did not necessarily share the popular perception that Canadian Indian policy was kinder and gentler than that of the United States, they also did not particularly challenge it.[5] The equanimity with which Canadian policy makers and implementers accepted starvation as normal among Indians certainly contradicts usual notions of kindness. The first complete critique of the bureaucratic dispossession of Indigenous Canadians came with Sarah Carter's *Lost Harvests* in 1980, fifty years after the first such critique, Angie Debo's *And Still the Waters Run*, was published in the United States,. It is not, perhaps, surprising that both books should have been written by women who were the granddaughter and daughter, respectively, of Prairie pioneers and who both introduced their own work with their sense of disjunction between their pride in their family stories of pioneering and their later discovery of the genocidal treatment of Native peoples that had been part of the framework—but not the rhetoric—of pioneering. In a sense, the twentieth-century narrative of Native people is "deficient" because it is part of the history of colonization. The people "should" have been uplifted by Christianity and civilization, and "should" have assimilated gratefully into whitestream society. Instead, they stubbornly maintained a separate identity despite (and perhaps because of) abusive policies and actions intended to force them into conformity. Isaiah Berlin was not talking about residential schools in the following excerpt, but he described their pathology perfectly:

> If the facts—that is, the behavior of living human beings—are recalcitrant to such an experiment, the experimenter becomes annoyed and tries to alter the facts to fit the theory, which, in practice, means a kind of vivisection of societies until they become what the theory originally declared that the experiment should have caused them to be.[6]

Angie Debo finished her dissertation during the 1930s, a time when there were still relatively many women in academe (though usually without the PhD), but a time in which almost no women were being hired for tenure-track university positions, a state of affairs that would continue into the 1970s and longer in Canadian Prairie universities. Unable to get the position to which her credentials and achievements should have entitled her, Debo found herself able, instead, to undertake extensive research projects such as that which resulted in *And Still the Waters Run*, which required the intensive mining of virtually every county courthouse and archive in the old Indian Territory. What she amassed was such a devastating portrait of the corruption used for the mass dispossession of the Five Southeastern Tribes that the University of Oklahoma Press turned it down for fear that the press would be sued or even shut down by the powerful interests of the state—who had acquired their power by their graft against the Indians. The editor soon moved to Princeton University Press, where he was finally able to publish the book in 1940, following a positive review by John Joseph Mathews and the strategic removal of the names of some politicians.[7]

In general, the record of the bureaucratic displacement and mistreatment that started before the ink was dry on the treaties is so ridiculous that it would be funny if not for the generations of human suffering that it produced. Contemporary Canadian political advisors and politicians such as Thomas Flanagan see the increasing recognition of Indigenous legal rights during the 1990s as perverse and counterproductive; they call for the nineteenth-century solutions of assimilation and a decent cloak of charity to cover whitestream society's continued acquisition of land and continued repudiation of treaty obligations. Twenty-first-century American politicians such as George W. Bush have been able to ignore Indigenous peoples and their legal rights altogether. Both strains of politicians see both Indigenous peoples and revisionist scholars as leftover hippie-pinko bleeding hearts who need to grow up and embrace competition. The rhetoric of freedom during and after World War I helped bring about voting and other citizenship rights for women and, in the United States, for Indigenous people. Revulsion at genocide in the aftermath of World War II helped open North America to Jewish refugees and brought about the recognition of human and civil rights of North Americans of Asian and African descent. Similarly, the rhetoric of justice and the right to defend one's borders in the aftermath of 9/11 ought to prompt countries like Canada and the United States to live up to their treaties with domestic Indigenous people or else to acknowledge frankly that violence, as the terrorists argue by their actions, is the only real arbiter of justice.

The story that Debo told about Oklahoma was certainly one of the efficacy of violence and the state power employed by law makers and courts. How did the people she described, who were pleasant family men, not psychopaths, justify the bureaucratic dispossession of Indians? As we saw in our discussion of Custer and Riel, it was necessary for both governments to demonize Native peoples in order to rally public opinion to support wars or other policies of eradication. But in order to convince citizens that they were being "humane, just, and Christian," both governments had to have some policy that was not explicitly genocidal. As Jill St. Germaine points out, in its treaties, the United States was very explicit about its "civilizing" mission, while in Canada, it was Indigenous peoples themselves who demanded agricultural assistance and education in Euro–North American wage-earning skills.[8] Canadian and American Indigenous peoples were often curious about and even drawn to Christianity, but none of them ever consented to deculturation or, except for a few who had been completely converted by their schooling, the criminalization of their religious observances and marriage customs, or the proscription of their languages. Although individual leaders, such as Joseph La Flesche of the Omahas, supported the allotment of land, that was mainly because both whitestream popular opinion and bureaucracy so favoured private property that it seemed a hopeful tool for holding onto at least some of the Omahas' land.[9]

From the point of view of the Crees in Saskatchewan in the 1880s and 1890s, Hayter Reed and his policies must have made no practical sense at all. As Sarah Carter has so painstakingly pointed out, the government forced the Crees to adopt farming methods that were completely and deliberately anachronistic and that were *bound to fail*. The Canadian prairies were converted to large-scale agriculture only in the context of technological, capital-intensive world market conditions. Before the development of Marquis wheat in 1911, the short growing season meant that even experienced wheat farmers with access to the newest time- (and labour-) saving machinery often failed to harvest a wheat crop before it had been damaged by frost. The farm instructors who were supposed to help the Crees were rarely experienced farmers, and none had prairie experience. Reed and Scott believed in an anthropological theory quite as baseless and farfetched as the rain-follows-the-plough theory: Indians would have to pass from "savagery" to "civilization" by way of an intermediate stage of "barbarism," in which they would perform all agricultural tasks by hand—even if that meant the custom manufacture of such implements as flails and scythes that were no longer used by Euro–North American farmers. The Blackfoot were told to *make* their own harnesses, hay rakes, and so forth. If Indigenous people, through their own initiative, were able somehow to purchase or obtain the use of up-to-date machinery, they were not to be allowed to use it on the reserves. They were also prohibited from selling their crops commercially. Unlike the homesteaders and the purchasers of railway and other land, Indian farmers could not mortgage their land to gain capital. Nor did the diminutive reserves contain enough land to support farming into the children's generation if the large families, needed to make any kind of success of such labour-intensive farming, were to grow up. Should, by any chance, any Crees or Blackfoot somehow manage to persevere, the Department of Indian Affairs frequently failed to provide seed in time for planting, thus dooming the entire year's effort. In areas including almost all of the Treaty 7 reserves, where herding was a more reasonable use of the land than row crop agriculture, people were discouraged from cattle raising by the requirement that all animals must wear the DI government brand.[10]

As short-sighted and counterproductive as these policies were, they did have certain advantages to the government. It was easier and much cheaper for Ottawa to insist that Indians make their own harnesses than to provide them access to capital to buy up-to-date machinery. If they were hampered in growing grain and forbidden to sell any product commercially, they did not compete with the homesteaders whom the government so dearly wished to settle on the Great Plains. Given that Canada was still an agrarian nation and given that both the government and enthusiastic promoters praised the fertility and ease of farming the land, central Canadians were willing to believe the government line that crop failures were the Indians' own fault because they were careless, lazy, incompetent louts who refused to do an honest day's work. Similarly, if rates of disease and child mortality were high on the reserves, central Canadians were willing to believe the government's assertions that Indigenous women were filthy and incompetent housekeepers, not that chronic malnutrition and the insistent refrain that everything Indian was wrong created sickness, death, and despair.

These attitudes had, of course, begun in Europe long before Europeans had ever dreamed of the Americas and had been practiced against Judaism and Islam since the codification of Christianity and the rise of the prophet Mohammed. When the *Mayflower* arrived at Plymouth Rock, its Pilgrims were ready to praise their God for having sent the small-pox to the Indians, leaving cleared fields and caches of dried maize that enabled the Anglos to gain a foothold on the shores of what they called New England. The Pequot War and King Phillip's War finished the effective depopulation of Indigenous New England, though survivors with great tenacity and courage have managed to remain to the present day. Throughout colonial and revolutionary times, the French, the English, the Dutch, and the other European powers that settled North America from the East were fairly straightforward in their treatment of Native peoples. They were partners in the fur trade, and each European nationality cultivated Indigenous allies against other European powers on the continent. Despite individual exceptions on both sides, Euro–North Americans interested in permanent agrarian settlements mostly saw Indigenous people as vermin to be exterminated, and Native people, who had at first welcomed and traded with the Europeans, fought back with arms, by acquiring the supernatural powers of their own priests and incoming missionaries, and by accommodating their economic systems to those of the newcomers. Of all the various Indigenous peoples, those of the southeast United States were most successful in accommodating commercial agriculture, private property, and European political traditions into an Indigenous world view. But their very success told against them. The discovery of gold in Georgia lit the spark, but it was quite clear by the 1820s that Euro–North Americans would no more accept a rich planter class of mostly mixed-blood Cherokees and Choctaws, Creeks and Chickasaws than they would accept Indians living in traditional subsistence patterns or the Seminoles who formed alliances with the daring Maroons, people of African descent who had escaped from slavery. The removals that became known as the Trail of Tears, especially for the Cherokees, were based firmly on the principle that whatever the Anglo-Saxon wants, the Anglo-Saxon takes, a tradition Mark Twain noted with disgust at the time of the Spanish-American War, and part of the Age of Empire that underlay Victoria's reign around the globe.[11]

Despite the blood feuds that had divided Cherokees and to some extent the other tribes between Treaty Party and non-Treaty Party factions, after removal the southeastern peoples deliberately discontinued the feuds to build successful mixed-farming method, mixed economies in Indian Territory. Traditional usufruct rights allowed anyone who cultivated land the right to hold that land as long as cultivation continued. Former rich planters, using the labour of enslaved and freed African-Americans, rebuilt their holdings, while traditionalists hunted and cultivated small gardens. No one was hungry. The American Civil War, however, once more split the Nations, more or less along the old treaty/anti-treaty lines. Although tribal governments supported the Union, the Confederate factions were prominent enough that after the war, the federal government levied heavy land cessions against the Nations and forced the enfranchisement and citizenship of all the formerly enslaved persons—something that of course did not happen in communities of Euro-American slaveholders. With infinite patience, the Nations rebuilt for the third time in three generations, again developing a distinctive blend of commercial and subsistence living with a vibrant political tradition that combined European and North American attitudes and institutions—though unfortunately one of the European elements was the disfranchisement of women.[12]

But the Five Nations of Indian Territory did not live in a vacuum, and what was happening to other Indigenous peoples of the Great Plains affected them. The western part of what would become the state of Oklahoma, then called Oklahoma Territory, along with western Dakota Territory, was to become the home of the remaining "savage" Plains tribes. The Five Nations hoped that this concentration could develop into the state of Sequoyah, named after the inventor/popularizer of the Cherokee syllabic alphabet, but this was not to be.[13] If the Five Nations had survived removal and the disruption of the American Civil War largely because of the strength of their system for internal conciliation and the flexibility of their land-use patterns, the Euro–North American system of allotment in severalty would prove to be the trickster device that disabled the regenerative powers of the people.

Allotment was first tried out among the Omahas in 1885. Joseph La Flesche, Iron Eyes, a mixed-blood of Ponca origins (though fixed tribal identity in the European sense was probably not a feature of Plains life before US annexation in 1803) was chief of the Omahas and a progressive, a leader of the "Make-Believe-White Man's Village." He had witnessed the removal of both the Pawnees and the Poncas from Nebraska to Indian Territory, and he feared that the Omahas, located not far upstream from the large newcomer settlement that had, ironically, been named after the tribe, would be next. Although the southeastern removals had proven that Euro–North Americans would not necessarily respect private property rights held by Indians, La Flesche and his advisors listened to the preachers of private property and determined that their strong rhetoric, whether or not it was useful or relevant, was the strongest defence against removal. Two years later, in 1887, Congress passed the *Dawes Allotment Act*, which applied to all reservation land outside Oklahoma. Henry Dawes, the Senator who designed allotment and worked it skilfully through Congress; Alice Fletcher, the anthropologist who befriended the Omahas and tried to move them "forward" through allotment; and other members of the society they called "Friends of the Indian" maintained that once each Indian family was settled on its own homestead and "surplus" reservation land had been distributed to incoming homesteaders so that no "unimproved" land would remain for hunting and gathering, all Indians would have to settle down and become Christian American farmers, earning their bread by the sweat of their brow.[14]

It seems likely that the reformers were sincere and believed their own rhetoric. Like the Ontario annexationists, whose experience and point of view were so narrow that they literally could not imagine that the Red River Métis and the various Indigenous nations of the North West would not rejoice in slipping the yoke of the Hudson's Bay Company to become a colony of the free and democratic Dominion of Canada, the Friends of the Indian had never questioned their belief in Christianity and private property. Such questioning, of course, was not widespread in whitestream North American society, but it was not entirely unknown. By the 1880s, Dostoevsky and Tolstoy were starting to become known to Americans, and Tolstoy's pacifism and collectivism offered an alternative model to

Manifest Destiny. Edward Bellamy's *Looking Backward* (1888) was published only a year after the *General Allotment Act* had been passed, and it suggested a complete turn *away* from private property. It became one of the bibles of the Social Gospel movement on the Prairies. Intellectual Christianity itself was undergoing a marked change in the larger society. French scholar Ernest Renan had published his *Vie de Jesus* in 1863, a biography of a remarkable but thoroughly human man that was translated into many languages and that, along with archaeology of the "Holy Land," served as one of the intellectual anchors of late nineteenth-century rethinking of Christianity. In 1902, William James published *On the Varieties of Religious Experience*, which served as psychological validation of religious experience as "real," but also moved it beyond anything remotely resembling orthodox North American Christianity. While the ideas of both of these books certainly affected the academic understanding of Christianity and had a trickle-down effect at least in those parishes with intellectually inclined pastors, a far more popular book was Charles Sheldon's 1896 *In His Steps*. Written by a Topeka pastor who had been raised in Dakota Territory, this still-influential bestseller asks its readers to follow a "what would Jesus do" model that involves solidarity with working-class and unemployed people and African-American civil and economic rights.[15]

As Charles Eastman noted during his period as a YMCA preacher, the Christian gospel's message of sharing sounded more like traditional Indigenous belief and custom than like the hard-edged, profit-oriented, bureaucratically administered system of private property that dominated those Euro-American settlement societies in contact with Native peoples. He recorded the response of one older Native man to his teaching about the life of Jesus:

> I have come to the conclusion that this Jesus was an Indian. He was
> opposed to material acquirement and to great possession. He was inclined
> to peace. He was as unpractical as any Indian and set no price upon his
> labour of love. These are not the principles upon which the white man has
> founded his civilization. It is strange that he could not rise to these simple
> principles which were commonly observed among our people.[16]

The intellectual countertrends in their own society seem to have been truly invisible to the Dawesites and the more pragmatic questers after Indian land—just as they are to many of today's Christian right that holds that Native people should not be entitled to any community rights or citizens-plus status in North America.

Just as the doctrine of barbarism and handmade hay rakes, however, was a great deal cheaper for the Canadian government than making sure the intending Cree, Assiniboine, Blackfoot, and other Indigenous Prairie farmers had, as the treaties seemed to promise, well-watered, fertile land and state-of-the-art equipment and seed, allotment was politically palatable in the United States. Western senators and representatives, particularly, liked the idea of "surplus" land that could then be distributed to enfranchised homesteaders—or to railroads and other large-scale landholders. As with the various Homestead Acts, general allotment paid vigorous attention to individual private property but served to commodify and capitalize the land. However earnest and sincere Dawes, Fletcher, and their fellow reformers may have been, the actual vote for allotment did not come out of concern for Indians but concern for non-Indians, especially rich and influential ones, who would have a better chance to acquire Indian land.[17]

And acquire they did. The declaration of land as "surplus" was only the beginning. Although allotted land was supposed to be inalienable for twenty-five years after allotment, unscrupulous land speculators, mostly non-Indians and mixed-bloods, put pressure on allottees to sell. Often allottees themselves wished to sell the land to raise money because they were forced to live in a cash economy. Since they could neither mortgage nor sell their land, it was difficult for many allottees to raise the capital to "improve" the allotment. They could not buy farm equipment, stock, or seed. Many poverty-stricken allottees needed money for food, since commons land for hunting and gathering was no longer available. Since most traditional healers were prohibited by government and church from working for the people—and in any case, introduced diseases did not respond well to traditional medicine—allottees often needed money for doctor's fees. Thus, there was a strong pressure, both licit and illicit, on the Bureau of Indian Affairs to allow allottees to sell or lease their land. The first relaxation was to let widows and orphans lease their land, on the ground that they had no one to work it for them. In many cases, land had already been leased to non-Indians before allotment, often for ninety-nine years—which seemed like perpetuity at a time when the reigning mythology was that Indians would die out as a distinguishable group within a generation or two. Once the land of the widows and the orphans was out of the hands of original allottees, the next group to be "emancipated" were the mixed-bloods of less than half Indigenous descent, who were often determined to be "competent" to patent and sell their land.[18]

The whole issue of "competence" derives from racial and gender stereotypes. Widows were certainly "competent" to manage their land, especially if they came from nations where horticulture was the responsibility of the women. While "blood quantum" could be a marker of assimilation and access to Euro–North American education, it was no marker of intellectual competence. Native persons who were not familiar with Euro–North American traditions of private property and market economies might have been naive about what the conventions allowed, but they were not stupid. Despite the "help" of the governments, some northern Plains peoples were succeeding in ranching and even farming before World War I, and the example of the Five Southeastern Nations demonstrates both individuals and social/cultural structures that were intelligent and adaptable, pioneering highly effective alternatives to the simple and often rapacious and wasteful market system of the late nineteenth and early twentieth centuries. As we shall see, the authoritarian squelching of land use in diverse and complex patterns of polycropping and wild land in favour of technological monocropping had been pioneered in Europe and would be repeated in the Soviet Union and Africa in the twentieth century. In those cases, everyone was equally dispossessed, but in Oklahoma, patronizing "competence" commissions of various sorts not only marked some people as "competent" and thus to be relieved of legal restrictions on the sale, lease, or mortgage of their land, but also marked others as "incompetent"—and thus to be subjected to trustees. Again, the process began with orphans. In some cases, all full-blood Aboriginal people were deemed legally incompetent. Trustees could then lease, mortgage, or even sell their allotments and pocket most or all of the profits.

Allotment was only extended to Indian Territory in 1906 (though western Oklahoma had been allotted earlier) to phase out reservations and the Indigenous governments guaranteed by treaty in order to clear the ground for Oklahoma statehood. The Five Southeastern Nations and the Osages, who in 1871 had purchased part of their old territory from the Cherokees with the proceeds from selling their Kansas reservation, had all developed workable land-use patterns in the context of the reservation. Allotment of more recently settled reservation peoples was the final act of a continuum of violence and destruction. For the Five Nations and the Osages, allotment was the third and most destructive of the US assaults on the successful blending by settled and prosperous peoples of Indigenous and newcomer economic, political, and cultural traditions. As Angie Debo wrote, "the general effect of allotment was an orgy of plunder and exploitation probably unparalleled in American history. . . . Personal greed and public spirit were almost inextricably joined. If they could . . . create a great state by destroying the Indian, they would destroy him in the name of all that was selfish and all that was holy." Debo's *And Still the Waters Run*, the title of which refers ironically to the terms of the treaties, is a blow-by-blow description of that "orgy of plunder and exploitation," particularly as it affected the "Five Civilized Tribes."[19]

It is not surprising that Euro–North Americans continued to "plunder and exploit" Indigenous people until there was nothing more that they wanted to steal. It is not surprising that prominent Oklahomans tried to squelch Debo's book and delayed its publication by four years. It is not surprising that Canadians had to wait more than fifty more years before a comparable study of bureaucratic dispossession was published, given Canada's belief that its Indian policy was humane and far preferable to that of the United States. It *is* surprising that the United States and Oklahoma were gifted with a historian as able, patient, and honest as Angie Debo. On the other hand, it is surprising and disheartening that neither the Osage Reign of Terror nor the "Still the Waters Run" saga make it into the general histories of the Great Plains. The decision of Richard Maxwell Brown and the editors of *The Oxford History of the American West* to omit all discussion of the Osage Reign of Terror from their chapter on violence is at best peculiar. The usual omission of all mention of Indians in most US texts for the period between Wounded Knee and John Collier and the Indian New Deal in 1934 considerably distorts the history of the Great Plains by implying that the end of the Indian Wars was the end of dispossession, and that brave cavalrymen defeated brave warriors in honourable battle to clear the way for brave pioneers eager to feed a hungry world.[20] Even revisionist historians who know that Wounded Knee was a massacre and that the Homestead Acts did not—and could not—create a heartland full of yeoman farmers still skip over the disruption, loss, and general infamy that was allotment.

Debo sets the scene by quoting Senator Henry Dawes and his apparently schizophrenic approach to Indian landholding. Speaking in 1883 to the Friends of the Indians meeting at Lake Mohonk, he described the Cherokees in glowing terms: "The head chief told us that there was not a family in that whole nation that had not a home of its own. There was not a pauper in that nation, and the nation did not owe a dollar. It built its own capitol, in which we had this examination, and it built its schools and its hospitals." Instead of concluding that the Cherokee nation was a society that the mainstream should endeavour to study and emulate, Dawes maintained that the Cherokee system was defective because "there is no selfishness, which is at the bottom of civilization. Till this people will consent to give up their lands, and divide them among their citizens so that each can own the land he cultivates, they will not make much more progress." A lot, of course, depends on one's definition of "progress." As we have seen, "progress" included the commodification of land by ownership in severalty through purchase, homestead, or allotment, and the establishment of capital-intensive, commercial, monocrop agriculture. We have also seen that this economy is inherently unstable on the Plains because extremes of climate mean that sufficient water for crops—or even, in some cases, for cattle—cannot be anticipated and because most production is for an export market that flourishes only during international catastrophes or with support from the American or Canadian government. The scene described by Pleasant Porter, last executive officer of the Creeks before the United States unilaterally abolished the Five Nations' governments in Oklahoma in 1904, is certainly not progress in Dawes's terms; rather, it is an image of biodiversity that may be more crucial for the environmental health of the region. (Ironically, it is what The Nature Conservancy has established—without the people—on a large section of what was once the Osage Reservation—not without the resentment of some Osages.)

> If we had our own way we would be living with lands in common, and
> we would have bands of deer that would jump up from the head of every
> hollow, and flocks of turkeys running up every hillside, and every stream
> would be full of sun perch. . . . That is what we would have; and not so
> much corn and wheat growing and things of that kind.

Porter believed that many of the Creeks would die off for "the want of hope" because their institutions were being destroyed too fast for them to make the transition to American individualism. Too much had been destroyed for the Creeks to regain their own ways, he believed, so the transformation had to proceed quickly to reach the stability of the other side.[21] But there would be no stability until the remnants of land, timber, and minerals left to the Indians were worth too little for the grafters to continue with the trouble of drowning out the voices of those who opposed allotment and graft.

Not surprisingly, the Indians with the most valuable properties were most in danger of the grafters. The luck of timing, the astute political sensibilities of the leaders, and the skill of their lawyers enabled the Osages to purchase both their land and its associated mineral rights in Oklahoma. (The people of almost all American reservations retained the rights only to the surface of the land, not to the minerals, including petroleum, underneath. Canadian reserve peoples have been more successful in maintaining claims to subsurface rights.) When allotment was forced upon the Osages, they had to parcel the land out equally among all tribal members, but they did not have to give up the "surplus" land. And they were able to keep the mineral rights in common for the tribe, with any profits divided up among all living members of the Osage Nation in 1906 with an addition of babies born in 1907. These "headrights" were inheritable, a circumstance that would bring a wave of murders to Osage country. Oil was found under the Osage lands during the 1910s, and as it was brought into production in the early 1920s, thousands of dollars per year fell into the laps of headright holders. Although most Osages tried to live as normally as possible in the midst of the boom, many died under mysterious circumstances, murdered by or at the behest of non-Natives who, through marriage, insurance policies, fraudulent wills, or other devices, stood to benefit from the deaths. One particular rancher and his nephew systematically murdered an entire extended Osage family to funnel the headrights down to the young woman who had the misfortune of being married to the nephew. Most estimates of the murders put the death toll at about sixty—out of a total population of about 2,400—but the number may be higher.[22] Ironically, had the Canadian *Indian Act* been law in the United States, the women murdered by their husbands or in-laws would not have been in danger—for they would have been dispossessed of all Indian claims for marrying a non-Indian.

Nor were the Osages the only people who suffered fraud, kidnapping, and murder, particularly if their allotments included or were thought to include surface rights to oil wells. Freedmen were at particular risk. Since the victorious Union government had required that the Five Nations adopt their freedmen as tribal citizens after the Civil War, the freedmen had been allotted on the same terms as Indigenous tribal citizens. According to Oklahoma's unusual social construction of race, however, all Oklahomans who were not of African descent, including Indigenous full-bloods, were "white," while all who were of African descent, in whatever mixture and whether or not considered by the Five Nations as citizens by blood and descent, were "black." Oklahoma courts were not at all efficient at protecting Indians, but they were particularly loathe to prosecute "whites" (in the ordinary understanding) who had defrauded "blacks" (by any understanding). Some freedmen were able to swindle the swindlers, but others were less successful. "Some spectacular crimes occurred, such as the dynamiting of two Negro children as they slept, in order that the conspirators might secure title to their Glenn Pool property by forged deeds," and other apparent murders, kidnappings, and other crimes were never solved. Some wealthy allottees were forced to flee the state for self-protection. While Debo found that most federal Indian officials were guilty of "general inertia and indifference" rather than "downright dishonesty," those who were dismissed for dishonesty tended to become grafters themselves.[23]

Although murder, kidnapping, extortion, embezzlement, and other crimes against allottees were rarely prosecuted successfully, legal ways of fleecing Indians were more popular and even less risky. The problems had begun with allotment. Dawes and other Friends of the Indian had long complained that innocent full-bloods who clung to subsistence homesteads in the hills where they could hunt needed to be protected from the elite and usually mixed-blood fellow tribal citizens who had established large cattle operations in the grassy valleys. (In this, the Friends sounded a good deal like the Ontario annexationists who had promised to deliver the Métis and Crees from the iron hand of the Hudson's Bay Company in the North West, but they also sounded like the AIM members who would arrive on Pine Ridge in the early 1970s to protect traditional families from the mixed-blood elite of the elected tribal government.) When it came time for allotment, however, the traditional hill dwellers were concerned to protect their homesteads and not concerned about the bottomlands assigned to them by the Dawes committee, who were excellent at the mechanical division of acreage, if less perceptive about the actual needs and wants of the Five Nations people. Speculators easily relieved traditional hill families of this "surplus" land through leases that automatically turned to sales when the federal rules ended restrictions on the alienation of "surplus" holdings. Lessors also managed to control timber lands and to strip them of trees before letting them go back to original allottees. (The same sorts of problems had plagued the Métis who received land scrip in Manitoba. Although the procedures were scrupulously "fair" in their operation, they were set up in such a way that it was almost impossible for the people to secure the lands they most wanted and needed.) The extremely high level of personal honesty among the Indians made them particularly vulnerable to fraud at the hands of unscrupulous or self-deluded speculators, but when the Nations did attempt to protect their citizens, the Indian office and the president disallowed it. "The Choctaw government made repeated attempts to deal with the whole allotment problem in a statesmanlike and constructive manner," asking that a commission with maps and field notes visit each township to help Choctaw citizens select their land. President Theodore Roosevelt vetoed two versions of the commission. "It is difficult to see why this intelligent plan was not adopted," writes Debo. Creek attempts at gaining protection also met with vetoes.[24]

"The most revolting phase of the grafter's activities," notes Debo, "was his plundering of children." Since allotments were made to each citizen of the Nations, young children often became owners of land that had great value to the grafters—especially if the land sat over oil. Because parents, expected to compete in a materialistic economic system, often knew no more about using and protecting their children's land than their own, legal guardianships at first could have served some useful purpose. Instead, they soon degenerated into a very lucrative process of farming children, particularly orphans, to gain the use of their property. Courts turned the guardianship over to grafters who were under no contract to insure a fair accounting or to abstain from swallowing up all their ward's profits. According to Debo, virtually all Indian children in the Territory could have been supported by the lease of their allotments, but hundreds received nothing at all, and many resided, destitute, in orphans' homes. During the territorial period, actual title did remain with the child; statehood, however, allowed alienation, and children whose parents died or who were otherwise deposed from guardianship—by being placed in a "state institution for the feeble minded," for instance—quickly had their land sold from under them.[25]

Not all Euro-Americans supported the grafters, nor did all Indians suffer. Those Indians who had already accepted Euro-American landholding customs and economic ways usually continued to prosper and were accounted founders of the state of Oklahoma and even elected to office. Guardianship laws had been genuinely intended to protect children and non-English speakers who were threatened by the grafters, and some Indian Affairs and court officials tried honestly and sometimes effectively to protect the people they were supposed to protect—even as other judges fattened their campaign finances by selling guardianships. For the most part the newspapers favoured the settlement of Oklahoma by Euro–North Americans by any means possible. A notable exception was the *Wevoka Democrat*, under the editorship of Dan Lawhead, who fought a pitched battle with the other two Seminole County papers in 1908, opposing the dispossession of Seminoles and Seminole freedmen. Within a year, however, Lawhead was silenced, apparently when one of the land dealers who "held a note" against Lawhead managed to have the *Democrat* plant repossessed by the sheriff and the paper began coming out under a "custodian." Whatever had sparked Lawhead's objections to the land transactions, it was not catching, and apparently nothing came of grand jury indictments of the Seminole County grafters.[26]

According to Debo, "The only serious attempt upon the part of the state to correct the situation came through the efforts of a remarkable woman, Miss Kate Barnard of Oklahoma City." The story shows the intersection between maternal feminism (though Barnard herself was not a mother) and Manifest Destiny—and the willingness of even the grafters to allow a little window dressing of piety and philanthropy to interrupt the manly business of fleecing Indians. Barnard, an active philanthropist, and the women's organizations that supported her successfully lobbied for the creation of a "Commissioner of Charities and Corrections," and when it was established, Barnard easily won election to the post. According to Debo, "The Constitutional Convention seems to have created the office as a gallant gesture that might please the women and would do no particular harm." The office had little money and little power, although as something of an afterthought (apparently), the first legislature authorized the commissioner as "next friend" to appear with minor, institutionalized orphans before the probate court. Barnard herself was not even aware of the exploitation of the Five Nations peoples, but as soon as she began to meet the dispossessed children, she took up the cause and worked with a will with the meagre instruments at her disposal. Co-operative judges allowed her interventions to succeed, though the attorney who served as her assistant was often high-handed and antagonistic. Many guardianship abuses were completely out of her control, and although Oklahoma congressmen boasted that the state was protecting the helpless, huge numbers of children were still being robbed.[27]

Exploitation of the Osages and the Five Nations of the old Indian Territory did not stop until the rich pickings were exhausted. When the murderers of Osage County finally struck a prominent white attorney who had been investigating the murders and the Osages themselves paid the FBI to investigate, the reign of terror came to an end.[28] Most of the Osage oil discoveries had been leased and were beginning to play out, so headrights were no longer as valuable. The Osages had already been swindled out of their "surplus" land, as well as of some of the homesteads, when they had become alienable after statehood. Similarly, by the mid-1920s, the Creeks, Choctaws, Cherokees, Chickasaws, and Seminoles had already lost most of their timber, oil, and "surplus" land. The orphans who had

received original allotments had already come of age and were of no further use to their "guardians."

And so the plunder stopped, not because of the courageous opposition of people like Dan Lawhead and Kate Barnard, or the incorruptibility of
the justice system, or the sense of fair play of the American people. In thirty
years, the people of the six Indigenous nations had lost their schools, their
government, their land, and sometimes their lives. The people had had a
comfortable and vibrant existence with a mixed economy and political and
social institutions that were not only assimilating to whitestream culture on
their own terms but even pioneering land-use and social-structure models
that might have been more advantageous to the Ozark Plateau and the Great
Plains than the mandatory commercial agriculture and petroleum industries that did grow up. And all of that was systematically destroyed. That
many people became demoralized and impoverished, and that others left
to develop lives where they would no longer be stigmatized as Indians is
not surprising. That most did manage to survive and to pass on some type
of Indigenous identity is almost miraculous and speaks volumes for the
personal and cultural tenacity of the people. While the grafters themselves
in some cases laid the foundation for subsequent Oklahoma fortunes and
political dynasties, they left behind them a tradition of public corruption and
land titles that were entangled in contested claims. Not surprisingly, they did
not openly celebrate that part of their heritage. Debo points out that at the
establishment of statehood, *all* Oklahomans adopted Pushmataha and the
Trail of Tears as their cultural heritage, and Oklahoma state license plates
still proudly sport feathers and the slogan "Native America." In 2000, the
Oklahoma Humanities Commission proclaimed Angie Debo as the writer
they wanted to represent their state. Newspaper stories of the 1920s talked
about how oil-rich Indians squandered their wealth—but not about how
they were murdered for it. A little pamphlet entitled *Oklahoma's Poor Rich
Indians*, written by Matthew K. Sniffen, Gertrude Bonnin, and Charles H.
Fabens, caused a furor when it was published in February 1924.[29] But its
fame was not long-lived, and although it contributed to the winding down of
graft, it had little institutional impact. The myth of the frontier completely
trumped the myth of American fair play. And so Custer remains in the history books—and Kate Barnard does not.

The 1920s marked not only the gradual tapering down of Indian exploitation in Oklahoma—and the exhaustion of anything left to exploit—but also
the gradual depopulation of the Great Plains (absolutely in some areas and
overall in relation to the rest of North America), which began in 1919.[1] The
extreme variability in moisture from year to year in a complex system of
greater and lesser precipitation cycles had developed the Great Plains ecosystem of grasslands with enormous species variety to be able to withstand
rain, drought, and prairie fire. Gophers and locusts harvested the grass to
protect the roots when there was little rain. Buffalo, elk, and other ruminants followed predictable migration patterns but ones that varied greatly
with rainfall and other climatic patterns. Prairie fires, bison ripping up
grass—roots and all—and pawing and creating wallows, and even the exca-
vations of prairie dogs and gophers exposed soil to blowing. At least during
the days of dog transportation, it seems as if Indigenous peoples mostly
lived on the verges and the riverine oases of the Plains. Travel by foot and
dog travois was slow. People in the southern and middle Plains, such as the

Pawnees, Mandans, and Hidatsas, maintained corn villages, while people north of the Missouri for the most part confined agriculture to the ceremonial growth of tobacco and hunted buffalo, using pounds or jumps perhaps in co-operation with the wolves. Despite their utilization of such stationary features as gardens, pounds, and jumps, and the slowness of dog travel, the people such as the Blackfeet were able to be mobile, to anticipate the cycles of the buffalo, and to move away from drought.[2]

The rapid reintroduction of horses back onto the Plains from the south increased the mobility of the people. They could range nearly as far and as fast as the buffalo, and they could carry food and tools with them. Horticulture became less necessary (lessening the significance of women's work and hence the prestige of women), and eastern peoples such as the Siouan confederacies and the Crees came (or came back) onto the Plains. The Kiowas came from the northwest and the Apaches, among the Athapascan peoples who had migrated through the Plains around the 1300s, returned from the south.[3] Although droughts, unusual bison movements, regional overhunting or overgathering, and, increasingly, deadly raids on small family groups menaced the people, theirs was a sustainable way of life. And a satisfying one. Historians and ecologists have not yet agreed on when and what the "climax" population of buffalo was or exactly when it began to decline, but sustainability was certainly hampered when the Great Plains became a hinterland to the fur and hide trades. The market for pemmican to feed the northern fur brigades certainly raised hunting pressures on buffalo in the Red River area, while a growing industrial demand for bison robes and bison-hide belts for steam machinery supported a hide industry largely carried out along the Missouri from St. Louis to Fort Benton. Not until the coming of the transcontinental railroads in the United States, however, did the buffalo vanish as a subsistence resource.[4]

The very mobility of the buffalo meant that their demise was necessary for the establishment of commercial agriculture on the Great Plains. Pronghorns gracefully feeding among cattle were no problem. A shaggy brown river flowing for days through fences and across ploughed fields was another matter. The demise of the great free-ranging buffalo herds, the agreements to the numbered treaties in Canada, and the abandonment of the treaty system and the retrocession of the Great Sioux Reservation in the north and Indian Territory in the south of the US Great Plains all occurred during the 1870s and set the stage for large-scale Euro/Afro/North American settlement of the Plains. The 1870s also saw the first bust for settlers who had come into Kansas and Nebraska during the relatively wet 1850s and 1860s, only to discover drought, the economic impact of which was magnified by a recession in the 1870s. A similar economic depression and drought in 1893 hit all the Plains states and territories and saw disillusioned settlers leaving the Plains and heading either west or back east. The 1890s was the decade of the Populists. Despite a few itinerant rainmakers, the Populists could do little for drought, but they could attempt to loosen the grip of the railroads, elevators, bankers, and mortgage companies on the farmers. Returning rains and prosperity by 1907 introduced the age of parity (1910–14)—the rate of return, in purchasing power, per bushel of wheat or corn or hundredweight of cattle or hogs that would be the benchmark for farmers seeking support programs for decades to come. The end of the Depression of the 1890s and the beginning of another prairie wet cycle initiated the extraordinary wheat boom of the Canadian Prairies that lasted until the outbreak of World War I in 1914.[5] As we have seen, the economic basis of both the wheat and beef booms of these years was speculation, investment, and, in the case of wheat, the subsistence work of women and children. What was happening in the Prairie Provinces was similar to what was happening in the western Dakotas and Nebraska, and on into Wyoming and Montana, helped on by the irrigation promoted by the 1902 *Reclamation Act*—though that had been directed more specifically at California, the Southwest, and the Great Basin—and the Enlarged Homestead Acts, such as the *Kinkaid Act.*[6] Neither farmers nor speculators nor the general public, however, saw rate of return as being the result of excess investment rather than the inherent productiveness of the land and the "scientific" farming techniques that had been invented to tame it. The Great Plains was the Last Best West, the home of the bonanza farms, where golden wheat to feed the world would make everyone's fortune.

Settlers poured in. Canadian cattlemen lost their leased and public domain grazing lands to homesteaders. Indigenous people disappeared from the public consciousness—except for spectacles like the Calgary Stampede—but the survivors of the starving years of the 1880s and 1890s found ways to combine subsistence, grazing, teamstering, and the sale of crafts and traditional foods such as berries to survive. They even began to reverse the population decline that had continued since 1492. Railway completion to the north, however, brought Euro/Afro/Canadian settlers to Peace and Athabasca River country, further marginalizing Indigenous hunting and trapping, subsistence activities, and even agriculture. The Canadian government's "barbarism" theories and the extreme niggard-liness of both federal governments in providing seed, draft animals, and implements had the ironic effect of protecting Indigenous peoples from the excesses of the wheat boom—though government expropriation of north-ern Plains land and herds during World War I, supposedly to produce more food to help make the world safe for democracy, saddled reserve and res-ervation communities with the ecological if not the economic results of the bust and substantially reduced land retained by the reserves.[7]

The development of Marquis wheat in 1911 did make the wheat bonanza plausible, but it took World War I to make it real. The virtual destruction of European agriculture and the insatiable demand of the allied armies for bread, beef, horses, and men raised the prices of all these Prairie products. Even with Canadian government price controls on wheat, Prairie farmers could pay off all their debts with a single harvest—which encouraged them to mortgage everything and buy more land at any price. The weather was not exceptionally good for most of the war, but it was good enough to make a crop. As a farmer says, cynically, in Edward McCourt's novel *Music at the Close*, "Matt, if them Huns can just hang on for two more years, we'll all be able to retire."[8]

By 1919, there were no more armies to feed, European agriculture was producing again, and the men that Prairie farmers had learned to do without returned home, looking for jobs and homesteads. And so the bottom fell out of wheat and land prices. The roaring of the 1920s on the Great Plains was the sound of banks failing and farmers losing their land. David C. Jones's *Empire of Dust* focusses on the dry belt of southeastern Alberta and especially the town of Carlstadt—which became Alderson during the anti-German days of World War I—but this area is simply an exaggerated version of most of the High Plains from New Mexico up. Alderson is in the western base of the Palliser Triangle, that area that the first expansionists had deemed part of the Great American Desert but that the optimists from Clifford Sifton and Frank Oliver down through the moneylenders and dry farming experts to the boomers and would-be town fathers had declared open for agriculture. People poured into the country. If many expected to get something for nothing, some did make speculative fortunes. The ones who believed in hard work and the decent life of farm and small town, however, mostly got nothing for years of trying to make a living. "Experts" were sure the region was destined for glory, as if willing it into agricultural land could make it agricultural land. "'Hard times will never affect Southern Alberta' [mortgage company co-treasurer Kingman Nott] Robins quoted Canadian paladin of the soil, Professor James W. Robertson [in 1910]. 'The interests of this district are now so diversified that there is no possibility of a pronounced depression.'" But diversification is no guarantee against drought. Nineteen fourteen brought crop failure; 1915 and 1916 brought bumper crops to coincide with the war demand. By the agricultural census of 1916, 45 percent of all farms and 75 percent of all wheat farms in Alberta were in the dry belt. Between 1918 and 1922, wheat prices dropped by more than half, though the prices of manufactured goods, inflated during the war, did not drop as quickly or as far. Meanwhile taxes leaped—in the Nobleford area from $2 per quarter section before the war to $36 by 1922. Land values dropped with wheat prices—from an average of $12.89 per acre in 1914–19 to $9.58 in 1920–21 to $7.51 from 1925 to 1929. The 1920s also saw record or near record low precipitation. After 1916, the grasshoppers, gophers, rabbits, and mosquitoes turned out in vast numbers. The pale western cutworm was largely responsible for crop failures from 1917 to 1920. And many farmers were paying off land purchased at high prices with high interest rates during the war years.[9]

Except for the municipalities, who were aware of the disaster facing their citizens and were generous with aid—resulting in debt and higher taxes—governments did not help the floundering farmers. Mrs. Reinhard Frerichs was one of many who wrote to Herbert Greenfield, the first premier in the United Farmers of Alberta government, asking for relief: "It eats all them years the dear seed and never gives it back." But the premier and the others in positions of power regarded Mrs. Frerichs and anyone else who complained as "anticapitalistic scaremongers of the worst order."

Euro–North American settlers were meeting the same ideology that had denied the suffering of the Blackfoot and Crees, Lakotas and Cheyennes, and others in the 1880s. The dry farming experts propounded their "ten commandments" of dry farming and insisted that no farmer who followed the rules could fail. They believed, on the grounds of their own high opinion of themselves, that since the land *was* occupied by farmer-settlers, it *could* be occupied, and Nature would have to obey the experts and nurture the farmers. Finally, even the experts acknowledged that in some years, no crops were possible. In 1926, the Lyman school board voted to paint the schoolhouse yellow so it wouldn't show if anyone relieved himself against the wall, and to paint a white elephant on the front. But it could have been no more than a gesture of defiance. Everything was kaput, and there was no money for paint. The out-migration that followed World War I in southern Alberta was as dramatic as the in-migration that had preceded the war, although most settlers stayed, at least during the 1921 census period. The whole southeastern Alberta area lost 21 percent of its population, but 48 southeastern townships lost 75 to 100 percent of their population. Southwestern Saskatchewan also lost population, but not as drastically.[10] The situation was similar in Montana and, to some extent, all over the Great Plains.

Why did the "Empire of Dust" see such a spectacular build-up and decline? As in the Vulcan area, just to the west, much of the original settlement was speculative, spurred by government hype, easily available mortgages and other money, and the desire to make a fortune while living a Wild West adventure. As with Vulcan, the people who suffered the most were those who really intended to make a living farming in the region and reaped few of the speculative benefits but had to pay for them in higher taxes and depreciating land prices. The wishful thinking of government, experts, moneylenders, and intending farmers is important. All concerned seem to have felt *entitled* to have rain follow the plough, to have technology vanquish the desert. Part of the problem was with the social construction of "desert." Just as Indians who did not wish to divide land into private property were deemed deficient and needed to be changed—even if the change primarily demeaned, demoralized, and impoverished them—land that would not produce dependable crops of European grains was also deemed deficient and in need of reform through conversion to private property and "breaking" to the plough. But if we remember that *no* ecosystem is deficient, that *all* ecosystems are sufficient for the organisms that have co-evolved with them, blaming the land is clearly paradoxical.

At the same time that the various Homestead Acts and the immigration propaganda was creating the belief in an entitlement to farm (and also an entitlement to make profits in land speculation) on the Great Plains, the railroads and the federal governments were creating contrasting economic uses for the mountains and the American Southwest as tourist attractions. Mining and irrigated agriculture would also play their parts in these arid lands, and the intensive irrigation around Cardston, Lethbridge, and the other parts of southern Alberta owes much to the Mormon experience of creating irrigated agriculture in Utah. It is instructive, however, to compare the propaganda for the corridor of national parks just to the west of the Great Plains to that for the Plains. The Y2Y (Yellowstone to Yukon) Conservation Initiative, currently supported by environmentalists, operates in the tradition of constructing the mountains as beautiful, fragile, and full of environmental diversity and splendour. While grasslands conservationists are now trying to apply similar imagery to the Great Plains, a century ago, the region was constructed as utilitarian, its diversity much better sacrificed to monocultures of corn and wheat, something we have already noted in Pleasant Porter's comments. It is also instructive to compare the CPR's tourist posters, for which they quite deliberately recruited artists, showing the mountain splendour of Banff, with their settlement recruitment posters of cornucopias and sheaves of wheat and a land transformed.[11] The railroads had received land as part of their payment for construction of the actual railways, and they had chosen *prairie* land, not mountains or the Canadian Shield. It was part of their economic role to construct the prairies in the picturesque tradition of homes and herds, while the mountains were the sublime of untamed peaks. The Santa Fe did the same for the desert Southwest.

The cowboy aesthetic, expressed by the Calgary Stampede, painters like Frederic Remington and Charlie Russell, and writers like Owen Wister and Will James, also created and bridged a dichotomy. The rodeo and Indian Village aspects of the Stampede to some extent mask(ed) and to some extent enhance(d) its utility as an agricultural exhibition, showing off wheat, barley, and tame forages like brome, timothy, and alfalfa. The relentless square survey also reinforced the idea that the Plains was a monotonous monoculture and the mountains—wild, varied, and unsquared—were sublime rather than deficient. Yet the one aspect of prairie restoration that has proven most difficult is the extraordinary complexity of prairie flora and fauna. Any given acre of tallgrass prairie regularly supports about two hundred different kinds of plants, and every slight slope, exposure, or soil variant supports a different mixture of plants, which shift again in response to drought or wetness or different kinds of grazing pressure.[12] From termites and gophers, to voles, to buffalo and elk and grizzly bears, the prairies are infinitely varied and variable. There is a good deal of irony that the great ruminants and predators of the Great Plains found refuge in the mountain parks of Banff and Yellowstone, Glacier and Waterton Lakes.

The settlement of the Plains and the preservation of the mountains was not inevitable—as one can see by the cattle and petroleum exploitation in the Kananaskis area just outside Banff, or the pressure for mines and wells outside the mountain parks, or even the reservoirs constructed within the parks themselves. The huge lake in Rocky Mountain National Park in Colorado is a reservoir for a trans-basin project bringing Colorado River water through a tunnel to irrigate wheat on the plains to the east, while Barrier Lake, at the north end of Kananaskis Country, was constructed to provide hydro power to the city. The flowery mountain meadows of July are no more beautiful than the flowery prairie meadows of May and June, except that the prairie flowers have almost all been ploughed under. The desert of southeastern Alberta is not the Great American Desert that Palliser believed he saw. Instead, it is a desert of wheat, created by a particular economic system. That does not negate the suffering of the intending farmers who came, but it does suggest that it was not the land that was at fault but the socially constructed belief in entitlement to farm European crops.

The continuing disaster of the 1930s on the Great Plains resulted from the same conditions that had plagued the 1920s, complicated by a more widespread drought and an international economic disaster that, like the collapse of wheat prices after 1919, was a result of the Great War and, in the case of the Depression, the shattering reparations imposed on Germany afterward. Donald Worster has called the Dust Bowl the greatest ecological disaster ever to hit the United States—possibly the worst in the entire world—and he blames both it and its coincidence with the Depression on the nature of capitalism.[13] He does, however, also acknowledge that even the Soviet Union followed a similar predatory policy in ploughing up land, if not in policies leading to the ecological disasters around Lake Baikal and the Caspian Sea. Yet the frenetic ploughing of the Great Plains did not happen in other parts of the capitalist world. The North American East Coast and Maritimes simply does not have the expanse of level land and deep soil that the Great Plains has. The tallgrass prairie of the American Midwest, sloping briefly into Canada, suffered ploughing even more intense than the Great Plains because it was less arid.

In general, the attitude of North American agriculture and public imagery focussed heavily on the idea that all grassland was deficient and had to be reclaimed—as if it had declined from some earlier, better use. While the clearing settlements in the eastern parts of North America and around the Great Lakes had the same attitude toward trees that Prairie pioneers had toward grass, timber was at least acknowledged as having value. The grasses were not. Even in the twenty-first century, I have had Greenpeace recruiters and other environmental activists argue against the validity of a grass/grazing utility for the land, opposing all meat production, even the range production of grass-fed beef, a far less ecologically damaging alternative to ploughing the grasslands and planting soybeans, most of which are now genetically modified. Grasslands can best produce protein for human use by serving as pastures for large ruminants—whether bison and elk or domestic cattle—as the Lakotas, Blackfoot, and other Indigenous Plains peoples have always recognized. Because the extermination of the buffalo herds was essential to the dispossession of Indigenous peoples and because the Indian Wars have been valorized in American popular culture and Walsh and his Mounties peacefully subduing Sitting Bull in Canadian popular culture, the very grasslands themselves seem to have become abhorrent to whitestream cultures on the Plains—just as Germany had become abhorrent to the victorious World War I allies, who imposed impossibly strict reparations upon Germany in the Treaty of Versailles. Although, as Worster says, we should have learned from the Dust Bowl that something was fundamentally wrong with the way capitalism used the Great Plains, even some conservationists have made only minor adjustments, not ones that would lead to a totally different and more appropriate relationship with the land.

The Depression was both a worldwide phenomenon and a sequence of experiences that varied by place, time, economic class, gender, Aboriginal status, and so forth. Both of my parents lived through the Depression in the small city of Calgary, now a major metropolis that is still distinguished by domestic architecture that was almost all built either before 1914 or after 1947. The Glenmore Reservoir, still the source of domestic water for downtown and the south side of the city, was, however, excavated and its dam built as a municipal relief project during the Depression. My maternal grandfather, a lawyer, held onto a middle-class existence during that grim decade. My father's family was less prosperous, especially during the last illness and after the death of my grandfather. Family legend has it that they escaped starvation and relief (a worse fate) during the winter my grand-father lay dying only because my uncle, teaching school at Morley on the Stoney Reserve, went hunting with his students and brought home a moose that fed the family through the winter. Yet Calgary was better off than the southeastern part of the province, where those settlers who had survived the 1920s slowly succumbed to the 1930s.

In the United States, the true "Dust Bowl" area of northeastern New Mexico, southeastern Colorado, western Kansas, and the panhandles of Oklahoma and Texas suffered from the tenure of "suitcase farmers" who had entered a particularly dry and windy part of the Great Plains, mined it for wheat for a few years, and left when the land began to blow away.[14] The dust storms further east, from Texas through Saskatchewan, hit communities that had been settled with Euro–North American farmers for three generations and had survived earlier droughts of the 1870s and 1890s. Some thought it was the end of the world. Yet climatologists tell us that, in terms of millennia, the drought of the 1930s was not a particularly harsh one. It was only that the people could not pick up and move with the dirt that made it a human disaster. In fact, the 1930s was the *least* mobile decade on the Great Plains since the Euro-Americans had come, and the one in which the Euro-Americans most resembled the Indigenous people in their subsistence methods. My uncle's hunting experience was not unique. Nor was the return to the "home place" and the support of the extended family. With the collapse of the market economy—despite the efforts of the Farm Holiday and other farmers' associations, the Agricultural Adjustment Administration (AAA) and Prairie Farm Rehabilitation Administration (PFRA), and other federal programs in both countries—women's subsistence activities and the ingenuity of both women and men had to replace the market until rain and war came again.[15] Rain and war were always the twins that seemed to make the Plains most commercially attractive.

The 1930s redefined politics in North America, as unemployment mounted to levels never seen before or since, except on reserves and reservations, and, more recently, in inner cities. Nowhere was the response more dramatic than on the Great Plains. The Populists had attained prominence in the 1890s, the Non-partisan League during the teens, and the Progressives during the 1920s. The thirties brought many new theories and leaders.

Mitigating but Not Rethinking: George W. Norris, 11
Tommy Douglas, and the Great Plains

The careers of George W. Norris of Nebraska and Tommy Douglas of Saskatchewan, two extraordinary Prairie progressives, cover nearly a century of political activism and tell us something about both what was possible and what was never even considered in the Great Plains. That their seemingly different heritages, one a dyed-in-the-wool Republican from Ohio and one a Scots Labourite, should result in similar solutions to the problems of European-style agriculture on the Great Plains illustrates the significance of geography, independent of ideology, in determining the lifestyles that will work for a region.

Richard Lowitt, Norris's biographer, describes Norris as a nineteenth-century liberal, but one who became a Progressive and then a New Dealer, developing his ideas to fit the exigencies of the twentieth century but maintaining his basic beliefs in the fundamental goodness of human beings, the value of honesty and hard work, and the role of government in helping people who are in trouble through no fault of their own. Although Norris carefully researched all the legislation he proposed and supported during his long career in the United States House (1902–12) and Senate (1912–42), he charted his economic and political course on experience rather than on readings in history or theory. Like many highly successful individuals, he had a few big ideas and he stuck to them, winning most of his greatest battles. He believed that government should be efficient, economical, and accountable to the voters. His institutional reforms included curbing the power of the Speaker and the caucus in the US House and in instituting a unicameral legislature in Nebraska. Norris started his career as a Republican but eventually became an Independent and one of the most intelligent backers of Franklin Delano Roosevelt's New Deal. Deeply moved by an international peace conference he attended in Belgium in 1905, he consistently supported a world body for arbitrating disputes, the disarmament of aggressor nations, and the restriction of munitions and armaments on the part of democratic nations. Norris, though originally not sympathetic to organized labour, came to see farmers and labourers as a necessary coalition. He believed that "big business" tended to exploit them both and that the role of government was to protect the people from monopolies by ensuring fair competition, bargaining rights for workers, and mortgage relief for farmers beset by bad weather or low markets. He also believed that electric power was a basic but transformative necessity that could serve people best if it were generated and distributed publicly, supporting both municipal and federal power.[1]

Tommy Douglas has so far attracted more adulation, more vituperation, but less painstaking scholarly analysis than Norris. There is no work on Douglas that compares to Lowitt's magisterial three-volume study of Norris. Nonetheless, the outlines of Douglas's career and beliefs are also clear. His intellectual tradition was that of British Labour of the Scots variety, deeply affected by the twentieth-century Social Gospel of Salem Bland and J.S. Woodsworth. Like Norris, his program was based on experience rather than theory—he quickly dropped the repugnant eugenics theory that had formed his master's thesis, and it never influenced any part of his public policy.[2] While Norris had had his pragmatic training as a lawyer and judge in southwestern Nebraska during the 1890s, Douglas's lessons in the world came from his own recurring bouts of osteomyelitis, his witnessing of the police riots against strikers in Winnipeg in 1919 and in Estevan in 1931, and his experience as a pastor and graduate student among the unemployed and desperate in the early 1930s. Unlike Norris, Douglas was satisfied with the parliamentary systems he worked in as both a Member of Parliament (1935–44, 1962–79) and as Premier of Saskatchewan (1944–62). He successfully advanced his beliefs, whether he was in the opposition or the majority. Like Norris, Douglas favoured arbitration for settling international disputes, though he was more willing than Norris to use force if he thought it necessary. During the 1930s, he was strongly opposed to Canada's selling any materials that might be used as munitions against Canadians or their allies. The Co-operative Commonwealth Federation (CCF), the party that, as CCF or later the New Democratic Party (NDP), would be Douglas's throughout his career, was founded as a farm and labour party: Douglas always saw farmers and urban workers as a natural alliance against what the Regina Manifesto, the founding document of the party, had denounced as "capitalism." Both Norris and Douglas came to see an alliance between farmers and urban workers as necessary if farmers were to overcome their increasing minority status as farm populations dwindled.

In *The Making of a Socialist*, the closest thing to an autobiography that Tommy Douglas left, he said that the notorious promise to "eradicate capitalism" contained in the Regina Manifesto really did eradicate the capitalism of the 1930s that had been a disaster for the people of Saskatchewan, leading to farm foreclosure, farm abandonment, and massive unemployment. The old capitalism had been banks calling loans and causing general social collapse—and the CCF had indeed done away with that. Condemning giant monopolies that had an unhealthy power to ruin the entire society was not the same as condemning private ownership. The "Pocket Platform" of the CCF in 1944, when it actually came to power in Saskatchewan with Douglas at the helm, called for home security and debt reduction; increased old age pensions, mother's allowances, and disability care; medical, dental, and hospital services; equal education; free speech and religion; collective bargaining; and the encouragement of economic co-operatives. The party wanted a mixed economy, including public, private, and co-operative sectors, with a strong role for private ownership in innovation and competition.[3]

Planning was crucial to the kind of economy that Douglas and the CCF envisioned in 1944. Because of Saskatchewan's relatively small population base, its remoteness from larger population centres, and its lack of large sources of capital, the CCF had to manoeuvre reasonably carefully in order to avoid scaring away potential investors who would have to fund most innovations. Social ownership would have to be small and experimental, and taxes could not rise above the norms in the rest of Canada.[4] Douglas argued that "those Industries that were vital to the life of the community, and were monopolistic in character, ought to be publicly owned." This included power generation and automobile insurance. Crown corporations based on natural resources could process primary products, create employment, raise production and provincial revenue, and return profits to the people. Even anomalies, such as the government's acquisition of a box factory, made sense when private ownership could not or would not provide investment or abide by labour laws.[5]

According to Richards and Pratt, planning kept the CCF from foundering like the "Nonpartisan League in North Dakota," but slowed down innovation, created friction with workers who wanted to share in management decisions for their plants, and discounted the need to take and reward risk, perhaps because, by definition, planning is intended to diminish risk. Nonetheless, Richards and Pratt found that the CCF Crown corporations were, as a whole, economically successful, but after 1948, the CCF's ability for economic intervention was dampened by anti-communist hysteria against any state enterprise, overstated opposition by the business community, and striking, if misleading, comparisons to Alberta's oil wealth after the first Leduc discoveries in 1947. Both Douglas himself and the more removed Richards and Pratt judged the CCF redefinition of capitalism in Saskatchewan as basically successful in diversifying the provincial economy, maintaining investment levels while capturing revenue streams for the province, and developing both a government bureaucracy and a managerial labour force by attracting highly qualified individuals from outside the province and educating those within. By contrast, Richards and Pratt see Alberta, much more richly endowed with petroleum and hence with wealth during this time period, as having produced only a "striking failure" at "nurtur[ing] a powerful class of Alberta entrepreneurs united with populist farmers in hostility to a takeover by external corporate and political interests."[6] Albertans, they suggest, merely counted the oil money as it rolled in instead of attempting to enhance provincial and local control of the industry.

In their study of Saskatchewan's economy after Grant Devine pronounced that the province was "open for business" in 1982, James Pitsula and Ken Rasmussen conclude that the free market economy, operating as it should, milked resources, capital, and people from economic hinterlands such as Saskatchewan. Devine and his Conservatives, working from an ideology derived from Reaganomics in the United States and Thatcher's policies in Britain, operated contrary to CCF policies, and they failed. "Driven by necessity but trapped by ideology [Devine's Conservatives] ended up entering into highly questionable deals that contradicted their own precepts about how the economy should work." Far from rescuing the province from some kind of "nanny state" regulations, Devine's failure illustrated that Douglas's prescription for Saskatchewan of a modified capitalism that used government planning had been correct. "The new right believes in the free market, but the free market judges Saskatchewan harshly."[7]

Norris, working in the federal arena, could not back such sweeping reforms for one state. His focus was narrower, but equally determined. He backed public electricity generation and distribution above all and, in times of war or crisis, other public ownership as seemed necessary for society to function. Both Norris and Douglas were successful in bringing rural electric power to their polities, but Norris's sustained campaign for public power ownership really had no analogue in Saskatchewan, as public power had been a norm in Canada since the 1910s. Douglas's role was more one of organizing parts into a coherent province-wide system. His most personal and deeply felt cause was universal hospital and medical insurance, which took twenty years to secure in Saskatchewan and helped bring about Douglas's 1962 defeat in Regina in the federal election and the 1964 defeat of the CCF-NDP government in Saskatchewan. Ironically, medicare has become Tommy Douglas's most enduring legacy and one of the enduring—though enduringly embattled—touchstones of Canadian society. There is no analogue for Norris or for the United States. Even contemporary "Obama-care" is far less inclusive and was developed nationally rather than in a particular region.

While Norris moved directly from local to federal office, Douglas began and ended his elected career as an MP but spent the most productive part of his political life as premier of Saskatchewan. Norris served only as a Nebraskan, while Douglas was elected MP from British Columbia after his federal defeat in 1962. The actual overlap in the careers of the two men was relatively short, from Douglas's election to Parliament in 1935 to Norris's defeat in the Senatorial election in 1942—or at most, from Douglas's first campaign in 1933 to Norris's death in 1944. Year-by-year comparisons, then, are not always relevant. Both Norris and Douglas worked in the context of other people: for Norris, the Progressives and the New Dealers, as well as his own circle of friends and supporters in Nebraska, and for Douglas, the CCF and later the New Democratic Party. They were not isolated prophets howling in the wilderness, but they were such significant leaders and shapers that one can attribute ideas to them without being misleading. Each has usually been discussed in the context of other Canadians or other Americans. For instance, Douglas is often compared to (and contrasted with) William Aberhart and the Saskatchewan CCF to the Alberta Social Credit. Norris is most often discussed as a Progressive and compared to Robert LaFollette, or as a New Dealer who was not a Democrat. Comparing Norris to Douglas, however, allows us to compare and contrast US and Canadian procedures and solutions, and to look at the ways in which the particular environmental and historical conditions of the Great Plains enabled the rise to power, the long and influential careers, and the distinctive arguments and successes (as well as those things misconceived or overlooked) of these two remarkable men.

The people whose anguish Norris shared in the 1890s and who commanded compassion from him and Douglas in the 1930s were those who had taken the promise of the *Homestead Act* seriously, whether they had homesteaded their land or purchased it. For them, leaving the land was neither emotionally nor economically sensible. They had followed all the rules to turn "free land" into farm homes, and they had failed because of forces they could not control—the climate; the international economic downturn; the pressure of outside financial, manufacturing, and transportation corporations; the workings of grain marketing boards; and the tax, tariff, and relief structures of municipal, provincial, state, and federal governments. Now they, like the farmers of southeastern Alberta in the 1920s, were being judged deficient. Norris and Douglas would do their best to change the conditions under which their constituents laboured, from government policies, to regional economics, to the very relationship of sky, land, and water. Although both men and their allies would be attacked as socialists and enemies to the market system, they, like the farmers whom they wanted so deeply to help, were concerned with making a conventional humid-culture market system work on the Great Plains. As the CCF and NDP would discover, public ownership was not nearly as much of a departure from market economics as theory would have it, and planning did not change the parameters of a sparsely populated hinterland. Neither Norris nor Douglas undertook reforms that looked to previous means and ideologies of using the land (such as riverine agriculture based on usehold rights, seasonal family-based migration to utilize different ecosystems, or a pastoralism based on the buffalo) that related directly to the particular ecosystem of the Plains or that moved outside the basic patterns of the market. The policies Norris and Douglas chose to develop, however, suggest both what their particular relationship with the Great Plains was and how economic development on the Plains might have happened during the twentieth century, and still might happen.

As a district judge in Nebraska from 1896 to 1902, Norris was frequently called upon to foreclose farm mortgages and to order sheriff's sales of the properties. Again we see the central trope of losing the farm, and Norris was in the thick of it. The southwestern corner of Nebraska (like south-central Saskatchewan, where Tommy Douglas would find himself in the 1930s) is a semi-arid region that receives an average of less than twenty inches (500 mm) of precipitation in a year, coming in alternating cycles of wet years and dry years as it does on most of the Great Plains. It is mostly cropped land rather than pastures and cattle country. Farmers who had come to Red Willow and the surrounding counties in the 1880s had arrived during a period of good rains and good crop prices. The decade of the 1890s was drier and featured the spectacular 1893 economic crash following the overbuilding of the railroads. Southwestern Nebraska, like south-central Saskatchewan and most of the territory in between, was oversettled—people moved in as if the land were suitable for humid-culture agriculture. The area was also overcapitalized. The rich soils coming under production were an irresistible magnet for eastern and European investors. Money, in the good times, fairly chased farmers. It was not the sale of wheat that produced the Prairie booms but the lending of money on the expectation of the production of even more wheat.[8] And farmers, during the 1880s (and later the 1900s, and particularly the great boom of World War I), were more than willing to borrow money to buy more land; to buy the machinery, such as reapers and threshers, that was necessary for increasingly large-scale agriculture; and to build improvements such a fences, drains, and irrigation works. The drought and contraction of the 1890s meant that farmers in Norris's district were producing less wheat per acre and receiving less money per bushel of wheat than they had a few years before. Making things worse was an agricultural economy that had been basically deflationary since the Civil War, meaning that each year the farmer needed more bushels of wheat to pay off the same amount of debt. No wonder the Populist Party rose out of this mess and called for railroad and elevator regulation and the free coinage of silver to inflate the dollars of the debtor farmers.

Judge Norris was an ardent Republican, not a Populist, but he was as concerned about foreclosure as anyone. Because his background as a lawyer was in working for lenders, he could see better than most people that selling out a hardworking farmer was a lose-lose proposition. The farmer and his family lost their home, and all the lender gained was a hard-scrabble ruined farm that no one wanted to buy and that had no one to work it, unless the former owner were willing to stay on as a disillusioned and angry tenant. Nebraska had no mortgage foreclosure moratorium law, so Norris simply stayed foreclosure and sale if he thought a farmer would be able to make it when the good times returned.[9] Only if he thought an individual were too shiftless or too heavily in debt to work his way out did Norris allow a sheriff's sale. At first, creditors were furious, but they soon came to see that Norris's solution was the most likely to repay their investments. For Norris, this was a pragmatic and humane solution to "the agony of these cycles of crop failure, heavy indebtedness upon the land, and ruinous farm commodity prices," and perhaps more important in the long run, it preserved both capital and democracy. For Norris, "national welfare and progress are stimulated by any system of capitalism which provides for the widest distribution of the natural resources of soil and its use by the largest number of legal owners."[10] Like the early modern bureaucrats who imposed square surveys on European commons and the men Worster believed had shaped the Dust Bowl, Norris did not consider other forms of land use than those of fee simple agriculture.

Tommy Douglas would not be in a position to deal with farm mortgages for another half-century or so, but his response was not dissimilar. Capital was safer if good farmers kept their land. Like Norris, Douglas believed deeply that one of the most essential roles of government (especially, for Douglas, in a Christian society) was to protect those who could not help themselves. The CCF fought the election of 1944 on the promise of farm security, and one of the first measures that the CCF government introduced was the *Farm Security Act*, designed to provide absolute protection to the farmer's home quarter section and to prevent foreclosure in years of poor yields. Although the war had brought a large measure of prosperity back to Saskatchewan, many farmers were still in debt. The *Farm Security Act*, like similar measures passed in Alberta, was eventually declared *ultra vires*, but the period during which the issue was tied up in court gave at least some farmers the breathing room that they needed. Neither Norris nor Douglas intended to interfere with capital's right to a return on its investment,[11] though the *Farm Security Act* did propose that investors be required to forgo interest in years when a farmer could not make enough crop to repay the loan. Sharing the risk is part of the investment process, however, and the higher the rate of return, the higher the shared risk is assumed to be. The object in both Nebraska and Saskatchewan—as well as the other polities that introduced or considered foreclosure moratoriums— was to make capital more flexible and capable of creating both a healthy rural economy and a satisfactory rate of return in the long run. Farmers forced to repay debts to distant eastern investors before buying local goods and services depressed the local economy and were less likely to succeed in the long run. Creditors who waited would get their investment back over the length of the climatic and economic, if not annual, cycles. Both Norris and Douglas approached farm foreclosure from an experiential rather than an ideological point of view. Republicans in the 1890s did not advocate for foreclosure moratoriums. Actual conditions on the Great Plains were the stimulus for the responses of both Norris and Douglas. A truly ideological response (and one that undervalued the ingenuity of actual farmers on the land) came much later, from Grant Devine, and, as Pitsula and Rasmussen demonstrate, it did not lead to more development but rather to more bankruptcies.[12]

Mortgage moratoriums, whether ad hoc or statutory, and the relief programs of the New Deal were not, however, sufficient to create or maintain healthy rural economies, and both Norris and Douglas continued to fight for changes to the way capitalism worked in the farm economy. Douglas was particularly interested in economic diversification, especially, as Richards and Pratt point out, into resource development and secondary manufacturing. Another piece of CCF legislation that was eventually declared *ultra vires* taxed mineral properties that were not developed. This legislation was intellectually akin to Henry George's ideas in *Progress and Poverty*.[13] Since economic value was created by society as a whole, landlords who held valuable properties out of production so that they could later reap speculative gains for themselves prospered at the expense of the rest of the society. Since the Canadian government had returned control of natural resources to the Prairie Provinces in 1930, the Saskatchewan CCF tried to spur development by taxing undeveloped mineral rights. Since much Prairie economic development has been premature in the sense that markets were not ready to support it—railroad building is the prime example—taxing undeveloped minerals was an ingenious attempt by the government, and hence the taxpayers of the province, to get the economic benefit of premature development even as owners waited for more economically viable production opportunities. Again, the courts ruled out this option, but it would have been an innovative way of frontloading the revenues that would eventually come to the province by way of royalties, back-ins, and other measures that Saskatchewan employed to share the revenue stream.

Norris's measures to mitigate the difficulties of raising humid crops in a dry environment were less innovative but actually attained the force of law. He was a great champion of dry farming, especially of the "Campbell method," and he pushed for federal support for agricultural experimentation with crops capable of withstanding Great Plains meteorological conditions. Both plant breeding and innovations in tillage succeeded in mitigating crop loss. Norris's greatest legacy to mitigation, however, was in the multi-purpose watershed projects that he championed throughout his federal career. Nebraska farmers were hindered by spring floods that inundated newly planted fields, swept away farm animals, damaged buildings, and, in the Republican River floods of 1935, resulted in the loss of more than one hundred human lives. Later in the summers, the lack of rain resulted in parched crops and diminished yields or even no harvest at all. Life on the farm in all seasons was lonely and labour-intensive. Dams could solve all of that, Norris believed, providing flood control in the spring, irrigation in the summer, and electric power and recreation all year long.[14]

Although Norris's response to the *Newlands Reclamation Act* of 1902 was to propose a reservoir along the border between Red Willow and Hitchcock Counties in the middle of his congressional district, his real introduction to dam building and public power came with the Hetch Hetchy project in California. As Norris saw it, the major purpose of the project was to create hydro power on public land and to make it available to the city of San Francisco, assuring cheap power for consumers and for the street railway. The dam would also provide flood prevention downstream and secure irrigation water for the farmers who were already using the stream. Although the dam would flood a wild and beautiful valley in Yosemite National Park, Norris thought a lake would improve the view, and the roads necessary for the project would make the area more accessible to tourists. Not surprisingly, private power companies as well as conservationists opposed the project, and while conservationists had to see the valley flooded, the private power companies eventually took over the distribution and sale of the hydro power. Although central California was far away from the Great Plains and the main purpose of the dam was electrical generation for a city rather than irrigation, it was Hetch Hetchy that introduced Norris to the "miracle" of dams and lakes.[15]

Norris's most famous multi-purpose river system development is also far away from the Great Plains. The Tennessee Valley Authority (TVA), as Walter Stewart has pointed out in both his 1987 study of Crown corporations and his 2003 biography of Tommy Douglas, is bigger than any of Canada's Crown corporations—thus somewhat complicating the assumption that publicly owned corporations are Canadian rather than American. After World War I, Norris saw an opportunity for the federal government to use the federally owned fertilizer plant at Muscle Shoals as the nucleus of a project to develop the entire Tennessee River watershed for flood control, navigation, irrigation, and hydroelectric generation. This time supported by conservationists (and not, in the beginning, by the Nebraska legislature), Norris deflected a private offer from Henry Ford, hung on through vetoes of public power by Herbert Hoover, and fought off the private power interests of the southeast. When the New Deal finally favoured public projects, Norris still had to hold on through Supreme Court challenges to the constitutionality of TVA before he saw its building and success. What marked TVA, the Rural Electrification Administration that Norris also sponsored, and the collection of dams and lakes (crowned by Kingsley Dam and Lake McConaughy, known as the "Little TVA") in Nebraska was their comprehensiveness. Norris made sure that each project even had a subsidiary that assisted farmers to purchase electric appliances so that demand would be ready when supply came on line and power would never go unused.[16]

If Norris's expertise in guiding dam building, irrigation, and power generation schemes did not begin with the Great Plains, it certainly lent itself to conditions on the Great Plains in the 1930s. John Wesley Powell had warned Americans since 1878 that water would control the economic development of the West, and that federal development of dams and co-operative irrigation districts was the most intelligent means to that development.[17] Although Powell's ideas were unpopular with western boomers, they were in many ways accurate harbingers of Norris's plans. The biggest problem with federally developed irrigation projects had always been that in most cases, irrigators alone could not pay for the cost of development. Hydro power could help subsidize construction, but not if private companies were able, as they were at Hetch Hetchy, to monopolize the sale and distribution of power. Navigation (not relevant to Nebraska, except on the Missouri) and flood control were federal concerns and could therefore command federal dollars that did not have to be paid back by the users of the water or electricity.

In putting together Nebraska's Little TVA, Norris had to fight New Deal administrators to make sure the state was awarded Public Works Administration funding in accordance with the disproportionate economic losses suffered by the Great Plains states during the Dirty Thirties rather than with average per capita US payments, and that farmers who had managed to avoid the dole were eligible to work on the projects. He had to cajole local backers of individual projects to work together instead of fighting among themselves for the primacy of their own local construction. And he particularly had to overcome the influence of his old adversaries, the "Power Trust," particularly the Nebraska Power Company, headquartered in Omaha. Their vituperation against anyone associated with Roosevelt or the *National Recovery Act* of the New Deal was just as scathing and more lethal than any of the opposition to the socialism of the CCF. Because the president of the Power Company was also the president of the University of Omaha board of regents, he was able to censure a professor for praising the TVA and to fire university president W.E. Sealock, a Norris and Roosevelt supporter. Three days after his firing, Sealock committed suicide. Norris persevered, however, and when World War II began to restore prosperity to Nebraska, farmers had the water and energy to increase production, while Nebraska's central location and plentiful, cheap electricity allowed it to land war-time production industries, though not as many as Norris desired. Once public power was harnessed into war production, it became patriotic rather than sinister and socialist. Ironically, though, Norris's scrupulous concern for public welfare may have cost both Nebraska and the TVA region postwar economic development. Both the TVA and the Nebraska projects directly hired local workers to construct dams, transmission lines, and other parts of the projects. In the Far West, however, the Bureau of Reclamation hired private contractors from San Francisco, Salt Lake City, and Portland, who were able to develop the corporate strength to grow even larger government industries during the war and to demand for both themselves and the region sustainable manufacturing and prosperity, unlike Omaha's, after the war.[18] One of them was Bechtel.

Saskatchewan's dams and hydro power lagged considerably behind Nebraska's. A Liberal government introduced public power to Saskatchewan in 1929, and the Conservative/Progressive government elected the following year endorsed it, but it was not until 1945 that the new CCF government began buying up all the private power companies in the province to gain economies of scale and to get rid of duplication. After that was completed, the CCF could move toward generation. Neither Nebraska nor Saskatchewan had to expropriate private utilities: they simply ceased to be economical after public power came in. In 1949, Saskatchewan Power (SPC) became a Crown corporation and began a rural electrification program for the southern part of the province, paid for by the farmers and SPC. Although the 1930s had moved Saskatchewan, like Nebraska, to look at damming major streams, particularly the South Saskatchewan River and its tributaries, the federal government dragged its feet on funding and authorizing such dam-building projects until 1958, when Prime Minister John Diefenbaker agreed to what would become the Gardiner Dam, holding back the waters of Lake Diefenbaker, which lapped against the shores of Douglas Park.[19] Not until the late 1960s did the South Saskatchewan plants begin generating power, and, as is the case in Nebraska, coal-fired plants provide much of Saskatchewan's power today. In both Nebraska and Saskatchewan, rural electrification was popular and uncontroversial. Isolated farm houses were not an attractive target for private power companies. The big difference was in the move to consolidate private power companies into a state- or province-wide public grid. In Nebraska, it was a hard-fought battle. In Saskatchewan, it was simply the model that other provinces already followed and the only surprise was that the CCF was able to hang on and get the job done in the vast and sparsely populated rural parts of the province. TVA and Nebraska Public Power are anomalies in the United States. Saskatchewan Power was formed by the CCF, but its public status was the norm—even highly market-driven Calgary hung onto its city power system, though not without controversy, during the recent rage for utility deregulation.

Neither Norris nor Douglas saw any problems with building dams. Norris, as we have seen, thought even very picturesque parks were better with lakes and access roads. Rivers that simply ran were, he believed, a waste of water. Yet dams on prairie rivers silt in rapidly and require dredging to retain their capacity to prevent floods and store water for irrigation and generation. The lack of flooding on the post-dam Platte means that the sandy islands characteristic of a braided prairie river are not scoured out, damaging the roosting habitat of sandhill and whooping cranes and the nesting and spawning habitats of various other species, some, like the whoopers, threatened or endangered. Just as the dams on the Columbia severely injured the native Pacific coast salmon, dams on prairie rivers have caused unforeseen ecological damage. Although Douglas opposed some provisions of the treaty governing Columbia River development, it was US control of the water, not habitat loss, that bothered him.[20] Yet renewable resources, such as water, may not be entirely renewable after all.

Dams and lakes have an adverse environmental effect that was not foreseen by their builders. Many of the dams on the Great Plains have also had an adverse effect on Indigenous people (as we shall see in chapter 13), who despite their articulate protests have in many cases received shorter shrift than the whooping cranes and snail darters. As F. Laurie Barron points out in his study of Tommy Douglas and the Native peoples of Saskatchewan, governments, particularly governments that explicitly set out to help the underdog, must be judged at least partly by their ability to perceive and to respond meaningfully to the most disadvantaged members of society, and on the Great Plains, that primarily means Indigenous people.[21] In Saskatchewan, the relationship between Native peoples and hydroelectric projects is not as contentious as it was in the case of the Great Whale projects in Quebec or the Oldman River Dam in Alberta. Power generation is mostly absent from the northern parts of Saskatchewan, the province with the highest proportion of Native and Métis people, although power generation and transmission was one part of the general disruption of Aboriginal societies in the vicinity of uranium-producing and pulpwood sites in the north. In Nebraska, dam building would mean substantial losses to Native people. While the Platte, Loup, and Republican river valleys had for the most part been "cleansed" of Native people long before the 1930s, Republican River dams did cause the flooding of one highly important Pawnee holy spring that was venerated by most peoples of the region. TVA dams also flooded Cherokee graves and other holy sites, long after the majority of the people had been removed from the area. The controversy over the snail darter and the Tombigbee River, long after Norris's time, obscured the Cherokee objections to the flooding of their ancient capital of Echota and other historical sites. The developments that caused the most damage both to living American Indian communities and to graves and holy sites, however, were those on the mainstem of the Missouri River. Although constructed long after the death of Senator Norris, these dams, like the project that flooded Echota, had been among his most cherished future projects.[22] As Michael Lawson notes, Missouri mainstem dams were consistently sited where they would not inconvenience many Amer-European settlers, but where they would inundate large percentages of the homes of Lakota and Dakota people settled on Missouri River reservations and in several cases, including the Santee in Nebraska and the Hidatsa in North Dakota, would flood whole communities. Although tribal leaders consistently testified that the dams would do considerable economic and social damage, their legitimate concerns were systematically denied at every level.[23] While Douglas was uncomfortably, if incompletely, aware of the failure of ccf policies to render substantial aid to Aboriginal communities, Norris, for all his tolerance and his disgust at racial hatreds, simply did not register Indian people. In the uplifting farewell chapter with which Norris ended his memoirs, thoughtfully suggesting his best hopes for a peace that would endure at the end of World War II, he wrote, "Never in its entire history has America coveted the lands and the wealth of other peoples," quite oblivious to all of America *being* the land and wealth of other peoples.[24] Building dams for irrigation and hydro power was a logical, even courageous, response to the conditions of drought and poverty, one especially relevant to the Great Plains. At the same time, it was a statement that both the land itself and the societies that had evolved there were deficient for proper human uses—so deficient as to be invisible. And it is hard to think of looking for guidance to something, or someone, that does not register on one's consciousness.

Dam building and the flood control, irrigation, and hydro generation that went with it were the main means of mitigating the climate that Norris and, to a lesser extent, Douglas implemented. Both, however, saw ways to change the structure of government itself so that it would better serve the particular needs of the Great Plains. Norris's lifelong goal of providing efficient and transparent government led to the formation of the only one-house legislature in the United States, Nebraska's non-partisan Unicameral. As was the case with public power, Norris expended a great deal of time and energy securing what Saskatchewan and most provinces already had, a one-house legislature, though of course Saskatchewan's is not non-partisan. The Unicameral was Norris's idea, although state governance was not part of his duties as a US senator. He organized the coalition that brought it into effect. While there was considerable public support for the idea, it would never have been raised except for Norris.[25] A sparsely populated, relatively poor polity benefits even more than a large and diverse one from a small, simple legislature. Fewer senators cost less. Non-partisanship allows for fluid alliances that change from issue to issue. While both Saskatchewan and Nebraska have more diverse economies than they had in the 1930s and both have developed urban centres with their own particular demands and concerns, neither has the diversity and polarity that might require two houses to protect. The major benefit of a unicameral system, to Norris, was the relative transparency that results when bills do not disappear into the mangle of conference committees and emerge with transformations for which no one is clearly responsible, something that could benefit any deliberative body. Norris brought the one-house idea to Nebraska because it was his home state, but voters may have been ready to accept it because it made particular sense for Nebraska and the Great Plains.

It is tempting but not actually useful to say that Norris's institutional innovations made government smaller while Douglas's made it bigger. Certainly Norris's support for the various New Deal agencies in Nebraska increased the presence of the federal government more than ever happened in Saskatchewan. The CCF government in Saskatchewan from 1944 to 1964 provincialized services that neither the federal government nor the private sector could provide. Co-operative marketing and purchasing boards were essential to farmers during hard times, though, like the New Deal agencies in Nebraska, they might come to be seen as impediments during the plush times. Provincial hospital and motor vehicle insurance were popular throughout Saskatchewan, but they were particularly helpful for farmers and their families. Because the farmer was self-employed, he had no employer to help out with medical insurance. And because farmers frequently owned valuable on-road vehicles such as pickups and other, larger trucks, in addition to private cars, they benefited more than the average urban driver from low-cost premiums.

One of the more striking parallels between Norris and Douglas during the years that they were both members of the federal legislature was in their reactions to the arms buildup before World War II. Norris, who had opposed US entry into World War I, had allies among the isolationists and the pacifists in American society, but he was neither an isolationist nor a pacifist. Douglas shared his attitude toward peace and armaments with his CCF caucus, but he particularly spoke out about munitions and war materials, especially after his 1936 visit to Germany. Norris opposed US entry into World War I because he thought Britain was as guilty of imperialism and the disregard of neutrality as Germany was, and because he believed arms manufacturers were stirring up a war hysteria to sell their goods. America could make the world safer for democracy, he believed, by staying out of the war. Before World War I, he had consistently pushed for smaller naval appropriations and more support for international arbitration mechanisms. Disputes were solved by the means at hand, he believed, and it was safer to make sure that one was supplied with agreed-upon international dispute resolution mechanisms than to be surrounded by warships. During the 1930s, both men consistently argued against selling materials that could be used as weapons to countries that might turn out to be enemies. In his maiden speech to parliament, Douglas pointed out that the federal government could scarcely talk of peace while selling Mussolini nickel and oil. In the spring of 1939, he made the same point (echoing William Jennings Bryan, another Nebraska statesman), calling on Canada not to crucify "a generation of young men . . . upon a cross of nickel." Norris, in the same year, similarly found it "heartbreaking" that the United States was selling scrap iron to Japan, airplane parts to Germany, and war materials to Italy. He proposed to keep materials out of the hands of aggressor nations by selling only on a "cash and carry" basis, which would have allowed England to buy—but not Germany and Japan.[26]

Although at first blush this opposition to selling war materials to potentially hostile powers seems like a combination of pacifist-tinged ideology with plain good sense, it also has a relationship to region. Sincere as both Douglas and Norris were in their opposition to munitions, it is doubtful if representatives with their outlook could have been consistently re-elected to the Senate or the Parliament from regions dependent upon war materials extraction or manufacture—and in this, a kind of "deficiency" may even be viewed as positive. Even after the outbreak of war, the Plains benefited far less from wartime manufacturing than did those regions already engaged in heavy manufacturing, such as the Ohio and St. Lawrence valleys, or the American West Coast and Southwest, where the companies who had built themselves up as federal contractors in the 1930s were most successful in landing war construction contracts. Montreal participated heavily in munitions manufacture during World War II, as it had in the Great War. Douglas, like all the federal members of the CCF caucus except for J.S. Woodsworth himself, supported Canada's declaration of war in 1939, and when Japan bombed Pearl Harbor, Norris joined in supporting the US declaration of war. Once committed, both men pressed their respective governments to provide adequate support for the troops. Norris pointed to Nebraska's central location and cheap, plentiful electricity to urge the siting of weapons and aircraft plants in his home state, and he was fairly successful. The *Enola Gay*, the plane that dropped the atomic bomb on Hiroshima, was fabricated in Omaha, Nebraska. Norris's defeat in 1942 came about for a number of reasons, but the effect of his pre-war pacifism or his wartime beliefs on the economy of Nebraska was not one of them. Saskatchewan received less in the way of manufacture, despite Douglas's greater degree of support for war. As a member of a minority party, he was not in a position to bring large wartime contracts to his province. Except for the training of Commonwealth pilots, the Prairies received little direct economic development during the war, but Douglas's long opposition to the sale of war materials was no hindrance to his successful campaign for a CCF government in 1944. During the Cold War, however, Douglas and his CCF government endorsed and promoted uranium production, even though it was being used for weaponry.[27] It was not sold to the Russians or other obvious potential enemies.

Both Norris and Douglas, then, mitigated the poor fit of humid-area cropping techniques with a semi-arid environment by supporting foreclosure moratoriums and direct relief for farmers, by championing irrigation and public power, by changing the structure of government to be more responsive to the people, and by articulating humanitarian and commonsense arguments against the excessive development of other regions, to the detriment of the Plains, through the sale of potential war materials. In what ways could they or did they try to change the humid-lands agriculture and economic system to fit the particular environment of the Great Plains or to mimic the past history of human land use there?

One could look at proposed variations to fee simple ownership and private property that seemed to have disproportionately affected Amer-European settlement on the Great Plains. Farm tenancy on the Great Plains has conventionally allowed new farmers to work themselves into the land and older farmers to work themselves out, rather than leading to permanent tenancy of the sharecropper version existing in the American South. Leasing is a somewhat different proposition that has primarily been applied to grazing and has allowed some approximation of the purposeful migrations of both the wild buffalo herds and the people who hunted them. Norris worked with fellow Nebraskan Moses Kinkaid to introduce a 640-acre homestead in 1908 to allow small ranches, especially in northwestern Nebraska's rugged Pine Ridge and Sandhills areas, but he later came to favour state ownership with leaseholds for cattlemen. Even a whole section was too small for a ranch, and the *Kinkaid Act* continued to result in violations of the law. Norris was sentimentally attached to the *Homestead Act*, and in 1935, he sponsored a bill to create the Homestead National Monument near Beatrice, Nebraska—the site, as noted earlier, of the "first" homestead claim in the United States. Norris does not seem to have been particularly involved in the withdrawal of all public lands from homesteading in 1934 and the substitution of the *Taylor Grazing Act*, which enabled ranchers to lease federal land.[28] By this time, Nebraska was no longer a public land state. Almost everything was in private or state hands and thus was not affected by the *Taylor Grazing Act*. Norris certainly recognized that private ownership of land did not always serve the farmer, or particularly the rancher, but his interventions were fairly limited and small scale.

Canada had experimented with leases for cattle ranchers, particularly in Alberta, before the mass influx of homesteaders to the Prairies. Public pressure had forced the opening of much of the grasslands to settlement and also expected ranchers to overgraze in order to be seen as more productive. During the waning days of R.B. Bennett's prime ministership, in the middle of the Depression, the Conservatives passed several acts intended to mitigate the effects of the hard times, including one setting up the Prairie Farm Rehabilitation Administration (PFRA). Although Douglas praised Bennett for this and other attempts to mitigate farm distress, as a member of an opposition party, he obviously could not participate in the

Conservative plans. During the 1930s, the CCF in Saskatchewan experimented with the idea of public ownership of agricultural land in two ways. The first, quickly repudiated, was to secure farm tenure through usufruct rights rather than through fee simple. The province would hold title to the land itself, but the farmer could use, bequeath, or even sell the usehold rights. Agnes Macphail, from the Ontario Farmers' Union, blocked the inclusion of such an idea from the Regina Manifesto, arguing that farmers would never support anything but absolute ownership rights to the family farm, and the proposal was attacked from the right as a precursor to the collectivization of farms.[29]

After that, the proposal was dropped altogether, but it is unfortunate that it did not receive more careful thought, because it had potential as a model for an environmentally sound land-use policy and it reflected both past and present Native land-use patterns. Forced collectivization of farming in the Soviet Union turned out to be a disaster, but it was not the only alternative to a fee simple title. According to Douglas, usehold title would be a voluntary option by which the province would pay off the mortgage but maintain the farmer on the land, and it was intended as an alternative to letting farmers slip into tenancy or lose the land altogether to foreclosure. The proposal's whole purpose was to guarantee owner-occupiers access to the land, the opposite of collectivization—though government programs do not always turn out exactly as planned. Usehold was the norm for the riverine horticulture practiced by plains peoples before the nineteenth century. Gardens belonged to individual women or coalitions of sisters or other female relations, who maintained specific plots as long as those met their needs. Swapping up or down as the situation changed seems to have been fairly easy. The Five Southeast Tribes who were moved to Oklahoma had proved that such useholds could work in commercial agriculture, as big holders and subsistence holders neighboured with each other, allowing wild game habitats to be interspersed with small fields and larger areas of monoculture until the system was destroyed by forced allotment.[30]

Could Saskatchewan have developed a successful usehold system? It is doubtful, given the excessive deference to private property that had developed on the Plains. Owning one's own land was a visceral response to the insecurities of European land tenure for peasant farmers as well as to the reputed (and real) overcrowding of European and North American cities. Useholds that could be sold and thus had a cash value would not have provided the flexibility of the riverine farmer women or the Creek pastoralists to change fields according to circumstances. Of course, many twentieth-century Saskatchewan farmers owned their land only in conjunction with their friendly neighbourhood banker or moneylender, so ownership may have been more in name than in fact for the very farmers whom Douglas was trying to help. But the legal system was not based on usehold, and therefore security was indeterminate—not a very reassuring way to control your home and means of livelihood. Usehold had another economic pitfall that also bedevilled the communally owned land on First Nations reserves. The reason farmers were facing foreclosure in the first place was because they had borrowed money against their lands. A mortgage was the cheapest and usually the only way a farmer could raise the cash for farm machinery, buildings, fences, and other improvements, or for buying more land to make the farm more viable. Once land title was held by the province or, as with the reserves, by the federal government, it could not be mortgaged, a circumstance that has often retarded economic development in reserve communities. Theoretically, this problem could be addressed, as it has been in some parts of Asia, with revolving development funds administered co-operatively, but this was not part of the CCF scheme, nor has anything like it yet been undertaken on the Great Plains. Usehold, which might have provided an innovative response to living with the normal climate fluctuations of the Great Plains, never got a hearing because it was discussed only in sentimental odes to the family farm or in terms of capitalism versus communism, neither of which system is particularly relevant.

By the early 1980s, the cry of collectivization rose again, this time in response to the Land Bank established by Allan Blakeney's NDP government to buy land from retiring farmers and allow young farmers to rent it in order to build up equity and to purchase the farm. When Devine's Conservatives abolished the Land Bank, farm foreclosures rose and tenancy increased—another indication that fee simple land policies in a pure market economy were not sufficient to support Saskatchewan farmers. In 2006, as Prairie farmers debated whether their spreads could ever be profitable in the age of globalization, the farm community was split between those who advocated a government-subsidized living wage for farmers and those who wanted more market control for farmers. Ironically, selling land to First Nations bands and leasing it back is allowing some Saskatchewan farmers to retain the use of their land.[31]

More successful than the abortive usehold proposal were the community pastures that are still part of Saskatchewan ranching, put in place by the Conservatives in Ottawa under the PFRA and implemented, also from Ottawa, by the Liberals.[32] Again, there was an Aboriginal prototype, but this time closer to home and probably familiar at least in theory to some in the Saskatchewan government. The river lots of the Métis settlements had included community pastures behind crop and hayfields. Anglo settlements as well had employed community pastures and a herdsboy in the early days of settlement before individual farmers had the time or money to fence their fields. By including relatively large areas of land extending across various microclimates, community pastures can allow ranchers to utilize range more rationally than if each spread had to feed all its own cattle, especially in areas where there were separate summer and winter pastures. Because Saskatchewan does not have the variations in altitude of the mountain west nor, in the south, the oil and gas deposits of Alberta or American portions of the mountain west, community pastures present a useful alternative to leased, multi-use pastureland that may encompass grazing, recreation, and gas and oil exploration and extraction.

Both Norris and Douglas introduced structural changes to their own polities that have remained in place and that distinguish Nebraska and Saskatchewan from surrounding states and provinces. Even though the CCF was less inventive in terms of the structure of government than United Farmers of Alberta and early Social Credit governments in Alberta, it did initiate many innovations that have stayed in effect. The Unicameral legislature has not proved as impervious to lobbyists as Norris had hoped, and it sometimes works on partisan lines, but it is efficient, effective, and economical (though recently imposed term limits seem to make it less so). Most Nebraskans are quite proud of it and even its detractors oppose its non-partisanship more than its unicameral nature. Douglas's changes involved the structure not of the legislative assembly but rather of the bureaucracy and its relationship to both individuals and industry.

Saskatchewan is still a sparsely populated province heavily dependent upon agriculture and to a lesser extent upon mineral extraction. According to Richards and Pratt, the CCF principles of social control and planning were at least as useful for the province as private ownership, but the CCF itself lost the nerve required to take the risks that would make social ownership as successful as it might have been, while the privatization attempts of Ross Thatcher and Grant Devine watered down some of the CCF heritage. Despite its mistakes—detouring local entrepreneurship into small secondary industries that really had no chance of long-term survival, or allowing northern fish, fur, and timber Crown co-operatives to undercut small private entrepreneurs such as sawmills—Douglas's first CCF government, especially in its first two years, did more to rationalize a sparsely populated, staple-producing province within a market economy than any other Plains government before or since. Provincial education, hospitalization, and medicare itself are deeply entrenched, and even enemies of the NDP admit that no government in its right mind would consider tampering with them.[33] Once the discipline of twenty-five years of privation and war had worn off, and once large farmers had become part of the business elite instead of floursack-wearing populists, however, people from Saskatchewan responded the way most North Americans, particularly westerners, responded to the slow patient slog of reinvestment, social equity, and the gospel of comfort rather than riches. They repudiated it. Like the casinos Saskatchewan would eventually erect, jackpots in the economy and the appeal of being a "have" province capable of flashing its overflowing billfold in front of Quebec and the Maritimes and even Manitoba was definitely appealing.

Neither Norris nor Douglas nor their supporters, however, has ever really dealt with the implications of a grassland ecology and the kinds of economic and social structures that might be most complementary with it. Much of Saskatchewan's resource economy is north of the Great Plains. Norris was never attuned to either the dispossession or the strengths of the land knowledge of the Indigenous peoples of Nebraska or the United States in general; Douglas and the CCF-NDP tried valiantly but mostly unsuccessfully to deal with dispossession issues, but his lack of recognition and use of Indigenous strengths tended to doom, or at least to blunt, reforms.

The legacy of Norris and Douglas is one of honesty, peace, goodwill, and successful mitigation of the grassland ecosystem to fit Amer-European norms of land use and participation in the market system. Their emphasis on altruistic co-operative handling of the environment worked for whitestream society on the Plains as long as it was not overwhelmed by prosperity itself. Their legacy also displays the great loss to both Natives and newcomers that resulted from their inability to "walk in Indian moccasins." Farming is always a gamble with the weather, the land, and the markets, and in North America, gambling is always supposed to pay off with a jackpot. Both Norris and Douglas believed that what most people wanted was freedom from want, a decent level of comfort, and security for themselves and their families. Perhaps that was not enough.

One can scarcely fault Norris and Douglas for not working completely outside the paradigm of market-based humid-lands society, yet it seems somehow a waste that, since they were challenging the status quo anyway, these leaders did not have access to a frame of reference that would have allowed them to plan reforms that started out with the great fact of the land and the thousands of years of history of its use by humans. But the Great Plains is always in transition. Unlike redwood forests that last for centuries, grasslands change from month to month and from metre to metre. Mad cow, drought, and the melting of the glaciers that feed the rivers of the Plains are all forcing change right now. The experiences of Norris and Douglas illustrate the limits of mitigation and could challenge us, the Plains dwellers of the twenty-first century, to look to what they missed—the ecology of the grasslands, the adaptations of native flora and fauna, and particularly the land wisdom still miraculously resident, despite over a century of suppression, in the Indigenous communities of the Great Plains. To rewire Plains whitestream societies in this way would require planning and government intervention in ways Norris and Douglas could not have dreamed. Unfortunately, however, the theory of planning is more useful in hindsight, to explain what has already happened. As we shall see in the next chapter, planning only works well when it proceeds fairly gradually and honours the land knowledge of the people—as it did in Tommy Douglas's Saskatchewan.

The crisis of the 1930s, particularly in Alberta and Saskatchewan, spawned new political responses to living on the Plains. The ability of the market to deliver unalloyed progress seemed questionable wherever it had been applied, but nowhere more questionable than on the Plains in the 1930s. Yet in 1934, during the so-called Indian New Deal, when Franklin Roosevelt's new commissioner of Indian Affairs, John Collier, safely shepherded through Congress the *Wheeler-Howard Act*—which, among other things, ended the process of land allotment in the United States—he was criticized as being too "Red," a supporter of communism. Although these charges seem to have stemmed from pro-assimilationist Indian groups, especially the American Indian Federation (AIF), other Collier opponents were not loathe to use them, and Collier responded by tarring the AIF as fascists. As a result, both sides in this battle of reformers lost face, and the quarrel may have helped lead to the later policy of "Termination," which was a disaster for those tribes—and most of the individuals—who were terminated. The Communist Party was active in both Canada and the United

States in trying to use agrarian discontent to its own purpose, but it was generally more effective in ethnic communities with a socialist background and in Canada than on the US Great Plains, and it had little effect on Collier or the Indians.[1] The skirmish, however, indicates the fear that existed of any kind of collective development, a fear that still animates the Great Plains, especially in the United States in this year of the Tea Party.

In any important sense, "planning" preceded "settlement" on the Great Plains. The various peoples who entered, lived upon, or left the Great Plains before Coronado or the fur traders studied the land for subsistence. They moved, for the most part, incrementally, paying careful attention to available resources, such as buffalo and buffalo jumps, berries and root crops, water for domestic use and horticulture, shelter, sacred places, materials for home building, and all of the other inarticulable elements needed to make space into a homeplace. As they moved onto, across, and out of or permanently into the Great Plains, they ingeniously adjusted their behaviours to get the most out of the Big Sky grassland country. The Pawnees developed riverine corn villages, while the Lakotas specialized in highly mobile buffalo hunting. For Amer-Europeans, on the other hand, the land was literally already mapped out—into 160-acre plots intended for fee simple ownership by individual homesteaders who were to "improve" the land into commercial grain farms operating in a global market. The railway lands, held out of the homesteading process, had in a sense already been "improved" by the simple passage of the rails and the telegraph. As James C. Scott has eloquently demonstrated, the square survey (or cadastral survey) and individualized land ownership were part of the centralization of power by early modern European states that wanted to be able to calculate and thus tax and otherwise control the products of land and labour. Farmers and other small holders resisted the loss of the commons and the loss of informal systems of usufruct rights embedded in the older land systems, but gradually Europe was surveyed and parcelled out. On the Great Plains, however, the federal governments treated for land rights, pushed aside the inhabitants, and laid out the land in square fields that paid no more attention to topography than to the people already living there.[2] Amer-Europeans, then, moved not onto a blank page but rather onto a colouring book created by an Etch-a-Sketch, where clear black lines

with square corners were superimposed on the bleached remains of a swirl-
ing and multi-coloured landscape that, no matter how blanched, could
never quite be subsumed within the lines. Planning for Native people on
the Great Plains has mostly meant ethnic cleansing and decollectivization,
while planning for whitestream settlers has been more incremental but has
still resulted in questionable gains for either humans or the environment.
The two strands of planning continue to interweave and overlap.

As we have seen, in terms of the market economy, the Great Plains is
a sparsely populated, resource-producing hinterland. If the purpose of the
free market is to allocate resources most "efficiently" in terms of producing the greatest amount of wealth, regardless of where that wealth ends up
or who benefits from it, then the proper role of the Great Plains is that of
being an area of exploitation.[3] In a sense, its deficiency is an asset. It *should*
export its minerals, its agricultural products, and the best and brightest of
its children. This is its inevitable economic destiny. In Scott's terms, the
Plains is an area where the laboratory experiment of huge-scale monocrops,
easily controlled by governments, agronomists, and agribusinesses, worked
better than they ever would anywhere else: "In a given historical and social
setting—say wheat growing by farmers breaking new ground on the plains
of Kansas—many elements of this faith [in monocrop technology] might
have made sense," Scott writes, but that sense was only temporary and
partial.[4] In the context of inevitable failure, Frank and Deborah Popper's
Buffalo Commons makes a good deal of sense. Frank Popper carefully studied a number of economic development plans for hinterland regions and
concluded that they did not work. And so he developed the idea of Buffalo
Commons, partly as a tongue-in-cheek commentary on the flaws in economic development modelling and partly as an exercise in planning for the
already occurring depopulation of a region with which he was not familiar.
He noted, correctly, that population in most Great Plains counties had been
decreasing since the end of World War I and concluded that the best thing
to do would be to move most of the people out and to re-establish the buffalo herds. He could number crunch, unhampered by personal experience
or sentiment that might contradict theory, to construct an *un*-development
model for an impressive swathe of the United States. (Like most American
scholars, he did not notice Canada.)[5]

We have already looked in some detail at the successful CCF experiment in Saskatchewan. The purpose of this chapter is to examine economic development models in order to search for other ways of conceptualizing what happened on the Great Plains. Although twentieth-century economic development theory is useful in hindsight for telling us why people behaved as they did, it has little predictive force. Scott's main point, in fact, is that the simplification implied by any linear theory that depends on breaking down incredibly complex interactions into a series of test tube experiments is *always* lacking because it cannot predict or even describe the myriad complexities and relationships of even a seemingly simple wheat field. Nor can it mimic the ongoing series of "gut" responses that an experienced animal, including a human, makes about real subsistence in a real environment.[6]

On a theoretical level, one can see the entire settlement of the Great Plains as a reaction to the revolutions of 1848 in Europe and the enormous fear of communism that developed in the middle classes. Although there was certainly a public demand for "free land" and for the removal of Indigenous peoples from any land deemed suitable for agriculture long before 1848, the monomaniacal tone of insistence on private property came, in part, from a fear of the revolutionists and a fear that they might be correct—that a perfectly workable market society could exist without the concentration of power required by monopoly capitalism. According to Scott, European governments had spent the last century perfecting their control with things like land surveys and systematized surnames. Canada and the United States simply copied France and Britain—Francis La Flesche tells how boys coming to the Presbyterian mission school on the Omaha Reservation were named for American statesmen and generals, and later, at the end of the nineteenth century, the United States would even sponsor a project for the systematic "renaming" of the Indians.[7] The revolutions of 1848, however, also sent exiles to North America who would be significant in forming a critique of nineteenth-century monopoly capitalism, influencing such writers as Edward Bellamy, whose 1888 novel *Looking Backward* would become an important source for the Social Gospel movement that flourished across the continent but, on the Great Plains, particularly in Winnipeg. In Bellamy's novel, the monopolies are the agents of their own downfall, expanding until they form a seamless web that then becomes the basis for a planned society affording comfort to all.

One way, then, to understand the Great Plains is to look at it as that part of North America that was settled and formed into a free market hinterland as a kind of trope of the discussion between the two sides of the debate over the meaning of 1848 and, more generally, controlled economies. This is certainly an implicit part of its intellectual history. But Plains people are not just ciphers without agency of their own in a national debate, and agrarian discontent and regional revolt have certainly occurred. The Grange, the Populists, the Farm Holidays, and so on have marked the history of the American Great Plains, although they have mostly been forgotten in terms of content and remembered only as a kind of resentment of government and used to fuel movements like the militias of the sort that seem to have animated Timothy McVey in his part in the bombing of the Oklahoma City federal building in 1995. Similarly, the Prairie Provinces, in their resentment of the National Policy, railroad and elevator rates, and particularly their lack of control over their own public lands before 1930, led to numerous protest movements, from the United Grain Growers and the Wheat Pool to the Social Credit party and the Co-operative Commonwealth Foundation. More recently, protest, as in the United States, has turned to the right, as we see with the Reform, Saskatchewan, and Wild Rose parties. In the United States, public lands have remained with the federal government except in the original thirteen states and Texas. This led to the Sagebrush Rebellion of the late 1970s and early 1980s, when conservative western states legislators—primarily in the mountain and desert West, rather than in the Great Plains, where most federal land had passed into private hands—mounted an attempt to get control over energy-rich lands for the states. Since Congress had just passed a law entitling local governments to payments in lieu of taxes for federal landholdings, the land stayed in federal hands.[8]

One must pay attention to the explicit intellectual history of the region as well as the implicit one. Regional economic development theory turns out to be an effective tool for this discussion. Regional planning is a paradoxical field, for one can only plan effectively for things that will continue on more or less as they are, and, as Scott points out, only on the basis of a very simplified version of how those things work. Most massive planning experiments have also been massive failures. The Land Ordinance and the various Homestead Acts were not failures in that sense, though they certainly caused harm to the land and the people who lived upon it, but even their successes were somewhat different than their proponents had intended. Planning can, however, solve specific problems. Thus, if the Canadian Plains are expected to be a wheat-producing region, agricultural scientists can develop a new, rust-resistant, short-season wheat—Marquis—and make it widely available to farmers. Or one can breed milo for areas of Nebraska too dry for dryland corn and without cost-effective means of irrigation. Or one can develop an oil seed that is low in acid and high in heart-healthy properties, give it a catchy name such as canola (instead of an unfortunate one like rape seed) and have a new crop capable of replacing wheat as a "king" crop. One can plan dams and irrigation works, and the utilization of new technologies such as center pivots and the older reapers and headers and combines, or specially constructed technologies such as the Noble blade.[9] One can also plan for freight rates, crop insurance, support payments, and tariff, tax, and other government policies that affect agriculture. Soil conservation measures such as contour ploughing, shelter belts, trash mulching, or even various no-till options are plannable. Even genetically modified organisms (GMOs), from Roundup Ready seed (which withstands herbicide) to far more exotic variations, are fairly easily planned.

Although the particular innovations may be unknown—Marquis was the usual mixture of trial and error, inspired guess, and sheer plod— the context, or paradigm, if you will, is perfectly clear. Once one accepts the basic premise of privately owned, monoculture, commercial agriculture, it is reasonably easy to see how it needs to be changed and to plan and innovate. Even less predictable innovations are easy to include, such as software to calculate exact profit and loss per acre, or online, real-time market quotations. Planning even accommodates changing the product stream to, say, pulses and sunflowers, or developing niche markets for buffalo or ostrich ranching, or switching to organic production, or encouraging ecotourism and farmstead bed and breakfasts. One can even plan, quite rationally, to take land *out* of production—either temporarily, as in soil banking or hedgerow habitats, or permanently, as in sales or gifts to private conservation groups such as the Audubon Society or The Nature Conservancy. Yet in all cases, this is conservative planning. One cannot plan for economic paradigm shifts such as the sudden appearance of e-commerce—though one can accommodate them as they occur. This kind of cumulative planning also complements, rather than replaces, the experience of real farmers on the ground, at least when the farmers are whitestream, commercially minded farmers, if not necessarily organic farmers or Native-style horticulturalists.

Planning is far less effective when it tries to become social planning. Tommy Douglas, to his credit, completely abandoned his eugenicist theories when he came into government. More inspired versions of planning for humans, such as the Saskatchewan government's attempt to implement the Carter Report in the 1970s and 1980s, and to provide meaningful legal aid foundered on professional jealousies and the huge changes in society that meaningful social reform would entail. In 1972, the Saskatchewan Legal Aid Committee, headed by Roger Carter, looked at legal aid issues in Saskatchewan and came up with a plan that was "perhaps the most far-reaching in North America" to revamp legal aid so that it would move out of the traditional adversarial and individualistic legal practice and serve as a systematic way of advocating for the poor and for Aboriginal populations, who were proportionately most likely to be charged with crimes, in such a way as to redress social injustice.[10] All of the western provinces embarked on Aboriginal justice inquiries, as did the Royal Commission for Aboriginal Peoples. All came up with intelligent, workable, thoughtfully articulated plans for truly re-forming the relationships between Native and non-Native society, but almost none of their suggestions have been implemented except in relatively small matters that do not involve reformulating Prairie (and Canadian) society. Even critically important social changes, such as the Canadian government's apology to the survivors of the residential schools and the "Truth and Reconciliation Committee" approach to reparations, are still within the context of capitalist, whitestream society. We are not talking here of massive relocation and development schemes of the sort that Scott critiques in the Soviet Union, Tanzania, and other places, though the ethnic cleansing of Native peoples and their replacement with Amer-European agriculturalists is in many ways a forerunner of the twentieth-century projects.

The economic development theory that planners have developed—particularly the theory that comes from close observation and trial-and-error interactions with real people in a particular physical place (rather than theoretical economic units in a homogenized theoretical space)—turns out to be an extremely useful tool for analyzing past development. Of course, this exercise is also partly tautological because development theory is based on history as well as experience and because the idea of the frontier as a special case of regional development is pretty much a given for the theorists. And the planners envisage a very Turnerian frontier, with little sense of the cost of the frontier to Indigenous peoples and without much probing into the nature of economic development as it swept from the Alleghenies or the Shield across the Plains to the coast. Nevertheless, when one reads of a "developmentalist state" that uses power to increase production by "supportive private *capital accumulation*" and by establishing "*state enterprises . . . on profit-making principles*" and that claims to represent all people through national development, we see a particularly clear picture of John A. Macdonald, the *Dominion Lands Act*, and the CPR, and a slightly less clear vision of the early dreams of US transcontinental railroads, Manifest Destiny, and the *Homestead Act*. Similarly, when we read Joseph Schumpeter's evaluation of the importance of entrepreneurship and of cumulative causation theory, it is easy to see why the destruction of the first generation of Cree entrepreneurs and the destruction of the distinctive entrepreneurial style of the Five Nations in Oklahoma had negative economic and social consequences that multiplied for generations.[11]

There is no dearth of literature on economic development theory, but boiling it down briefly to apply to the recent history of the Great Plains is not particularly easy, especially as much of the more recent anti-development theory, like Scott's, deals with a far more authoritarian and grandiose compulsory development. Even though some of the theory was developed from bonanza-style farms on the Great Plains, the real mistakes in terms of massive relocation of people and destruction of place-specific knowledge that Scott describes did not occur for whitestream farmers on the Great Plains.

Let us start by looking at "dependence theory." Although this is foreshadowed in Marx, the classic statement comes from Gunnar Myrdal:

> That there is a tendency inherent in the freeplay of market forces to create
> regional inequalities, and that this tendency becomes more dominant
> the poorer a country is, are two of the most important laws of economic
> development and underdevelopment under *laissez-faire*.[12]

Both capital and labour are mobile, but capital is more mobile than labour because it has no attachments to home and family. Nor does capital require passports or visas to cross international borders. Once a region starts to develop, it attains its own momentum. Banks siphon off savings from poor regions and invest in rich ones, and small local industries and even farms cannot compete with the economies of scale (or in contemporary agriculture, the export subsidies) of the richer region. The money that flooded into the Prairies during the various pioneer booms was lent on extremely high interest rates, and one could argue that had Prairie farming been successful, the farmers would have lagged in gaining prosperity simply because it would have cost so much to pay off their loans. As we have seen in Voisey, it is precisely those farmers who did succeed and who did stay on the land who made less pure economic gain than the speculators who skipped out, with or without defaulting.

Human capital follows financial capital. Soon many of the best and the brightest young people from the poorer region begin to migrate to the richer region, leaving the poorer region even less competitive and with a social structure that disproportionately includes older people and very young families The market thus creates and perpetuates the economic and political dependency of the poorer region, and the outflow of people cannot keep up with the outflow of capital. Thus, the market *cannot* cure regional development disparities. If one wants, for political or sentimental reasons, to promote development of hinterlands, the market must be tampered with. But dependency theory holds that the economic elites, most of which are the former colonial powers, do not want the economic development of the dependent regions, most of which are former colonies: they need to have them in a dependent position in order to be able to continue to exploit their resources as if they were still a colony. The metropolises expropriate the "economic surplus" of the hinterland for their own life. This theory can also apply to internal colonies, such as the Great Plains.[13]

A less ideological statement of similar principles is the theory of cumulative causation. Growth starts somewhere—perhaps because of rich and accessible resources, perhaps because of a particularly inventive entrepreneur, perhaps because there are a lot of people and nothing clearly exploitable and people have to live somehow. Once started, it keeps going. Similarly, decay is cumulative, as one can see by looking at many small towns in the Great Plains. Loss of business on Main Streeet means the loss of population, which results in the closing of the school, which brings about more loss, and so on. This process has been happening on the Plains since the end of World War I. According to classical economic theory, this population loss is a good thing. As *Empire of Dust* shows us, southern Alberta, like many parts of the Great Plains, was grossly overpopulated in terms of any conceivable kind of crop agriculture. Even had the short growing season allowed high-intensity truck gardening, there were no urban markets to support it. It is at this point, as the Poppers propose, that traditional economic planning would suggest the management of an orderly depopulation and the return of the land to Buffalo Commons. Although the Poppers' work addresses only the US side of the Great Plains, it applies even more to the Canadian Plains, where the shorter growing season makes agriculture even more problematic.

One of the problems with all economic development theory is that it tends to deal in terms of space rather than place. Scott suggests that this has in part to do with the nature of scientific research, which must narrow down the variables to be examined and thus is more effective in a generic laboratory than in a specific field on a specific farm with specific weather conditions. Theoretical location is only relative, and concerns tend to be couched in terms of transportation costs and distribution issues across uniform and undefined space. But the Great Plains is a *place* gifted with certain distributions of soil, moisture, and climate as well as certain histories. The current basic theory of economic growth centres on the "growth pole," a place where economic activity takes off and that then defines the economic activity of the region. As usual, that theory is not terribly useful for examining the Great Plains because most of the theorists require a city of at least 300,000 to serve as a growth pole, and most Plains cities of this size are only on the fringes of the Great Plains. During the last fifteen years,

Calgary has, for a number of reasons we shall discuss later, become one of the growth poles on the Great Plains. Wichita, Kansas, during and after World War II, became a growth pole because of the aerospace industry that located there. Similarly, most western growth poles took off thanks to government spending, particularly during and after World War II. Seattle and San Diego are particular examples. Canadian Plains cities are justified in feeling that they were overlooked to favour sites in Central Canada. The revolt caused by awarding the CF-18 contracts to Bombardier and the province of Quebec instead of to Winnipeg in 1986 is a particular focus of western Canadian discontent. Peter Newman and others regard it as the proximate cause of the creation of the Reform Party, which was intended to register the West's outrage at being passed over (yet again) for Quebec.[14] Even Vancouver has not fared as well in defence contracting as West Coast cities in the United States, even taking into account Canada's smaller total defence spending.

Another aspect of growth pole theory that confuses its application to the Great Plains is that the important growth poles are often *outside* the region. This is implied in the metropolis-hinterland theory of Canadian economic growth as articulated by Harold Innis, Donald Creighton, and J.M.S. Careless. American historian William Cronon relied upon these Canadian models in his *Nature's Metropolis*, which focusses on Chicago but explains the economic growth of the Great Plains, one of Chicago's hinterlands, and Cronon, like Scott, points out the distortions of a bland, undefined "space" that does not consider the specifics of a given "place." A growth pole outside the region also distorts the theory by introducing variables of transportation and communication into the mix. Growth pole theory, as applied to regional development worldwide during its heyday in the 1960s and 1970s, for the most part gave regional development a bad name. Growth pole theory relies heavily on increasing exports while reducing dependence on imports. Thus began the practice of plopping factories down in Lesser Developed Regions to manufacture products for export or products to replace imports. In most cases, these initiatives were failures. I remember hearing in the late 1970s the story of how a light industry in snelling fish hooks—putting the short filament leader onto the hook so the fisher could attach it to the end of the line—was set up on the stunningly impoverished Pine Ridge Oglala Reservation in South Dakota. It employed a few women and went out of business as soon as the federal economic development subsidy ran out. It was the classic case of an enclave development. Oglalas, unlike, say Wahpeton or Sisseton people, have had until recently no tradition of fishing, no sport-fishing resources on the reservation, and little contact with the nearest sport-fishing opportunities as a minor tourist occupation in the nearby Black Hills. Pine Ridge thus offered neither a local market nor a local source of knowledgeable and enthusiastic workers whose involvement with the sport would bestow meaning on the repetitive task of joining hook to snell. If the hook factory had no existing connection to the reservation community, it had even less possibility of growing in the context of the community. Neither the hooks themselves nor the filament could be produced anywhere near Pine Ridge. There was no facility for printing the cardboard on which the hooks were packaged nor for developing associated products such as lures, bobbers, or even sinkers. The snelling enterprise is a perfect example of what *not* to do in economic development. Pine Ridge was also the home to other short-term manufacturing attempts, including an arrow factory. One woman who had worked there said she had to quit because not only was she bored by making two thousand arrows per day, but the glue was giving her increasingly severe asthma attacks. Eventually, many of these operations were moved to Mexico or other countries.[15]

Failures and successes, however, have refined the notions of economic development over the last forty years, and these more complex theories are useful for understanding what has and hasn't worked—and to some extent why and what to do about it—over the market history of the Great Plains. The most important effect of the fur trade on the American Great Plains was the Osage assumption of the crucial middleman role during the period of French and Spanish claims to sovereignty and the early years following the Louisiana Purchase. The northern US Plains were more a region that American trappers crossed to get to the beaver in mountain streams. As we have seen, American mountain men tended to trap beaver for themselves instead of trading with Indigenous trappers, and even the role of Indigenous women in curing furs was largely eliminated by the mountain men, thus providing relatively little room for Indigenous middlemen. On the Canadian Plains, however, the Cree and Assiniboine long played the middleman role, and when pemmican became the fuel for the Athabasca brigades, buffalo hunting became adjunct to the fur trade. Thus, the Canadian Plains indirectly entered the market economy as a staples producer—of pemmican, if not fur directly.[16] So far, the history fits nicely with traditional economic theory—the staples theory is a version of export-base theory (which underlay the hook snelling, creating an export commodity) and fits with the growth pole theory, the European metropolis driving hinterland growth. Import replacement—facilities for making things like guns and iron kettles—were not at this point a possibility. Trade with the metropolis would provide such goods.

Euro–North American settlement of the Great Plains adds more complications. Economic development theory, as Higgins and Savoie discuss it, indicates what was mistaken from an economic (never mind a human) point of view in the "development" of Indigenous peoples. Some current dependency theorists maintain that growth can occur in Less Developed Regions only under a socialist framework—in complete *opposition* to the free market beliefs of Henry Dawes and Hayter Reed and their contemporaries; Higgins and Savoie, however, argue from experience in numerous societies around the world that development can happen "whatever the socio-cultural framework" and without the enormous human cost of social and cultural disruption that Scott describes so accurately for the Soviet Union, Tanzania, and Ethiopia, Ferguson for Lesotho, and Patel for parts of India, Mexico, and Central America.

Allotment was a political and ideological decision that unnecessarily devastated the Indians to satisfy the land hunger of Euro-American pioneers. It was "justified" by the overly enthusiastic embrace of technology and the elite theory that both the people and the land were deficient without Amer-European management and markets. But development could have proceeded successfully without allotment. Pleasant Porter was right; development in the context of Creek and other Five Nations traditions made just as good economic sense and considerably better human sense than allotment in severalty. The elites of the Five Nations were certainly committed to development, and their success in twice rising from the ashes of profound disruption is proof positive of their ability. Nor do non-market "peasants" need to go slowly in embracing technology, as Hayter Reed contended, to the great disadvantage of the Prairie Crees. Savoie and Higgins instance a group of Malay fishermen who had to make the transition to motorized boats. They chose to go all the way to inboard motors, not to make the transition by degrees, starting with small, less-expensive craft powered by outboard motors. They astutely recognized that in the long run, the intermediate step was simply a waste of time and money that might hamper their eventual use of the most efficient technology.[17]

As Scott notes, "traditional" peoples are not static but extremely adaptive, quick to pick up the knack of any crop, product, or method that proves or even promises to be useful.[18] Thus, economic development theory bears out the conclusions of historians like Angie Debo, Sarah Carter, and Russel Barsh. All of the Plains tribes had Indigenous elites of one sort or another, from the Pleasant Porters and John Rosses of the Five Southeast Tribes, to the La Flesches among the Omahas, to Lakotas such as Black Elk who developed cattle herds, to Blackfoot, Cree, and Dakota farmers and herdsmen. Allotment was not necessary and it created enormous and completely avoidable human suffering, as did the similar villagization of Tanzania in the 1960s and 1970s. Hayter Reed's insistence on obsolete farm machinery—a warped variation on the value of technology (headers are too complex for any but the "civilized" to use them)—delayed the adaptation to the most useful machinery and left Indian-owned farms fatally behind their neighbours' farms in terms of economic takeoff. Throughout the Great Plains, official Indian policy worked to discourage and demoralize Indigenous entrepreneurs. Only those most willing to assimilate completely survived economically in Oklahoma, while the destruction of the cadre of first-generation economic entrepreneurs on the northern Plains of the United States and Canada crash-landed their economic takeoff. It also virtually guaranteed, through cumulative causation, the continuing economic failure of reserves and reservations and the cumulative demoralization of the people—and this without even considering the disheartening effects of residential and boarding schools, and the prohibition of culturally important practices such as the Sun Dance and the giveaways.

Although Euro–North Americans benefited in the short run by gaining title to Indian land and shielding themselves from Indian economic competition, the net loss to society as a whole was far greater. Even putting any human decency or humanitarianism aside, the creation of a demoralized, dependent, and rapidly growing population is a poor economic choice. What's more, as economic development theory teaches us, innovation does not necessarily come to the most "favoured" areas. Singapore is a relatively small and resource-poor island that has constructed wealth based on the ingenuity of its people. Higgins and Savoie note that non-irrigated areas in India tend to be more innovative than the more favoured irrigated areas. It makes sense—as Scott points out, a fisherman living by a river that always serves up plenty of fish in the accustomed places is less likely to innovate than the fisherman whose river demands the utmost in wile to land a steady catch.[19] Reservations on the US Great Plains were almost always set on parcels of land that Euro-Americans deemed least useful for farming, including the South Dakota Badlands and the Missouri Breaks and Coteau, land in most cases prized by the reservees, who cherished its diversity and opportunities for subsistence hunting and gathering. Although under the numbered treaties, Canadian bands had the right to choose their own reserve land, this was not always honoured in practice, and the small size of the reserves hampered their economic viability. Nonetheless, in both countries, the squalor and extreme poverty most North Americans today associate with reserves and reservations are not the result of the inadequacy of the land or the inability of the people to adapt to "civilized" economic and social behaviour. Instead, they are the result of the determined squelching of all successful entrepreneurial behaviour. Judging by the successes of the Five Nations in Oklahoma before allotment and of individual and small groups of Cree, Omaha, Lakota, Dakota, Osage, Blackfoot, Kiowa, and other peoples in the first reserve/reservation generation, the reserves and reservations could have been village-sized growth poles, innovative adaptations of introduced economic behaviour to an intimately understood place that newcomers persistently regarded as deficient. Unlike the projects that Scott describes that were doomed by an excess zeal for technology and mechanization at the expense of practical, place-centred knowledge, however, reserves and reservations were tacitly expected to fail by the white governments and negotiators who set them up. They were supposed to vanish, like the Vanishing American they housed, and not to flourish. That this idea of the economic success and even exemplary counter-narrative of the reserves and reservations seems romantic or nostalgic is merely an indication of how strongly what *is* conditions our idea of what *could have been,* and how much contemporary society has bought into just one tidy, technocratic idea of "success."

If theory suggests an unimaginable alternative to history, it also spawns explanations of success. As we have seen, the theory of the developmentalist state exactly describes Canada at Confederation. The National Policy of Macdonald's Tories, whether conceptualized in the nineteenth century or imposed on the actions of the past by twentieth-century scholars, fits perfectly with how one would establish a state. In both Canada and the United States, federal support of a transcontinental railway led to a government "trapped between the possibility of a legitimation crisis and a fiscal crisis."[20] As we have seen in the United States, this led to the Credit Mobilier scandal. Because the government had to be accountable for the costs, the railroad was too undercapitalized to succeed. Only by sweetening the pot through an elaborate kickback scheme could financiers gain a rate of return commensurate with the risk involved in this colossal and premature enterprise. Given Canada's small population and imposing landmass, most development is premature, thus demanding more government support and a much narrower margin of error. As we have seen, the CPR was on the verge of bankruptcy when the Northwest Resistance gave it the legitimacy it needed to raise more capital.

In both countries, the railroads were examples of "spatial integration" to improve transport and communication, and to increase capital by reducing transfer costs for merchandise. Compare the speed of shipment by rail to the old fur trade routes that required a year out and a year back for there to be any return on capital! Or look at what Cronon has to say about the cost of slow return to merchants and farmers alike when all consumer goods and all crops were shipped by water instead of by rail! The railroads allowed the commodification of agriculture (and the commodification of land) by providing the farmer with reasonable access to the world market. According to Gore, "Part of state policy in developmentalist states is thus directed toward *inventing* a nation."[21] Although he is speaking primarily about nations with arbitrary boundary lines created by the withdrawals of colonial powers in

Asia and Africa, the statement applies perfectly to Canada and to the nation building that incorporated the Prairies into Confederation. Even in the United States, the transcontinental railroads and the quick western settlement that those railroads allowed were a crucial part of reinventing and repairing the nation after the Civil War. The epic of "The Last Spike," then, is a function less of human imagination and courage than of the demands of economic development in an international market system.

As Janine Brodie points out, region is as much a political as a physical designation, and the term "region" has no particular meaning except in relation to nation, or province, or city, or whatever.[22] One can explain the relative lack of economic development in the Great Plains as opposed to Central Canada or to the East and West Coasts of the United States in a great many ways, from the use of tariffs, to the destruction of an Indigenous entrepreneurial class, to the lack of federal defence spending, to dependency theory/cumulative causation and the basic operation of a free market economy on a resource region, but it is always easier to explain why something *did* happen than why something did not.

As we have seen with Scott's *Seeing Like a State*, more recent works have criticized the whole "development" model in general. James Ferguson's 1990 *Anti-Politics Machine* looks at the context of development in Lesotho, a tiny, landlocked country entirely within South Africa that supports a huge, internationally funded development "machine" that continues to sponsor projects congenitally unsuitable to the land and the people. In *Stuffed and Starved*, Raj Patel, in examining the global food system in which the Great Plains region competes, details the ways in which it damages both farmers and consumers. He points out that food processing is more profitable than food production and that the processing and wholesaling bottlenecks now built into the global food chain keep farm prices disastrously low, while failing to produce comparable savings for consumers. He notes that the first phase of the Green Revolution in India did raise yields, but only by raising costs for seeds, fertilizers, and pesticides to farmers at rates that could not be repaid by harvests. The new farming structure also substituted monocultures for a biological diversity that had developed to fit various niches in growing areas and that had produced a well-balanced diet. The Green Revolution itself depended upon a kind of deficiency theory—the plants and tillage practices of people like small farmers in the Punjab were deficient, something Scott explores in considerable detail. People could not raise enough food to feed themselves, resulting in endemic malnutrition and periodic famines. But Patel points out that the deficiency was not in the plants or the tillage, but in the method of distribution of the products. By the time of the Green Revolution, India had spent more than a century under indirect or direct British rule. The old feudal system, which required landlords to put aside enough grain to feed all the people in time of want, had been deliberately replaced by a market system that required surpluses to be exported—mostly to Britain, to keep its food prices low and to stave off social unrest. It was politically more expedient to the Raj for poor Indians than for poor Britons to starve.[23]

Of specific relevance to the Great Plains are Patel's sections on food aid, wheat, and soybeans. Commodity distribution to Native Americans, though not something Patel discusses, is certainly a Great Plains issue. As reservation land was expropriated and Indians stubbornly refused to "vanish," the US Department of Agriculture introduced commodity distribution, which not only provided an outlet for food surpluses of grains, dairy, and other products, supporting overproducing farmers, but also dampened social unrest by feeding people who had lost their subsistence hunting-and-gathering grounds but had not acquired agricultural acreage or employment opportunities to replace them. Commodities and other cheap foodstuffs available in cash-poor and increasingly welfare-dependent reservation communities included large proportions of processed grains, sugars, fats, and dairy products, all unfamiliar in North American ancestral diets. Diabetes, which was literally unknown in American Indian populations in the 1940s, is now endemic in these same populations at rates several times those in any other North American demographic. Patel points out that the same sort of thing happened in the rest of the world—wheat imports in the 1950s and 1960s left recipients, mostly in the Global South, "hooked on the most expensive grain." His solution is disarmingly simple: instead of providing cheap food for poor people, development should instead focus on economies that allow everyone to purchase good and diverse foodstuffs.[24] Patel is particularly critical of the international politics of soybeans and corn, major Great Plains crops in the United States, if not in Canada.

After World War II, American soy growers had virtually cornered the market, controlling more than 90 percent of the world's soybeans in the 1960s. Brazil, however, embracing a positivist model of development, entered into soy production and, after a brief 1973 embargo by the Nixon administration, emerged as a serious competitor to the United States, moving soy production into the *cerrado*, which is, like the Great Plains, a grasslands ecosystem underlain by a huge freshwater aquifer. Patel sees hope for a more sustainable development model in Brazil through progressive, democratic co-operatives of landless rural workers and Indigenous people whose land is being invaded and destroyed. Meanwhile, back on the Great Plains, he points out that the soy farmers try to compete by talking about seeds, fertilizers, herbicides, and other inputs of corporate farming—as one can see by looking through the glossy advertisements in any farm magazine or by counting the seed and herbicide commercials on local television. But, quoting activist Emelie Peine, Patel notes that Great Plains soy farmers "are talking about ways to become more competitive but not about why we're having the race."[25] Certainly this is true for my Nebraska students whose families grow soybeans.

Corn, Patel shows, is even more insidious in its results than soybeans. High fructose corn syrup provides a market for corn that rewards processors and does not interfere with US tariff protections for domestic sugar. It appears in everything from soft drinks to spaghetti sauce and helps account for the obesity epidemic in North America. Worse, Mexican economists, educated in the United States and believing in market and developmental models, included food—most notably corn—in the North American Free Trade Agreement (NAFTA) despite US officials' concerns that "the economic impacts of free trade on farmers would cause such poverty that it might destabilize the Mexican countryside." In fact, farmers did fail in large numbers, and while prices fell for the Mexican food processors and wholesalers, they rose for consumers.[26]

In his 2010 book *Murder City*, Charles Bowden looks precisely at the horror of what the rural destabilization came to mean for Ciudad Juárez, a border city with a population somewhere between one and a half and two million people. And with the highest murder rate in the world, moving from an already stunning 48 murders in January 2008 steadily up to 324 murders in October 2009. And rising. Bowden's analysis is even more chilling than his figures.

> The trade agreement [NAFTA] crushed peasant agriculture in Mexico and sent millions of campesinos fleeing north into the United States in an effort to survive. The treaty failed to increase Mexican wages—the average wage in Juárez, for example, went from $4.50 a day to $3.70. The increased shipment of goods from Mexico to the United States created a perfect cover for the movement of drugs in the endless stream of semi trucks heading north.
>
> American factories went to Mexico (and Asia) because they could pay slave wages, ignore environmental regulations, and say fuck you to unions. What Americans got in return were cheap prices at Wal-Mart, lower wages at home, and an explosion of illegal immigration into the United States. This result is global, but its most obvious consequence is the destruction of a nation with which we share a long border.
>
> The main reason a US company moves to Juárez is to pay lower wages. The only reason people sell drugs and die is to earn higher wages.

Juárez, Bowden claims, is the logical future "of a religion called the global economy."[27]

The warnings that Patel and Bowden give us, as well as Scott's horror stories about forced rural relocation and devaluation of land knowledge, are closer to the removal of Native peoples from the Great Plains than the bland analysis of planning and economic theory. The model of the economy that Great Plains society, as a whole, has embraced is not serving anyone except for economic elites mostly outside the area. If none of the old methods of planning that serve to explain the past can work in the future, what might be a productive way for the Plains and the rest of the globe to visualize the economies in which we are all ensconced?

As we have seen, the Plains supported hunting-and-gathering economies, especially on the verges; a mounted buffalo-focussed economy; and a commercial pemmican economy associated with the fur trade. At the point of Euro/Afro/North American settlement, it supported a speculative cattle economy and a speculative wheat economy underlain by women's subsistence work. In both countries, the "sodbusters" focussed primarily on wheat, with a large admixture of corn in the United States and a smaller admixture of barley in Canada—and with enough oats to support draught horses in both countries. As Voisey notes, although experts urged mixed farming and farmers agonized over taking their advice, it was not economically sensible.[28] Wheat gave the best return and was the most easily portable crop, and the droughts of the Great Plains were so much more intense than the droughts of Ontario or Ohio that the pastures and ploughed fields alike dried up and were given over to gophers and grasshoppers in the scorching summer sun. More recently, federal support payments, especially in the United States, also helped focus farmers on wheat and other large commodity crops that were price supported. The high prices of oil in 2004–7 gave rise to a fervid boom in the ethanol industry and to planting corn "fence row to fence row," but the depression and rapid energy price declines of 2008 quickly ended that boom, and the farmers and ethanol plants alike await a rebirth.

Available moisture and length of growing season are the two crucial variables on the Great Plains, and they are associated with microclimates. Frost pools, and crops at the top of even a modest slope may be untouched while only a few metres below in an almost imperceptible hollow, a killing frost may settle. Moisture also pools, and the hollows are wetter than the ridge tops, even as the change in altitude is slight. This is easily perceptible on the tallgrass prairie, where little bluestem and stypas typically grow on the upper slopes and southern exposures of gently rolling land while big bluestem, switchgrass, and Indian grass thrive on north-facing slopes or slightly further down the incline. Contour ploughing, terracing, and grass waterways serve not only to conserve water but also to define and provide distinctive uses for the microclimates for farmers willing to work with this kind of diversity—often a luxury, given the size of contemporary farm machinery. Plains agriculture has the potential for on-farm diversification that provides some of the flexibility that mixed farming did in more humid climes, but this diversification is more theoretical than practical, especially since ag experiment stations and seed companies have not worked to provide the niche seeds, and experimenters like Wes Jackson have mostly been laughed out of the industry. For the most part, diversification has involved moving into other field crops such as canola, sunflowers, soybeans, and pulses. Or diversification has meant another kind of extraction from the land, particularly involving the oil and gas industry. Monocultures are still taken for granted. Even the Land Institute, which has consistently experimented with growing perennial grain crops that do not require annual ploughing and planting, can only work toward a diversity of a few mixtures of seeds and plants. When some of my colleagues at the University of Nebraska–Lincoln experimented with growing a traditional Omaha garden, their vari-coloured corn grew ten feet high, supported by beans and squash and watered by hand-built wooden pipes. This kind of diversity and polyculture does not seem to be in the cards, but its place-based systems are, at the least, worthy of serious study. We shall return to the concept of diverse sufficiency arguments in the last chapter, but first let us look at two twentieth-century planning efforts, for water and for oil, whose partial failures advance the tale.

Out of necessity, Saskatchewan developed a mixed economy that managed to buffer both agriculture and industry against the disadvantages of great distances and sparse populations. It lost it when a global way of thinking found momentary support—and proved again that the unbuffered market will not work on the Great Plains, especially not in the purest part of the Great Plains, where there is what W.O. Mitchell has called "the least common denominator of nature, the skeleton requirements simply, of land and sky—Saskatchewan prairie."[1] If the eighteenth and the first part of the nineteenth centuries began the transition of the Great Plains to hinterland through the fur trade, the late nineteenth and early twentieth centuries cemented that relationship through the railroads and homestead settlement. But, as we have seen, that expansive wheat economy reached its apex in World War I and never really recovered its strength or its importance in the national economies of the United States or Canada after the disaster of the 1920s. Yet keeping up the boosterism, the belief in progress and growth, continued largely unabated through the twentieth century and into the twenty-first. While growth in most of the rest of North America was first industrial and then service and knowledge oriented, the Great Plains for the most part missed out on the economic development promoted by the wartime munitions and aerospace industry—continued in the United States by the Cold War. Although the American Great Plains did get wartime investment, most states were unable to parlay that into a permanent industrial complex because the West Coast and Southwest were home to the giant contractors who were able to capitalize on and to keep the armament nuclei of industrial production. Canadian Prairie cities did not even have any plants to lose. Efforts to proclaim the University of Nebraska the "Harvard of the Plains" or to make the University of Saskatchewan the information technology centre of Canada have mostly lost out to the two old problems—distance and sparse population. Except for petroleum centres like Tulsa and Calgary, economic development on the Great Plains has stayed focussed on agriculture and its sine qua nons of land and water. Land and water cannot be created and cannot even, for the most part, be transported. But water can be saved up in reservoirs behind dams, and arid lands can become commercially productive with the application of water. So economic development on the Great Plains in the last half of the twentieth century has continued to focus on getting more land from the Indians and putting more land under irrigation, processes that, as we shall see, are often interlinked.

As shown earlier, the Amer-European settlement on the Great Plains started as a planned economy, denominated especially by the square survey. For the most part, classical development theory explains how the ploughed grasslands merged with a global market economy, though it does not deal with issues such as the loss of fertility and the growing dependence on hybrid seed, chemical fertilizers, and pesticides that mark monocrop agriculture, especially on the Great Plains. Planning for Native peoples on the Great Plains, however, had never been as neutral and beneficent. In fact, it looked a good deal like the disastrous farm schemes in the Soviet Union, Tanzania, and Ethiopia that Scott describes, though without the promise of equality and material abundance. From the point of view of the Great White Father or even the friendlier Grandmother England, reservations and reserves were temporary refuges, not unlike Buffalo Bill's Wild West (through which Black Elk would actually meet "Grandmother England"), where people could play at being Indian until they got the real message and gave up their heathen ways to become brown white men and women. Since the reservations were temporary, they did not need to be economically viable and indeed were founded with the promise of annuities and rations in lieu of the subsistence hitherto provided by the land that was, according to the written parts of the treaties, to be relinquished to the Amer-Europeans. Thus, it is not surprising that the modern states of both Canada and the United States, with their firm belief in better living through technology, would "improve" reservation and reserve lands by flooding them and moving the inhabitants from their chosen plots. Although the rhetoric of these moves stressed the benefits of dam and lake building, the benefits were almost all off-reservation and the unacknowledged losses almost all on-reservation.

In the 1950s, the United States came up with the twin policies ominously named Relocation and Termination. After the war, Native people, especially veterans who had seen the world and were determined to succeed in whitestream society, began, like other rural North Americans, to come to the cities. Relocation was ostensibly a plan to help them move, adjust to new conditions, and train for and secure new jobs. For the most part, it failed, marginalizing people in urban ghettos instead of merging them into the economic mainstream. Termination was the federal withdrawal of recognition from tribes, thus "freeing" their land and other assets for the good of the society as a whole or, in some cases, forcing them to relinquish valuable timberlands to private companies. In Canada, the reformulation of the *Indian Act* in 1951 also focussed on whitestreaming Aboriginal peoples. Despite clear testimony to the contrary from Aboriginal speakers, the new act still assumed that it was the united purpose of everyone for Native people to assimilate as quickly and fully as possible. The resource boom of the 1950s also pushed Euro-Canadian resource exploration further north, into what had hitherto been considered inhospitable and marginal land, deficient for economic use. Other federal programs in the 1950s and 1960s resettled whole communities, particularly in the North, for the ease of providing mandatory schooling and other "services." Since early contact, Native people had valued technological training in the European arts and crafts, including reading, writing, and arithmetic, which they had actually embedded in some of the treaties. They never, however, asked for or wanted training that was coercive or primarily assimilative. In some cases, at least partially for the sake of defending Canada's claims to the North from other countries, villages were relocated far from any resources that they were trained and equipped to hunt or gather, forcing them into welfare dependency. In both countries, the relocations were socially and culturally damaging to the people who were relocated, and the resultant social and health pathologies became excuses for the rapidly increasing practice of taking Aboriginal children from their families of origin and placing them, sometimes illegally, in non-Aboriginal foster or adoptive homes. All of these processes were ostensibly aimed at assimilating Native people into whitestream society but also had the added benefit—from a non-Native point of view—of decreasing reserve and reservation populations so that there would be more "surplus" land.[2] The self-serving and duplicitous nature of this sort of transaction was no more apparent to most non-Aboriginal people of the mid-twentieth century than it had been at any other period of North American history, especially as it was usually couched in terms of individual equality of opportunity.

Although reserve and reservation lands had originally been placed in areas that Euro–North Americans did not expect to want, lands that were seen as completely deficient for "civilized" uses, uses began to develop early in the twentieth century, especially for road corridors or for areas to be flooded behind dams. Indigenous people had frequently chosen land that included rivers, breaks, and riverine forests, the Great Plains habitats most conducive to subsistence lifestyles because they featured wood, water, and game. Euro–North American farmers commonly favoured level uplands instead. But these lands that Euro–North Americans had customarily defined as "unused" and "uninhabited" became prime sites for dams and reservoirs. A relatively early and small-scale project was Calgary's Glenmore Dam and reservoir, built during the 1930s, largely as a relief project, on land abutting the city and purchased from the Sarcees (Tsuu T'ina), who had preferred that area to the dryer regions further south where they had first been settled under Treaty 7. Land for the Ghost Dam and reservoir was leased from the Stoney people at Morley in 1929.[3]

The most egregious and most studied case of "dammed Indians," however, is the saga of the Pick-Sloan projects, built on the mainstem of the Missouri through the Dakotas. Michael Lawson's study, *Dammed Indians*, dramatizes the institutional imperialism of the Army Corps of Engineers and the Bureau of *Reclamation*, both of whom needed to build dams in order to justify their own existence. As we have seen, George Norris had envisioned such a project along with his TVA, and he and the actual planners shared an automatic acceptance that dams were an unmitigated good, providing flood control, irrigation, and hydro generation with no side effects. Rivers without dams, which flooded or simply ran away to the sea, were deficient rivers. All the government bodies concerned also automatically accepted the premise that reservation lands were expendable. Although some of the power of the Corps and of Reclamation was related to the sheer inertia of bureaucracy, their real power was bound up with ideas. The Corps had been involved in exploring, mapping, and "taming" the West. They were the bridge builders and road builders who brought civilization to the hinterland. Dam building, especially for flood control, was another aspect of "taming," one for which there was a good deal of public support, especially in downstream communities that had been built on flood plains and were frequently inundated during spring runoff. The very name of the Bureau of *Reclamation* stated its ideology. The arid and semi-arid West was implicitly not only in a deficit state but a deficit that represented a fall from some happier, Edenic time when, presumably, it had been a well-watered garden. It was not being "claimed," but "*re*claimed."[4] The Bureau of Reclamation, then, redeemed and restored what God or Thomas Jefferson or some other such venerated father had intended to be the homeplace of Crevecour's American Farmer. Once again, the land could no longer be declared sufficient to its own flora and fauna.

The main argument for dams made by the Corps was to govern the flow of the Missouri and to keep it navigable for barge traffic. The main function of barges was to haul wheat. Water transportation was a huge part of the hinterlanding of the Great Plains during fur trade times, and shipment of wheat through the St. Lawrence system or down the Mississippi was vital to the market economy of the Prairie Provinces and of the Cornbelt States, respectively. Moving cargo on the Missouri, however, especially above Sioux City, Iowa, was a different kind of proposition. From Nebraska to Montana, the Plains had been settled primarily by rail, not water, transportation, though steamers sailing as far as Fort Benton had been an important part of the trade in buffalo hides. By the 1940s, when Pick-Sloan was devised, a navigable Missouri was not a significant part of the transportation puzzle, and in the ensuing years, the cost of maintaining locks and a navigation channel has often outrun the economic benefits of barge traffic. Flood control and electric generation were obviously more popular and sensible arguments.

Perhaps the most salient effect of the Pick-Sloan projects was the siting of dams so that most of the flooding took place on Indian lands. This was not the case on the Great Plains alone. We have already looked at the TVA's flooding of Cherokee lands. The most ambitious dam-building projects in North America in the second half of the twentieth century were proposed or built on Cree lands in northern Quebec. Hydro projects throughout the North flooded or otherwise impinged on Native peoples' lands. Even in New York and Pennsylvania, the story was the same—dams, lakes, and expressways were somehow sited on the little bit of land that had been left to Indigenous people.[5] This was more than coincidence. The old imperial ideas were certainly important—Manifest Destiny involved pushing out the Indians in favour of Euro–North Americans—but economic ideas were even more important. Although dams and their lakes were not private property, the downstream structures to be protected from floods, the upstream crops to be shipped to market, and the homes, farms, and businesses to be served by electricity were. And recreational facilities, such as lakes for fishing and waterskiing, and parks for camping, unlike income-producing properties, seemed to be entitled to public ownership because they could be used for the private enjoyment of all—including Indians, even if they mostly did not like sport fishing and mostly could not afford speedboats.

Conversely, the things that made the riverine forests so valuable to the Dakota and Lakota people of the upper Missouri seemed to be of little or no value to the Euro–North Americans. Even those who might value the habitat in general did not value the particular land bounded by the reservations. Recreational hunting of whitetail deer is an important sport in areas like Ontario and New York—important enough that any big box electronics or sporting goods store in North America will feature a videogame or two on whitetail hunting—and perhaps bass-fishing and pheasant-hunting games, too. But whitetails have become very common on the Plains, so conservation of particular riverine habitats has little value even for whitestream hunters. Ducks Unlimited and Pheasants Forever are popular charities in the Great Plains, where they do extremely important habitat protection and restoration work. Pheasant and other upland game hunting is popular in the eastern Dakotas, while in the West River part of these states, hunters turn to antelope and mule deer and elk. Again, most of this does not depend on reservation land, though the Santee Sioux, for instance, do cater to hunters and fishers who like the open, unspoiled nature of the river where the reservation was relocated after the old town was flooded out by the dam. Rabbit hunting or snaring is of very little recreational value to Euro-Dakotans, except for those who practice some subsistence hunting themselves.

The whitetails and cottontails of the Missouri Breaks are an important part of subsistence living, and were particularly significant in the 1940s, but the value of Indian lands to be flooded was determined exclusively in terms of "fair market value." Wild fruits, herbs, and beans were of even less value than game to Euro–North Americans. Mouse beans are perhaps the best case in point. These wild seeds are gathered and stored by harvest mice. In Sioux tradition, one took the beans from the mouse nests, replacing them, handful for handful, with dried corn or an equivalent, and then cooked the beans in a soup or stew.

> When Lower Brulé representatives asked $6.00 per bushel for the value of their mouse beans, Richard LaRoche, Jr., recalled that the Congressmen "laughed like hell and said we never heard of such a damn thing." Thus the Indians were required to gather samples of this food source and submit them to a University of Maryland botany professor, who finally verified their worth to Congress.[6]

Although the Lower Brulé did get paid for their loss of existing mouse beans, the loss of the ongoing connection between beans, mice, and people had no economic function for Euro–North Americans. It is unlikely that Congress would have overlooked the economic potential of a commercial grain farm or proposed compensation for only the crop in the field and not the loss of future earnings.

The laughter of the congressmen at mouse beans is a parable of exactly what Scott terms "seeing like a state." In her novel *Waterlily*, Ella Deloria records the complex relationship of mice and humans, beans and corn. It is local knowledge that can be gained only by doing and that provides diversity in the diets of both mice and men. The exchange required women to remember the sites of harvest mouse hoards and to recognize the signs of new hoards as they were developed. Unused beans and corn in the hoards preserved seed stocks through times of drought. Ploughing or flooding out mouse habitat destroyed the value of local knowledge as well as the actual resource, and impoverished both the people and the land, not to mention the harvest mice. Because the mice did not abide by the square survey and could not be owned or farmed, they were of no use to a modern state concerned with orderly fields and deficient rivers that had to be reconstructed so as not to flood valuable commercial buildings. Along with the mice, the people were flooded out, with the Lower Brulé (whose representatives had spoken for the mice) and their cross-river cousin Crow Creek (where my claimed family hails from) flooded and moved twice. Crow Creek still uses the "temporary" school buildings built after the original townsite was flooded and still has difficulty supplying potable water to the Fort Thompson school. As in Tanzania, the local knowledge that allows a rich, sustainable way of life in a complex semi-arid place like Buffalo County, South Dakota (the poorest county in the United States in the 2010 census), was laughed out of court in favour of a technological fix that is not working.

Perhaps the last major dam project that will be built in the Great Plains is the Oldman Dam in southern Alberta. It was planned almost completely for irrigation, a final answer to the entitlement mentality that marked the boomers of the region just west of the one Jones discusses in *Empire of Dust*. In this case, the lake was designed to cover the lands of several Euro-Canadian farmers, but the main opponents to the project were the environmentalists and the Lone Fighters group of the Peigans, who pointed out that the dam would inundate some off-reserve ceremonial sites as well as diminish downstream flow and threaten the regeneration of the cottonwoods needed for Sun Dance ceremonies. Although the environmentalists won several important court battles, the dam was built, despite the rulings and the determined civil disobedience of the Lone Fighters and their leader, Milton Born-With-A-Tooth.[7] Flooding sacred sites on Great Plains rivers was not a new phenomenon. A relatively early dam on the Republican River in Kansas drowned a spring that was accounted particularly holy by the Pawnees and many other central Plains peoples.[8] Given how little land remained to Aboriginal people after the numbered treaties, the *Dawes Act*, Termination, and Relocation, it is depressing how routinely Great Plains dams after the 1940s continued the frontier pattern of taking Indian land in the name of the market, technology, private property, or newly discovered communal Euro–North American rights to recreation or, in other parts of the Dakotas and Alberta, for conservation and parks. Badlands National Monument was taken from Pine Ridge Reservation, while the Siksika Nation in 1960 filed a claim to the Castle Mountain area of Banff National Park, which had been promised to them as a timber reserve and then removed from their possession with no compensation. The Tsuu T'ina Nation near Calgary received back parts of their land that had been taken for use as a bombing range, but they had to clear the unexploded ordnance themselves.[9] At the same time, the needs of drought-stricken farmers are real, and as global climate change makes the Great Plains climate even more extreme, with periods of drought and periods of flooding, people want still more dams and transfers of water. As tourism replaces farming as a major revenue stream, parks and lakes for recreation will continue to be economically important as well as being amenities for the people who live in the region. While producing food *is* a noble calling and making two blades of grass grow where only one grew before *is* a kind of miracle, there is a reason miracles are few and far between. The assumption that monoculture is better than the diversity of the Plains and that an artificial lake surrounded by fields of wheat or corn or canola is better than a break alive with deer and harvest mice has been central to the idea of the Great Plains as hinterland, as we saw with Oklahoma. "As long as the waters flow" has many meanings.

Although most of our discussion has focussed on the Great Plains as a hunting-and-gathering or agricultural economy, the region has also supported extractive industries. The fundamental difference is that hunting, gathering, and agriculture, at least in theory, exploit renewable resources that may be sustained for centuries without an endpoint, while mineral deposits, once exhausted, are permanently gone. Other minerals may be discovered to be valuable and exploited in turn, but they, too, are finite. Clay, pipestone, and flint have been used for tens of thousands of years but are still relatively plentiful since they have mostly been used on a subsistence rather than an industrial scale. Gravel, cement, and limestone have been used more recently and more industrially, but they are still widely available. Gold in Montana sparked a brief rush and a wave of vigilantism, and gold led Custer into the Black Hills and the US government into the abrogation of its treaties with the Lakota peoples, resulting in, on the one hand, Custer's death and on the other, an extremely lucrative gold-mining industry. Surface potash was a valuable resource at the time of World War

I and, with the advent of deep-mining techniques and a growing demand for fertilizers, a significant international trade product through the end of the twentieth and into the twenty-first century. Soft surface coal deposits were a curiosity for Indigenous people and a useful fuel resource for early Euro-Americans. Wyoming, Saskatchewan, North Dakota, and other northern Plains polities produce significant quantities of primarily open-pit coal, and huge deposits still remain. But the resource that has had by far the most significant environmental and cultural effects on the modern Great Plains is the remains of fossil fish and dinosaurs and their ecosystems manifested as petroleum products, oil and gas.

The petroleum industries in North America started in Pennsylvania and Ontario, but the most valuable deposits turned out to be further west, including Texas, Oklahoma, Alberta, and to a lesser extent, North Dakota, Saskatchewan, and other parts of the Great Plains. What these oil deposits and their exploitation would mean for individuals, Indigenous nations, state and provincial governments, large oil companies, and federal governments differed radically in terms of time period, jurisdiction, and the power of the oil companies. In the United States, all title to the land not already distributed resides with the federal government, except for land in the thirteen original states and in Texas, which kept control of its public lands when it became a state. In addition, the federal government retains mineral rights to all Indian reservations except that of the Osages, who purchased mineral as well as surface rights. Private land owners usually, but not always, own both mineral and surface rights. Mining laws passed in 1872 and 1873 granted absolute mineral rights on federal lands for a nominal sum to whoever found and exploited them, throwing in the land itself as well, an aspect of the law that twenty-first-century developers are now learning to exploit to carve condominium communities out of federal land while going through the motions of "developing" mineral deposits. Not until 1920 did the federal government switch to a lease system for petroleum development, but the goal was regulation, not revenue, and large oil companies in particular welcomed the stability of regulations that enhanced their prosperity by making it harder for newcomers to undercut them. Regulation is still notoriously lax, however, and the US Geological Survey, which is supposed to monitor lease payments, has no way of determining if oil companies are actually paying what they owe.[1]

In Texas, the state or private individuals own the land and the mineral rights, which can then be leased to oil companies, big and small. In the early twentieth century, Texas anti-monopoly laws prohibited Rockefeller's Standard Oil of New Jersey or other established oil companies from achieving monopolies through "vertical integration"—any one company could choose only one aspect of production: extraction, transportation, refining, wholesaling, and so on. The Texas Railroad Commission regulated the industry, and the state received lease money and also profited from the general prosperity that oil booms brought to Texas throughout the twentieth century. Not surprisingly, Texas politics and Texas oil interests became inextricably combined, eventually influencing not only Texas regulation policy but moving up to influence, and even to some extent control, US foreign policy relating to oil and to those areas of the world where oil is produced, including Canada and the Middle East.[2]

We have already seen the deleterious effects of oil on the Osage and other Indigenous peoples settled in eastern Oklahoma. The attempt to suppress Debo's *And Still the Waters Run* indicates the power of oil and graft in Oklahoma in the 1930s, while Robert Sherill noted in 1983 that "if an oilman must stand trial for fraud [including false reporting on oil obtained from Indian land], Oklahoma is by far the best place for him to be; in that state, judges are notoriously sympathetic to anyone who handles crude." This is not to say that all oilmen were or are corrupt. John Joseph Mathews, whom we have already met at some length, was closely associated with the Oklahoma oil industry, both as a geologist and as the beneficiary of an Osage headright. His 1951 biography of E.W. Marland, *Life and Death of an Oilman* (apparently the only full-length book published by an American Indian author during that decade), provides a strong sense of the early oil industry as practiced by actual field geologists and of how that differed from the institutionalized industry that came to dominate the oil business. Mathews admired Marland for his ability to look closely at the land, the way an Osage might have were he exploring it for oil, and for his employment of young university students, also trained in a specific and rigorous kind of land knowledge. Mathews also honoured Marland's willingness to pay a good wage and to provide decent housing to his workers. Marland succeeded in building a successful oil company but, according to Mathews, lost it to the Morgan bankers who, instead of capitalizing his expansion, eventually forced him out of his own enterprise. (As we shall see shortly, the saga bears a remarkable resemblance to the later history of Jack Gallagher and Dome Petroleum, a similarity to ponder for those who blame the National Energy Policy for all that went sour in the Alberta oil patch in the 1980s.) Although Marland became a congressman and eventually the governor of Oklahoma, Mathews presented him as a thwarted figure who never lived up to his personal or professional potential.[3] Oil has been a mixed blessing for many of the people who have touched it.

The early twentieth-century oil developments in Alberta were, like those in Texas and Oklahoma, colourful mixtures of wildcatting geniuses, boom-and-bust wealth, and both co-operation and conflict with different levels of government, but there were vast dissimilarities as well. At Confederation, the *British North America Act* (the *Constitution Act of 1867*) apportioned control of natural resources to the provinces. But when the original North-West Territories were formed into Saskatchewan and Alberta in 1905, Ottawa retained control of the resources, supposedly because the new provinces were in a state, more or less, of tutelage. Control of natural resources was a particular sore point and focus for western alienation until 1930, when the *Natural Resources Transfer Act* finally transferred control over lands and resources to the Prairie Provinces. As in the United States, ownership of surface lands does not necessarily mean ownership of mineral rights. Depending on the year, homesteaders might or might not have obtained mineral rights, and Indigenous parties to the numbered treaties, which cover the Prairies, have consistently argued that their intent was to share surface use of the land, not to surrender exclusive use to all of the land and resources, a concept that neither the federal government nor the provincial governments have effectively acknowledged or negotiated.[4]

Although people had noticed and even made casual use of various oil and gas seeps for centuries, the western Canadian oil and gas industry began in Turner Valley just before World War I. Natural gas quickly became a household commodity, but the oil industry did not fully take off until the famous Leduc No. 1 well came into production in 1947, the first of the fields that would establish Alberta as a major world oil producer.[5] As we have seen, Saskatchewan also discovered and exploited oil and gas, though on a smaller scale. While the CCF actively developed petroleum policies that would subsidize social programs for the people of Saskatchewan, Alberta's Social Credit government charged relatively low royalties and gave the oil companies—domestic and foreign, but particularly American—more or less free rein. Wrote John Richards and Larry Pratt, "The striking failure of the post-Leduc resource boom to replicate the historical conditions of turn-of-the-century Texas and to nurture a powerful class of Alberta entrepreneurs united with populist farmers in hostility to a takeover by external corporate and political interests is one of the great puzzles of modern prairie development."[6]

Perhaps it can be explained in part by the different relationships between corporations and the federal government in Canada and the United States. As noted earlier, the transcontinental railways in both countries were examples of premature development. Both had federal subsidies, but neither subsidy was enough, so both groups of entrepreneurs attempted "creative financing" schemes to raise the rate of return. Credit Mobilier in the United States was successful in that the Union Pacific was built, the builders got rich, and the fallout from the scandal was minor and transient. Rather than being tarred by it, James Garfield was subsequently elected president with a reputation for honesty. In Canada, the Pacific Scandal toppled the government and delayed the completion of the Canadian Pacific Railway until after Macdonald could be re-elected. The federal government subsidized construction and guaranteed the repayment of monies that were borrowed. Still, both the railway and probably the country would have failed had not the Northwest Resistance shown the "need" for both a federal government and a transcontinental all-Canadian railway. The size of Canada's landmass and the forbidding nature of most of its territory (at least to European temperate-zone sensibilities) coupled with its small population base means that most major Canadian economic enterprises are "premature," whether they be the CPR in the 1880s or the Alberta oilfields of the 1940s or the oil sands of the 1980s to the present, and thus must be underwritten in part by federal or provincial governments, or both.

Most Canadian political scandals, historically and currently, involve skimming off government funds raised from taxes for the private gain of the skimmers and their allies. Most American political scandals, at least the directly economic or financial ones, on the other hand, involve the private taking of "public" properties, whether it be land or resources owned by Indigenous people, oil or other minerals discovered on public lands, or intangibles such as broadcast airwaves, although Enron broke the pattern by manipulating natural resources for private gain. Canadian taxpayers are, not unjustifiably, enraged to see funds drawn in whole or in part from direct tax dollars going into private coffers with no public return. American citizens, on the other hand, with the exceptions of various public watchdog organizations and their individual supporters, are remarkably blasé about the theft of land, minerals, air, and water, and are even likely to applaud the grafters, at least up to a point. One can see the continuation of these traditions in the reactions to Ad-scam and Enron around the year 2005. Ad-scam, the Canadian sponsorship scandal, led to an election in which a large Liberal majority was replaced by a whisker-thin minority Liberal government in 2004 and a 2006 election that yielded a Conservative minority government. Enron had almost no political fallout and led only to a few highly publicized trials, despite the far more widespread economic and social destruction wrought by the Enron collapse both on the community as a whole and on individual workers and shareholders and despite the close connections between leading Republicans and Enron. By the time the Democrats gained both the White House and substantial Congressional majorities in 2008, Enron was long forgotten by almost all voters.

Pratt and Richards are perhaps naive in judging Texas's success in withstanding external takeovers. To a large extent, the Texas oil magnates simply merged with New Englanders, such as the Bush family, successfully extending their influence to Texas and Florida. To some extent this is happening now in Alberta as corporate headquarters move from Montreal or Toronto to Calgary. On the other hand, Calgary, long the most American city in Canada, is becoming more and more an outpost of American oil. Alberta, like Texas, certainly had its oil pirates (some were my relatives) as well as its highly skilled and community-minded oil magnates, such as Eric Harvie, whose gift lay in sorting out and acquiring land titles confused by years of overlapping provincial and federal jurisdiction and by arguments over the siting of land claimed by the railways for successfully laying tracks across the continent. What Alberta lacked was the Texans' facility for mutually beneficial graft that allowed state governments, their regulatory bodies, and their regulated industries to move the most public domain into the fewest private pockets with the least public outcry.

The changes in the oil industry between the OPEC oil crisis of 1973 and the worldwide recession a decade later allow us to see both differences and similarities in the petroleum industries of the Canadian and American Great Plains, to evaluate their continuing impact on regional and local economies, and to consider how, or if, they could be managed for a more equitable and even sustainable future. In his provocatively titled 1975 book, *Making Democracy Safe for Oil*, Christopher Rand argues that OPEC (and its oil embargo of 1973, leading to long lines at the gas pump) was not the cause or even the beneficiary of the price rises and the reorganizations of the industry that followed the embargo. Rather, the major American oil companies brought about the shortage by refusing to invest in refinery capacity in 1972. (The rapid rise of oil prices between 2005 and 2008 was also the result of limited refinery capacity, and while oil companies blame environmentalists for the lack of new facilities, it is those very companies that have reaped record profits from the rapidly rising prices and have refused to build less polluting facilities.) In the early 1970s, the Nixon administration publicly agreed with the oil industry's contention that the embargo was working, in order to force US policy away from a pro-Israel slant to become at least somewhat more even-handed toward the Palestinians. According to Ed Shaffer, the shortage in the United States could have been avoided by taking Canadian oil, and eastern Canada would not have been dependent on imported oil had Canada not accepted the American and oil company arguments for not extending the cross-country pipeline all the way to Montreal in 1961, when it was extended to Ontario: accepting those arguments caused western Canadian producers to become dependent on US oil and gas markets. More than other Western industrialized nations, Canada has a split between producing and consuming regions. Before the 1973 crisis, Ontario actually purchased Alberta petroleum at higher than world prices, but after the oil shocks, made-in-Canada prices meant that Alberta oil companies were selling their crude to the rest of Canada at prices that eventually equalled only slightly more than half of world prices. Even though production costs had not gone up, both Albertans and the oil companies themselves resented the loss of profits—from the inflation of world oil prices—that could have been ploughed back into exploration and development.[7] The 1973 crisis and the events surrounding it entrenched the relationship between the Texas and US governments and the big American oil companies. It also increased oil company profits, and thus activity, leading to booms in Texas, Alberta, and other oil-producing areas of the Great Plains. And it set off a considerable power struggle between Alberta and Ottawa over who should keep the profits of the boom, a struggle that intensified in 1980 with the National Energy Policy and has continued through the first decade of the twenty-first century.

In the 1971 election, the somewhat sleepy laissez-faire Social Credit administration in Alberta had been defeated by the dynamic business-savvy leadership of Peter Lougheed, who was determined to capture oil revenue for the province. Meanwhile, Pierre Trudeau had to manage the fallout of oil prices rising rapidly in the less economically robust East while Ontario enjoyed Alberta oil that could certainly be profitably produced and transported for considerably less than the world price. Alberta resented what felt like subsidizing the rest of Canada, especially as the estimates of oil available in Alberta fell drastically between 1972 and 1974.[8] Whereas Lougheed had a substantial majority government, Trudeau was at the helm of a minority government dependent on the NDP, who consistently argued for using the revenue of natural monopolies for the people as a whole. Although Texas oil was nominally regulated by the Texas Railroad Commission, the commission was for the most part controlled by the American oil companies, while the Federal Power Commission under the Nixon administration began controlling gas prices to give incentives to production companies, not just to maintain fair prices.[9]

The US system was cozy, and the industry, state, and federal shares of oil and gas profits seem to have been satisfactory to all, though consumers endured a nasty shock in both oil *prices* and oil *accessibility*. A final source of contention in Canada in the 1970s was the commission headed by Thomas Berger to look at the feasibility of a gas pipeline down the Mackenzie Valley. Berger listened closely to the concerns of Indigenous people and returned a plan that would postpone development until an Indigenous infrastructure was in place to protect against the economic and social damages that had resulted from previous projects such as the Al-Can Highway.[10] The lack of a Mackenzie pipeline threat weakened Canada relative to the United States in the late 1970s, but along with oil sands development, it currently represents a huge, untapped Canadian source of energy of considerable interest to the United States. And the United States, both in terms of its oil companies and their intertwined relationship with the state and federal governments, was a major presence in all of Canada's internal arguments about oil. After all, Americans owned much of Canada's oil and gas infrastructure, whether companies were subsidiaries of major American oil companies or American-owned independents. Canadian- owned and -operated vertically integrated oil companies were simply non-existent, partly because provinces controlled resources but the federal government controlled commerce and hence Canadian pipelines, which tended to be partly publicly owned. Canadian independents excelled at exploration and production. Although there are oil refineries such as the Strathcona plant (Humble Oil) in the Edmonton area, and the city's National Hockey League team is called the Oilers in honour of the region's industrial role in the industry, much of Alberta's oil is still refined outside the province in eastern Canada or the United States. Arguments about economic diversification include urging Alberta to develop more "downstream" processing for its oil.[11]

In 1976, the United States elected Jimmy Carter, a most anomalous president, who, unlike his immediate predecessors and his immediate successors, did *not* come from a major oil-producing state or from an oil family. Particularly concerned with bringing peace to the Middle East, he was ironically rewarded with the toppling of the Shah in Iran and his replacement by a hardline, anti-American religious leadership in 1979. The Shah's fall was a case of chickens coming home to roost, but they were not Carter's chickens. American power had overthrown the Iranian nationalist government of Mosadegh in 1953 in order to displace both the Iranians and the British as the controllers of Iranian oil and to guard against any alliance between Iran and the Soviet Union. The Eisenhower administration's decision to topple a non-sectarian, nationalist, and popularly elected government in favour of the Shah continues to resonate around the world, but in 1979, the Shah's overthrow had the effect of discrediting Carter's peace policies and also sending world oil prices skyrocketing once again. Not surprisingly, this also shattered the accommodation on oil prices that Alberta and Saskatchewan had reached with the federal government.[12] Another collateral casualty of the Shah's fall was the brief Progressive Conservative government of Joe Clark, which fell over a budget that tried, unsuccessfully, to balance the needs of the oil-consuming and oil-producing provinces. Trudeau and the Liberals regained power, while in the United States, Carter gave way to Ronald Reagan, a nationalistic neo-conservative from a major oil-producing and -refining state, California. The stage was set for the National Energy Policy, the West's most hated and most debated program since the CPR.

Trudeau's National Energy Policy (NEP) was a political solution to a number of problems raised by increasing oil prices and the resultant increase in activity and prosperity in the Alberta oil patch particularly. Alberta had about 10 percent of Canada's total population but 80 percent of its oil revenues. As a federal leader, Trudeau really had no choice but to capture some of the "economic rents" from oil and gas for the nation as a whole and particularly for population- and industry-rich central Canada.[13] Although Trudeau opposed Quebec nationalism and chose world peace as his legacy for Canada, he was enough of a Canadian nationalist to be wary of the power of US oil companies, US investors, and their symbiotic relationships with the US government. Trudeau and his advisors believed that Canada's future economic and energy security depended on having Canadians owning and operating their own oil industry. These were not unreasonable goals for Ottawa to pursue, nor were they even goals most Albertans would challenge as a whole. Even a quarter century after the NEP, when oil prices *have* reached and far exceeded the $50-per-barrel prices envisaged in the early 1980s, the oil sands *are* economically (though perhaps not environmentally) viable, the Mackenzie pipeline *is* once more on the drawing board, and American ownership and management are even *more* strongly entrenched in Canada's oil industry, most Albertans are not averse to the goals of the NEP—even if they do recoil in horror at hearing those three successive initials. As Shaffer says, Peter Lougheed was not in favour of having the "eastern Canadians ['bastards' was the bumper sticker language] freeze in the dark." He wanted Ottawa to import more oil, forcing Alberta oil prices up to the world benchmark more quickly. Lougheed, as was evident in the negotiations to patriate the Constitution, was a federalist and a Canadian nationalist in many of the same ways as Trudeau.[14]

The actual effects of the NEP as they developed in the real world are much more difficult to calculate, let alone evaluate. As Tammy Nemeth points out, the unilateral arrogance of the way the Liberal government imposed the NEP has poisoned the discussion in Alberta forever afterward. Certainly, the various taxes that the federal government was able to impose shifted substantial portions of the economic rents of the oil industry away from the oil companies and away from Alberta, and keeping a made-in-Canada price for Canadian oil protected Canadian consumers from greater price shocks than they would otherwise have received, at the expense of oil company profits and Alberta provincial revenues. Federal incentives for oil exploration on federally, rather than provincially, owned lands did divert exploration away from Alberta, though the following quarter century suggests that most of Alberta's future petroleum reserves lie in the oil sands rather than in large finds still to be discovered by conventional exploration: the exploration rigs, therefore, would have moved out in any case. Nemeth shows, however, that the oil rigs began moving in response to the NEP, harming Canadian juniors and secondary and tertiary businesses in Alberta before the economic downturn of the early 1980s created the same sort of busts in the oilfields of Oklahoma and Texas.[15] Small finds are still being made in Alberta, showing that exploration did not stop altogether. New and environmentally controversial methods of gas production are coming into vogue, again suggesting both that exploration did not stop and that finds of the old sort were not there to be made, no matter what government policies were followed.

Petro-Canada, constructed as Canada's window on the oil world and as an alternative to the nationalization of the oil industry (as was happening in other countries), was able to grow by acquisitions, but it never served its purpose and has since, bit by bit, been privatized. Canadian ownership also meant the purchase by Canadian independents of American and other foreign-owned oil properties, but they were buying in a seller's market and in inflated dollars from the "stagflation" period of the late 1970s, with high interest rates. The saga of Dome Petroleum is instructive here. Dome, one of Canada's most innovative and successful independent companies, purchased Hudson's Bay Oil from the American company Continental Oil in 1981—with a $2 billion mortgage held by both Canadian and US banks. When the economic downturn came, Dome was in effect taken over by the banks—exactly as Mathews' friend Marland had been more than half a century earlier and half a continent further south. That some of the banks had strong connections to US oil companies was probably no coincidence.[16] Buying a national oil industry in the open market without any Daddy Warbucks was spectacularly unsuccessful.

As for the rest of the National Energy Policy, even Roger Gibbins, CEO of the Canada West Foundation, is doubtful. In an op-ed piece for the *Alberta Centennial*, Sydney Sharpe quoted him as saying: "It is difficult to disentangle what happened to world energy prices from the NEP itself. . . . I've always wondered whether we've loaded more on the NEP and whether that was appropriate given the other changes that were going on globally. Lots of things went pear-shaped in terms of energy markets at that time."[17] To be sure, the drill rigs left Alberta and their departure was hastened by NEP incentives to explore on the "Canada Lands" in the North and East— but they pulled out of Texas as well. The same bumper stickers appeared on trucks at both ends of the Great Plains: "Please God, if You let us have another boom, I promise not to piss it away this time." Oil mansions were foreclosed in Calgary and in Houston. The secondary industries (those who produced parts and services for the exploring crews, drill rigs, and pipelines) and tertiary industries (those like hotels and gas stations that served the entire economy) were harder hit than the oil companies themselves because they could not pick up and leave. In Saskatchewan and Alberta, it was the junior companies that were really the Canadian-owned and -operated part of the oil and gas business, the ones that any successful program for Canadian ownership needed to nurture and protect—but they were the ones that really suffered from the NEP.[18]

Neither the big oil companies nor the US government wished to see the NEP succeed, so it is likely that American companies that pulled out of Alberta and Saskatchewan, complaining that it was not fair to change the rules in the middle of the game, were as concerned to defeat Canada's goals of self-sufficiency and control as they were with their own short-term profit. The rules in the oil game were *always* changing as conditions changed, and American oil companies had been remarkably successful in getting the rule changes they wanted from bodies as different as the Texas Railroad Commission and the US State Department. Reagan began deregulating the American oil and gas industry soon after the NEP began operating, rule changes that were *not* opposed by the oil industry, and small Canadian oil operations who moved to the States did not actually find the US market system as congenial as rumour had it. If we look to Saskatchewan, we can also see that world economics were more important than attaining a free market system for extractive industries. For all of Grant Devine's proclaiming that Saskatchewan was "open for business," privatized potash was not bringing big prices any more than Alberta oil was selling for $50 per barrel.[19] The oil companies, the federal and provincial governments, and the banks had all believed that high prices had come to stay in 1980. They were all wrong—as they would be again in 2008.

Peter Lougheed was certainly right in believing that the federal government was moving to block Alberta's rise to power. A federation in which 10 percent of the population, concentrated in one of ten provinces, was becoming fabulously wealthy in contrast to and to some extent at the expense of most of the rest of the population (assuming that BC, Saskatchewan, and the Territories stayed relatively the same) would not have been stable. As was often the case with Trudeau and the West, the issue was not really so much what he did as how he did it. The West saw him as insufferably arrogant in general and particularly so in his attitude toward the West. Trudeau saw the Prairies not necessarily as deficient—but certainly as insignificant. Provincial control of natural resources, for which Alberta and Saskatchewan had had to fight so hard for the first twenty-five years of their existence as provinces, was a particularly sensitive subject, and the idea that Trudeau would come waltzing into the oil patch with a bevy of new fees deliberately designed to shift oil wealth from Edmonton to Ottawa maddened Albertans. They were legal in that they were neither exactly royalties nor exactly income taxes, but they were levied only against oil and gas, and were especially galling, since the feds called on Albertans to drop their royalties if the federal tax bite threatened the profitability of any of the oil companies.[20] The Liberals would not dare do something like this to Ontario or to the sacred cow of Quebec, Albertans reasoned, and if oil profits could be siphoned off, so could the hydro profits of Quebec and Ontario, which also come in part from the US market. And the Albertans were absolutely correct in this suspicion. But Ontario and Quebec each have a good deal more than 10 percent of the population of Canada, and hydro competes in a limited regional market, not the global market of oil. Hydro is also renewable, not a finite resource, and it has never undergone the extreme price spikes that petroleum did in 1973 and 1979. The federal government believed, correctly, that if it did not get its finger in the conventional petroleum pie in the 1970s there would be nothing but crumbs left in the future.

By 1981, Alberta and the feds had shaken hands. By 1984, Trudeau was gone for good, and Conservative Brian Mulroney was in office. He would abolish most of the NEP, begin the reprivatization of Petro-Canada, and negotiate a free trade pact with the United States. Free trade outlawed any kind of National Energy Policy and eased both American participation in Canadian oil and gas extraction and Canadian dependency on US oil and gas markets. According to Shaffer, the actual winners in the battle between Lougheed and Trudeau were the oil companies. Alberta had received 48.5 percent of oil revenue in 1979, dropped to 41 percent under the NEP, and then further down to 30.2 percent in the agreement Alberta supposedly *won* from the feds in 1981. Meanwhile, the federal share of the revenues rose from 12 percent in 1979 to 27.4 percent under the NEP before dropping slightly to 25.5 percent. The oil companies, however, dropped from 39.5 percent to 31.6 percent under the NEP before rising to 44.3 percent in 1981![21] Surely this was not what sent the drill rigs out of the province in 1982!

The price rises of 1973 and 1979 touched off state, federal, and industry conflicts in the United States as well, but they assumed different guises. Both Texas and federal regulators, as we have seen, were most solicitous of the oil industry. The trouble, such as it was, arose from alternative fuels or alternative technologies in other states, such as Colorado, where shale oil promised fortunes, or Montana and Wyoming, which sat over huge deposits of coal. For the most part, it was the states that wished to go slowly to avoid the social costs of overly rapid development and to conserve the land and water of the West. The bust of the early 1980s devastated the Texas and Oklahoma economies, as it did Alberta's, and slowed or halted energy development in Colorado, Wyoming, and Montana. Paradoxically, it also stirred up the Sagebrush Rebellion among conservative (not conservationist) westerners, who wanted the federal government to cede public lands to the states to speed up economic development. According to Richard White, the Sagebrush Rebellion failed because western urban and conservationist movements opposed it, and western governments, from school districts to states, could not operate without the funds that the federal government had begun paying in lieu of taxes in 1976.[22]

According to White, "Westerners, it seemed, agreed they were being abused; they disagreed on the nature of the abuse and the identity of the abusers." Even Albertans, convinced as they were that the Liberals were the abusers and the NEP their tool, were slightly unsure exactly what the result was, except for "the memory of Calgarians walking away from their houses, leaving the banks to repossess them."[23] But such strong images and deeply held beliefs on the Plains of both countries about the perfidy of the federal government and of government in general has fuelled both what Roger Epp calls "de-skilling" and an incongruous belief in the rightness of the political right and the free market. These images and beliefs have also convinced many rural westerners that they are helpless to do much but continue to vote for politicians whose policies do not favour them and for development, whether it be petroleum or sour gas or hog factories or even federal farm supports, that does not help them. As Epp writes,

> Farm families in many areas [of Alberta] exist in an uneasy, subordinate relationship with the energy sector, which is a source of lease income and off-farm wages, as well as disruption and environmental threat, and which (once pipelines are included) constitutes in some rural municipalities the largest, most influential source of tax revenue.[24]

The sour gas wells, the huge trucks barrelling along narrow rural gravel roads, and the exploration activities taking place on traditional northern Indigenous peoples' hunting grounds are not showing up in the neighbourhoods in Calgary where oilmen walked away from their houses in the early 1980s. But resentment against the NEP provides a good way for the current politicians to deflect anger away from themselves and their royalty regimes in the petroleum industry and to deflect thinking away from a far more profound and productive regional rebellion than we have yet seen.

In July 1990, my four-year-old son and I drove from Batoche to Kingston. All the way along, I listened to the radio and read the papers, following the Oka Crisis and believing that things were going to be different, that finally Canadians, and even North Americans in general, were going to see that after five hundred years of survival and resistance, Indigenous peoples were—had always been—entitled to control their destiny, as much as any humans ever can. And that controlling their destiny meant something other than assimilation and "vanishing" into the Amer-European whitestream culture. Then we took the ferry down to Wolfe Island and over to New York State, and as CBC faded off our radio, all news of Oka disappeared. Even the *New York Times*, it seemed, had never heard of Oka or Mohawks. It was a parable not only of American blindness toward everything Canadian but also of whitestream willed oblivion to anything Indigenous that did not fit stereotypes.

When Europeans began sustained contact with what they called the New World after 1492, they almost immediately put into place the rhetoric of deficiency and punishment that they had already developed for the "infidels" against whom they had been staging "crusades" for centuries. Despite the splendours of Inca or Aztec civilization, Euro-Christian rhetoric condemned the Indigenous peoples of the western hemisphere as deficient beings whose lands and possessions were automatically forfeit to the Europeans and whose lives and liberties, if not absolutely forfeit, were still so badly in need of improvement that they were only to be enjoyed at the pleasure of the civilizers. That this was both inaccurate and unfair, by the civilizers' own rules, was rarely thought and even more rarely expressed by the Europeans, save for those who, for whatever reason, threw in their lot with that of the Indigenous people. The basic attitude that "Indians" had no rights that a white man need respect really changed little over the next half millennium, though as Michael Murphy has shown, different nineteenth-century European thinkers had various theories about whether Indigenous individuals were inherently inferior to Europeans or whether environmental and cultural deficiencies had rendered their social relations inferior, even if individuals were capable of improvement and assimilation.[1] Only in the second half of the twentieth century, when the horrors of the Final Solution made inescapable the fallibility of the idea of racial superiority and even of Euro-Christian enlightenment and its imperialist attitudes to "lesser breeds without the Law," did Amer-Europeans begin to listen seriously to the intellectual arguments of global decolonizers, including Indigenous leaders in the United States and Canada, though those arguments took over half a century fully to register and are still not clear in all quarters today. Despite my experience with Oka's disappearance from my airwaves, July 1990 was approximately midway in a series of crises that reshaped Indigenous affairs in Canada and prompted a wholesale reconsideration of the relationship between Indigenous and whitestream Canadian laws and ideals. Although the United States is now lagging far behind Canada in recognition of Indigenous rights, the Canadian lessons apply to the States as well.

The purpose of this chapter is to review the discussion that has ensued over the last twenty years and to mine it both as a model for how deficiency can come to be understood as sufficiency and as a source for Aboriginal ways of viewing the contemporary world that provide us with a specific way to reconceptualize our relationship with that other deficiency, the Great Plains. To understand *The Terrible Summer*, as Richard Wagamese called the summer of 1990 in a collection of newspaper columns he wrote at the time, it is useful to examine the Oka Crisis in the context of other "Native News." What issues were Indigenous communities across Canada pursuing? How did Indigenous artists conceptualize decolonization? How can we use the continuing rhetoric of colonialism to defeat itself? What is the contemporary rhetoric of decolonization, especially as it relates to common, criminal, and constitutional law, themes that repeatedly arise from the news stories? And finally, how does this serve as a model for reimagining and justifying the Great Plains?

We can find a context for the Oka summer in texts produced by and for Native communities in western Canada in *Windspeaker*, an Edmonton newspaper that has survived tumultuous times to remain an influential Native voice. A central theme running through most of the discourse is that Indigenous issues are holistic, that one cannot isolate one strand from another. At the same time, it is impossible always to talk simultaneously about everything, so the human brain is forced to create distinctions even for the purpose of reuniting them. Although I will focus on events from the Prairies, many of these issues, like Oka, were national in scope but produced particular responses on the Prairies. One could start in many places, but let us choose the early morning of 9 March 1988, when Cree leader J.J. Harper was shot and killed by a Winnipeg city policeman with a reputation for being a cowboy and a racist. Manitoba empanelled an Aboriginal Justice Inquiry (AJI) the following April to look into Harper's death and also into a vicious murder in the Pas in 1971, in which a young Cree student, Helen—called Betty—Osborne, had been kidnapped and killed by four white men who remained invisible to authorities for more than a decade, even though they were well known in the community. Also in 1988, the Lubicon Lake Cree band organized a widely publicized boycott of the *Spirit Sings* exhibition at Calgary's Glenbow Museum during the Winter Olympics to call attention to their unresolved land claims in northern Alberta and to the physical and economic degradation of their territory by oil drilling operations.[2]

In 1990, matters accelerated. The Nova Scotia Justice Inquiry began the year on 26 January by exonerating Mi'kmaq Donald Marshall, Jr., for the 1971 murder of Sandy Seale, for which Marshall had been wrongly convicted and had served eleven years in prison. The province apologized, not only for the wrongful conviction, but for the statement of the judge who finally acquitted him, that Marshall had been the author of his own misfortune. In March, the federal government slashed funding for Aboriginal communications networks, including publications like *Windspeaker*, and for Friendship Centres and Aboriginal government organizations nationwide. The Province of Alberta persisted in building a dam on the Oldman River without fulfilling the environmental requirements that courts had demanded and without the permission of the Peigan people, whose land and river would be affected and whose people were divided over the dam. Mohawks in Kanesatake protested plans of the neighbouring municipality of Oka to enlarge a golf course onto an area the Mohawks called The Pines, which the Mohawks had reforested in the nineteenth century after the failure of European agriculture on the sandy soil. The Mohawks had used The Pines as a community sacred site ever since. A land claims suit was wending its way through a court in British Columbia. Wilson Nepoose was still in an Alberta jail for the 1986 murder of Rose Desjarlais, though his conviction would eventually be overturned. Brian Mulroney's window of opportunity for the ratification of the Meech Lake Accord to reconcile Quebec to the 1982 patriation of the Constitution was coming to a close when Elijah Harper, a Cree MLA from the North, refused unanimous assent to its passage in the Manitoba Legislative Assembly on 12 April 1990. The accord, Harper accurately pointed out, did not deal with Aboriginal rights, and Canada could not afford to conclude constitutional dealings with Aboriginal rights still in abeyance. Harper never gave assent, and Meech Lake ran out of time on 23 June. It had never been particularly popular outside of Quebec, but the image of Harper holding an eagle feather aloft as he refused assent would inflame Quebec's resistance to the Mohawks' claim to their land.[3]

On 11 July 1990, the Terrible Summer began in earnest, when Sûreté du Québec troopers attempted to clear out the Mohawks blocking the golf course expansion. A quiet and peaceful occupation turned immediately into an armed standoff, and somehow one policeman was shot and killed. All of a sudden, Indigenous issues were front and centre in Canada.

The newspapers, the radio, the television all devoted time and stories to the Oka Crisis, especially when the Canadian Army was dispatched to . . . do something—whether to bring peace or just to end the blockade is not entirely clear. Mohawks from Kahnawake blocked the Mercier Bridge and shut off the southeastern approach to Montreal. Aboriginal people felt that *finally* non-Native Canadians were paying attention to what they had been saying all along. Elijah Harper had stopped Meech—with a feather. A few determined Mohawks were holding the Canadian Army and the Province of Quebec and the City of Montreal at bay. Aboriginal groups enacted their own blockades—including a brief and symbolic stoppage of the Louise Bridge in Calgary.[4]

But as the Terrible Summer wound down, solutions did not seem near. Residents of Chateauguay stoned the Kahnawake Mohawks. Milton Born-With-A-Tooth and the Lone Fighters managed to dig a diversion ditch around the dam site and free the Oldman River on 3 August 1990. But they were surrounded by police and shots were fired on 7 September. Milton Born-With-A-Tooth was arrested on 12 September. Despite winning some battles against the oil companies, the Lubicon Lake Crees were losing the war. The federal government was angling to recognize a band, whom the Lubicons believed were malcontents organized by the government, with rival claims to the area, and the Japanese conglomerate Daishowa was moving ahead with a giant pulp plant for Lubicon territory. On 26 September 1990, the Mohawk Warriors left the treatment facility where they had taken their last stand at Kanesatake. Most were arrested.[5]

On 28 January 1991, in Prince Albert, Saskatchewan, Leo LaChance, an Aboriginal trapper, was shot and killed by Carney Nerland, the former leader of a white supremacist group. There was conflicting testimony as to whether this was an accident or a deliberate and racially motivated slaying. (At Nerland's trial and subsequent inquiry, despite police attempts to maintain secrecy, it became apparent that Nerland had been a police informant. He received four years for manslaughter and disappeared into a police protection program after his release.) On 8 March 1991, Justice Allan McEachern of British Columbia handed down a verdict dismissing Gitksan and Wet'suet'en land claims. Canada, he said, was under no obligation to accept oral tradition in a land claims case nor, really, to respect any Aboriginal claims not backed up by European written treaties. Indigenous people, like everyone else, could casually use Crown lands not otherwise under lease or contract. But that was all. Delgamuukw would appeal McEachern's decision—and the condescending manner in which it had been reached and written—to the Supreme Court of Canada. On 11 December 1997, the Supreme Court would overturn McEachern on every particular and grant oral tradition—be it story or dance—and the underlying relationship between the people and the land full standing in Canadian courts. On 25 March 1991, an Alberta judge sentenced Milton Born-With-A-Tooth to eighteen months in jail—apparently for firing a gun in the air in the presence of police. The Oldman Dam would be completed. On 29 August, the AJI of Manitoba presented the results of its three-year investigation. Like the inquiry in Nova Scotia and similar inquiries in BC, Alberta, and Saskatchewan (and in fact some thirty inquiries in twenty-five years, as counted by *Windspeaker*) the AJI would conclude that the Manitoba justice system consistently and across the board failed Aboriginal people who were victims of crimes, victims of police brutality, perpetrators of crimes, wrongly charged or convicted, or "criminalized" by child protection services or job discrimination. It could not have been more plain in its condemnation nor more precise in its suggestions for change. (Twenty years later, few of those changes have been made.) That same August, the federal government authorized a Royal Commission on Aboriginal Peoples (RCAP) to study all the issues raised by the Oka Crisis and the Nova Scotia and Manitoba justice inquiries.[6] While RCAP considered many issues in addition to the justice system and its failures, the revisions the commission and others suggested show the greatest capacity for immediate, effective system change—although implementation lags far behind inspiration.

The five hundredth anniversary of the Columbian invasion was in October of 1992—ironically, on Thanksgiving. No celebration for Indigenous peoples of the western hemisphere, the anniversary was, however, a chance to look back over five hundred years of survival and resistance, which Indigenous artists did in two travelling exhibitions: *Indigena*, mounted by the Canadian Museum of Civilization, and *Land, Spirit, Power* by the National Gallery of Canada. Both surveyed the past as a way of building a better future. And of course, life went on and was recorded in the pages of *Windspeaker*. Young Aboriginal artists won awards. People went to powwows and rodeos and hockey games. Students won scholarships and graduated from programs in nursing, law and corrections, and forest management. People started new businesses. Students wrote and performed their own plays. Tony Thrasher, the "Skid Row Eskimo," died in Edmonton in July 1989. AIDS was stalking Aboriginal communities, and Ken Ward, the first treaty person to be diagnosed with full-blown AIDS, went public in 1990 and began to visit prisons and schools and communities to talk about AIDS prevention. Although the "scoop-up" of the 1950s and 1960s was over, Aboriginal families still lost children to "welfare" agencies, and adult children who had been adopted out of Indigenous communities came back, struggling to find and know and be claimed by their birth families. In August 1991, a mother staged a hunger strike to protest the decision to have her children adopted by non-Native parents instead of by the family members to whom she had entrusted them. Prostitution and substance abuse ruined lives, and people got off the streets and sobered up and reclaimed their lives. Incarceration rates rose more rapidly for Aboriginal people than for other segments of the population, especially in the Prairies, and people fought to get a Healing Lodge for women started in the Prairies because Aboriginal women were dying in the antiquated Prison for Women in Kingston. Aboriginal birthrates were among the highest in both Canada and the United States, and while Aboriginal dropout rates were also high, more and more Aboriginal students completed high school and succeeded in post-secondary education. Videographers scrutinized the valour and the horror of Indigenous life—kids committing suicide and communities coming together to provide hope for the kids. In 1992, Wilson Nepoose's murder conviction was overturned and he was released from jail, but Alberta never ordered a new trial, leaving him in limbo, neither guilty nor exonerated. A year and a half after the RCAP reported its considerable findings and recommendations, which have never been implemented, Wilson Nepoose, in early January 1998 (or perhaps the very end of December 1997), died in the bush near his sister's house. His remains were not found until the following summer.[7] (The murder of Rose Desjarlais remains unsolved.)

In the United States, the pages of *Lakota Times* (now *Indian Country Today*) included many similar stories, but the events that, like

Oka, had shone a national spotlight on Indian affairs had happened about two decades earlier, and they had arisen from a somewhat different political and cultural context than the events in Canada. As we have seen, in 1932, Black Elk and John Joseph Mathews had articulated powerful visions of Siouan world views as exemplary, even as they worried about the possible disappearance of the people who understood those traditions. The thirties, however, were a hopeful decade for American Indians. In 1934, as we have also seen, the *Indian Reorganization Act* had de-outlawed the Sun Dance and other ceremonies, ended allotment, provided for tribal sovereignty—if not always on tribal terms—and in many ways ended direct colonization of American Indians. The federal and even state arts and other employment projects of the Depression did employ some Native people and included Native motifs as the "universal" heritage of the United States. One of the most striking examples is the Nebraska State Capitol, a singularly beautiful building with Indian themes developed by University of Nebraska professor Hartley Burr Alexander, including a senate chamber intended to inspire high-minded political thought with exclusively Native themes and images.[8] Although this was to some extent simply cultural appropriation, it did show a genuine willingness of whitestream power to learn from Indigenous philosophy. During the 1940s, Native American intellectuals could still retain some optimism. As had happened in World War I, Native men volunteered for the armed services in very high proportions, while Native women moved to the cities and took on Rosie-the-Riveter roles. World War I service had been rewarded with full US citizenship for Native people in 1924, and it was reasonable to expect social improvement after World War II as well. In 1944, Ella Deloria published *Speaking of Indians*, an eloquent explication of "A Scheme of Life That Worked" and a prescription for a better postwar society that would at least accept Native people's right to live by rules based on kinship and sharing, and perhaps inform whitestream society as well.[9] Unfortunately, that would not happen. Instead, the postwar years were marked by the federal policies of Termination and Relocation, which attempted to open up remaining reservation resources to non-Indians, to terminate the Indian status of tribes and individuals, and to move Indigenous people off the land and into the cities. Although couched in terms of assimilation, these programs for the most part resulted instead in marginalization, alienation, and increased poverty and welfare dependence. Public Law 280 replaced federal jurisdiction over reservations in some states—including Nebraska—with state jurisdiction, often leading to selective enforcement and a lack of protection for reservation residents. The ensuing disorder hastened the breakup of Indian families as children were taken away and put into foster care or entirely adopted out. As we have seen, the Pick-Sloan dam projects systematically flooded reservations on the Missouri mainstem, including Crow Creek, flooded out twice, leading to more social and cultural disruption and the apprehension of children for placement in often brutal foster and institutional settings. (Some were my claimed family.)

Not surprisingly, Native Americans fought back. Ella Deloria's nephew, Vine Deloria, Jr., was the most influential Native American intellectual from the late 1960s to the end of the century. Just the titles of his books—also bumper stickers—give the flavour of his discourse: *Custer Died for Your Sins*; *We Talk, You Listen*; *God Is Red*; *Red Earth, White Lies*. Unlike Black Elk, Mathews, or Ella Deloria, Vine Deloria was less interested in arguing or delineating the exemplary nature of Lakota or Native society in general—he pretty much took that for granted—as he was in pointing out the deficiencies of both government and academic treatment of Native Americans and how Indians themselves would organize to go about resolving their own issues. His *Custer Died for Your Sins* is a strong parallel to another book published in 1969, one that also used irony and humour to point out government mistakes and how Indian-controlled programs could redress them: Harold Cardinal's *Unjust Society*. While Deloria and Cardinal and others provided theory, during the 1960s and 1970s, thousands of Native people from all over the continent provided the specifics of a Red Power movement that took inspiration from the more general civil rights struggles, mostly in the United States, and from specific responses to Indigenous issues such as Relocation and Termination, the infamous "White Paper" put forth by the Trudeau government in 1968 that would have effectively terminated Indian status in Canada, ongoing problems with residential schools and education in general, fishing and hunting rights guaranteed in treaties but abrogated in practice, the needs of a new class of urban Indians, and the loss of subsistence and the resulting welfare dependency on reserves and reservations.

Cardinal was nothing if not blunt in his first paragraph, and his basic premises are similar to Deloria's:

> The history of Canada's Indians is a shameful chronicle of the white man's disinterest, his deliberate trampling of Indian rights and his repeated betrayal of our trust. Generations of Indians have grown up behind a buckskin curtain of indifference, ignorance and, all too often, plain bigotry. Now, at a time when our fellow Canadians consider the promise of the Just Society, once more the Indians of Canada are betrayed by a programme which offers nothing better than cultural genocide.[10]

Both Deloria and Cardinal saw the missionaries and their schools as failures and travesties, and both, especially Cardinal, demanded education—as the Indigenous treaty makers of the nineteenth century had—in the trade and professional skills necessary to earn a competence in whitestream farming, resource extraction, and other fields. Deloria particularly disliked anthropologists, who, he believed, tended to sentimentalize Native people and hinder their mainstream success by locating them in a romantic past. Both Deloria and Cardinal looked for peaceful solutions but noted that violence was possible if whitestream society did not acknowledge Native rights. Deloria distinguished between Indian nationalists—who "are primarily concerned with the development and continuance of the tribe" and were not much influenced either by whitestream assumptions or blacks' aspirations of inclusion in whitestream society—and militants, who were "reactionists." The nationalists might use violence if necessary, but militants used violence only to attract attention to themselves and thus had nothing but violence to provide. Cardinal noted that Canada's Indians were watching television to learn about the successes and failures of Black Power in the United States. They were doing their best to organize, despite generational differences, funding deficits, and divisions between status and non-status Indians, and urban and reserve populations. He warned Ottawa to honour its own words. The White Paper had been a particular affront because it contravened all the promises of consultation that the government had made to Canada's Indians and handed down a mandate that had nothing to do with what Native people in Canada wanted. He warned of a "Red Explosion" if Ottawa proceeded on its heedless way.[11]

Writing in 1968, both authors clearly envisaged the Red Power movement that began with the occupation of Alcatraz in 1969 and culminated in the Wounded Knee takeover in the spring of 1973. Robert Allen Warrior and Paul Chaat Smith painstakingly chronicle these events and the ideas behind them in *Like a Hurricane*, but their emphasis is on the rationales for the occupations and on the great significance of the occupations in arousing "Red Pride" and counteracting internalized racism. Although Warrior and Smith did not look at the Canadian participants in these events or at the occupations and other actions in Canada, Jeannette Armstrong's 1985 novel *Slash* is an excellent distillation of that Canadian story; she shows her hero in the BIA building in Washington as well as in a similar action in Ottawa and responding to Wounded Knee even though he cannot participate. Armstrong's focus is on Okanagan tradition—and other Indigenous cultures and ceremonies when necessary—as a way of healing the alienation of colonialism for her Okanagan and other Indigenous characters, and also for providing a model of how to live with the land for the ignorant white people, whom she acknowledges, as Cardinal does, are "here to stay." As Deloria notes in the introduction to the 1988 republication of *Custer Died for Your Sins*, AIM created a feeling of solidarity and pride in the 1960s and 1970s, but by 1988 it was virtually moribund.[12] He did not note, though he could have, that it had largely been silenced by the FBI's domestic terrorism through their COINTELPRO (Counter Intelligence Program) initiative and a determined program of selective prosecution that ate up precious funding and time. AIM leader Leonard Peltier still languishes in federal prison, doing two life sentences plus seven years for a crime even prosecutors acknowledge he did not commit and for which he was illegally extradited from Canada.

All of the stories of Black Elk and Mathews and Ella and Vine Deloria and Harold Cardinal and Jeannette Armstrong and many, many others are parts of the intellectual background for Native issues in Canada, and the Prairies in particular, during the 1990s. But let us look in a bit more detail at the ways both Deloria and Cardinal suggested Native society could be exemplary for whitestream society. As Deloria noted frankly, "The United States operates on incredibly stupid premises," so it could be influenced to more intelligent, peaceful ways "by any group with a more comprehensive philosophy of man if that group worked in a non-violent, non-controversial manner." Deloria saw hope in urban Indians who have access to libraries and night schools, and by 1988, was particularly optimistic about "an increasing number of young people" who, with "well-organized community support [could] greatly influence the thinking of the nation within a few years." Similarly, Cardinal saw the "rebirth of the Indian, free, proud, his own man." While co-operation between Native and whitestream societies has hitherto demanded that all the change be on the part of the Indigenous groups, "our older people think that it is part of the responsibility of the Indian to help the white man regain this lost sense of humanity."[13]

Though Deloria (Dakota, Denver) and Cardinal (Cree, Edmonton) were both from Plains tribes and were living in cities on the edge of the Plains in 1968–69, they spoke for national and to some extent pan-Indian movements, as did *Windspeaker* and the *Lakota Times*. The Wounded Knee takeover was on the Plains, on the Oglala Pine Ridge Reservation in South Dakota, but Alcatraz and Washington were on the west and east coasts. Similarly, the events around 1990 took place across Canada, and one would be hard put to identify a particularly Plains or Prairie view on issues such as Meech Lake. Elijah Harper was from northern Manitoba, and it was a federal mandate to bring Quebec into the Constitution that he stopped, but he held up his feather in Winnipeg. Local, regional, and national issues all morph into each other, which is not surprising: Native sovereignty cannot be meaningful if it is based on single reserves, since almost all important land-use and political decisions are made regionally, nationally, or even internationally. To make sense of the arguments for Indigenous customs as sufficient and even exemplary, we must see region in a global context. Let us look, then, at how these arguments are developed and sustained in the production of three artists who were working in Alberta at the time of the Terrible Summer and who were often featured in *Windspeaker*. Joane Cardinal-Schubert and Jane Ash Poitras are Albertan visual artists with international reputations—both took part in the *Indigena* exhibition and both were reviewed frequently in *Windspeaker*. Richard Wagamese, journalist and novelist, was an award-winning columnist for *Windspeaker* and the *Calgary Herald* during this period. These three artists and intellectuals represent a Prairie-based creative nucleus for discussing the ethic of survival and resistance that developed in Indigenous Prairie communities during this turbulent period.

At a June 2005 gallery talk/demonstration in Calgary, Joane Cardinal-Schubert showed a series of recent works in which she juxtaposed rectangular city spaces with organic images of horses. Her talk mixed technical discussion of colour theory—how different colours work together to create the illusion of depth, of shapes protruding or retreating from the surface of the canvas—with her concern about representing images that belong to her own lived experience, such as horses, without reproducing or even suggesting stereotypes of "the Plains Indian" as mounted warrior. For *Indigena*, Cardinal-Schubert prepared a complex installation piece, "Preservation of a Species: DECONSTRUCTIVISTS (This is the house that Joe built)," which combined painting, drawing, photography, sculpture, assemblage, and text, and was completed in 1990. In her artist's statement in the published catalogue of *Indigena*, she says that it "is an installation that visually discusses RACISM through an examination of labels and imposed stereotypes that I have experienced growing up in a non-Native society." She deals with the forced assimilation of children through the mission schools and the foster care system, the systemic categorizations of people by "status" and number, the fencing in of people in the reserves, and the necessity for resistance—her own, her father's (Joe), her grandmother's, her brother's (architect Douglas Cardinal). The installation includes the text of "Joe Cardinal's message to his children from his deathbed 'IF I HAD MADE A STAND—you wouldn't have to/you've got to stand up to them. Don't let those bastards get you. Just Stand up and Never give in . . . '" (ellipses in original). But her father she also associates with the land, which she includes in a "large painting of the lake." Her artist's statement continues, "We should be thankful that the Native people have become the barometers, the 'eco-meters' who point out the dangers of pollution to us." Part of the text talks about her father as hunter and "just really part of the forest." But she concludes that text with the words, "(Eventually Joseph came to believe nature's biggest enemy wasn't poachers, but his employer, the Alberta government, which seemed to be in league with the exploiters)."[14]

The nature of the assemblage forces the viewer to see all the parts as connected—but not whole. The peaceful painting of the lake and water lilies is behind a box labelled "FOSTER CHILD" and is littered with bottles, a syringe, money, a scrub brush, and "CULTURAL IDENTITY" locked in another, smaller box. Posts become women in head scarves, some wearing newspaper clippings attesting to daily realities and all with the bark on, still identifiable as trees, as part of the tree. In his 2001 book *A Feather Not a Gavel*, A.C. Hamilton, co-chair of Manitoba's Aboriginal Justice Inquiry, combines memoir with research and experience to explain why Canada's justice system is failing Aboriginal people and how it might be reconceived. He emphasizes that Native people repeatedly pointed out to the justice inquiry that no problem could be solved in isolation. Cardinal-Schubert says the same thing visually.[15] Using different media and enclosing part of her installation in a house, a box, so that it must then be viewed through different windows that break things up and prevent the viewer from seeing everything at once, she actually emphasizes the wholeness. We are perfectly content not to see "the whole picture" when our little vantage points present what appears to be a complete view, but we are frustrated when we are forced to recognize that part of what we are trying to see is blocked. Cardinal-Schubert forces us to put the pieces together—the mission schools against the foster child; the clear water (which she points out in her statement can also be used as a weapon) against the bottles of despair, the weapons of oblivion for the powerless; the artistic survival and resistance of herself and her brother against her father's defiance and his recognition that the state as keeper of the game is more destructive than the poacher.

In the same *Indigena* exhibition, Jane Ash Poitras also used mixed media, but in the form of two-dimensional collages. Her three-panel compilations, "Shaman Never Die" (1990) and "A Sacred Prayer for a Sacred Island" (1991), both foreground Native spirituality, which she further discusses in her artist's statement. As does Cardinal-Schubert, Poitras combines glyphs—horses, bison, bear—with words, photographs, newspaper clippings, and chalk overdrawing. Her clippings deal directly with issues such as Oka and Meech Lake, but her glyphs and historical photographs link these closely to five hundred years of cultural survival. While her imagery is not as explicitly land-based as Cardinal-Schubert's—perhaps because she grew up in Edmonton rather than on the land—her third panel in "A Sacred Prayer for a Sacred Island" includes "A New and Accurate Map of the World" from about the seventeenth century, which of course does not show the northern Great Plains nor what we now call Alberta at all. Although it clearly shows the continents as islands (though not as sacred), it also represents the continuing misperception of Europeans and Euro–North Americans of what is here. The question is not whether the map is an accurate projection—though the fact that by twenty-first-century standards it is not definitely *raises* the question—but whether such scientific mapping can ever be commensurate to a sacred place. Above the map, as in the other two panels, are historic photographs, overdrawn with both glyphs and crosses. Despite the deliberate depthlessness of the collage—everything is melded onto the same plane—context of place remains in the backgrounds of the photographs and in the map. Poitras does not explicitly recognize "city" as "place" in the way Cardinal-Schubert paints lake and forest as "place."[16]

Like Jane Ash Poitras, Richard Wagamese was removed from the land by Children's Aid when he was still a small child, but unlike her, he spent an uneasy childhood with several non-Native foster and adoptive families in several different towns and neighbourhoods. Land and memory of land became a constant for him. "For Indians," he wrote in one of his columns, "the single most important element that defines them as individuals, bands, clans and nations is the land." Later, looking back on the summer of 1991, which he saw as a summer of hope following the Terrible Summer of 1990, he talked about experiencing the beauty of the land.

> For me, as an aboriginal person forced by circumstance to be a city dweller, it's a vital reconnection to what my people refer to as the heartbeat of the universe.
>
> . . .
>
> It's the foundation of everything, because the land is the teacher and the tool which allows us to continue to define ourselves mentally, spiritually, philosophically, and emotionally. All things are tied to it.
>
> It's not difficult to understand . . . And in this, we are all Indians.

In the penultimate essay in his collection, Wagamese writes, "I believe we become immortal through the process of learning to love the ones with whom we share this planet."[17]

A Quality of Light, Wagamese's second novel, set partly during the Terrible Summer, talks in more detail of the land, both in itself and as a trope for the kinds of human knowledge protected by tribal traditions and crucial to the survival of all those "with whom we share this planet." The present of the novel is a fictional hostage taking at the Harry Hays federal building in Calgary by Johnny Gebhardt, the militant childhood friend of the Reverend Joshua Kane, the protagonist and narrator of the novel. Johnny, though of German descent, is staking out a tribal position of solidarity with the Mohawk Warriors of Kanesatake, while Joshua, an Ojibway, seems comfortably assimilated into the Euro-Canadian mainstream. Wagamese thus vividly illustrates that loss of touch with the land and tribal ways of orienting oneself, as a human being, to the land and the universe are as destructive to non-Native as to Native North Americans. Johnny begins his healing at a traditional camp in the mountains, where he spends a winter by himself, in a tipi. *"Above it, the sky is a tremendous bowl, like a pipe bowl, the universe gathered within it. . . . The land veritably pulses with energy"* (emphasis in the original). In the manifesto, which is to be read on live television as the price of his surrender and the freedom of the hostages, Johnny writes, *"Tribalism is an expression of the needs of one honored by the whole. We are all tribal people. We all have, within our genes, the memory of tribal fires. Some of us have distanced ourselves from that memory . . . But it lies within each of us like a latent hope"* (emphasis in the original). Responsibility to the land and to the people with whom we share it is the central motif of *A Quality of Light*, though Wagamese explores many variations on it. In particular, he deals with the specific nature of imperialism and oppression of the Indigenous peoples of North America through death, displacement, and the systematic debasement of Indigenous cultures for generations of Indigenous people. On one level, *A Quality of Light* can be read as a primer on the texts and struggles of decolonization, of those who have fallen and those who have survived, and Wagamese refers specifically to writers like Harold Cardinal and his Métis contemporary Howard Adams. In allowing his Teutonic character to fulfill the role of warrior for the people, Wagamese points out that although the problems the Columbian invasion created for Native North Americans are real and distinct, they cannot be resolved without a spiritual change of the whole society to honour the needs of one segment of that society. Land is central and it provides inspiration, but it also requires human interpretation, which the book attempts to fulfill.[18]

Oka was about land. Cardinal-Schubert's lake and Poitras's islands were about land. The Columbian quincentenary was definitely about land. Wagamese's Joshua makes a joke of it, at one point introducing himself as the pastor of "St. Geronimo's parish of Our Lady of Perpetual Land Claims."[19] The significance of Columbus was not that he "discovered" America or even that he brought it to the attention of Europe. Both of those had been done long before. The significance was that he and his backers, the Spanish monarchs, began the land grab. Other discoverers had either settled, married into the people and become at home, or visited for a while and returned whence they had come. The Columbian invasion used physical force based on a supremacist ideology that granted entitlement; the invaders used systematic methods of seemingly neutral activities such as mapping and accounting that allowed for the bureaucratic stripping of both individual and community identities, and, sometimes unwittingly, brought vast armies of microbes that devastated the previously healthy peoples of the western hemisphere. Land, justice issues, and Columbus did not just happen to coincide in the pages of *Windspeaker* and in the work of the three artists discussed above. They have always been connected.

The artists and the newspaper respond acutely to Native issues. I would like to end this section, however, by focussing on three texts that are comparative overviews of the large relationships between Indigenous and whitestream philosophy and practice in Canadian society. Let us begin with a text that explicitly works from a deficiency model of Indigenous thought, examine its shortcomings, and then look at the exemplary models given of Indigenous society and how they may benefit whitestream people and their institutions. This, I believe, gives us a working analogy to the Great Plains, rooted in its millennia of occupation as a humanly satisfying environment. The flaws in the deficiency model of Indigenous North America suggest analogous flaws in the deficiency model of the Great Plains, while the actual strengths of the sufficiency models that Indigenous people hold of themselves and their land suggest analogues for "reading" the sufficiency of the Great Plains and thus potential ways of healing the woes we have been discussing. They also show us the need for Indigenous people and their philosophy in any humanly satisfying future for the Great Plains and, indeed, for all of Turtle Island and its blue-green globey earth.

Reading accounts of nineteenth- and early twentieth-century Canadian Indian policy is always troubling because it deals with immense human loss and seems to rest on a questionable premise: that Christian, Amer-European principles of economy, society, and culture are inherently superior to Cree, Blackfoot, Dakota, or other First Nations principles of economy, society, and culture. The twenty-first-century reader is apt to ask what Hayter Reed or Duncan Campbell Scott or even John A. Macdonald would say were he writing today. If we read Thomas Flanagan's *First Nations? Second Thoughts* and take it at face value, we are likely to conclude that Reed, Scott, Macdonald et al. were on the right track all along. They should have changed nothing in their beliefs—they should simply have been more consistent in carrying out their work of civilization and not indulged in the paternalism of trying to treat First Nations (or Métis) peoples any differently than any other Canadians. While I contend that Flanagan works from an interlocking series of untenable premises, I believe that the rhetorics of his presentation are very effective—even seductive—and thus help us to understand why the deficiency models of both Indigenous societies and the Great Plains still seem to make sense.

Flanagan's involvement with Native issues began on the Prairies with his careful and intelligent translations and editions of the writings of Louis Riel and with an essentially positive biography of Riel. More recently, he has changed his point of view to one more critical of Riel, which may be why the federal government contacted him to carry out research relating to Métis land claims. That research led to Flanagan's publication of his own summary of the issues in *Métis Lands in Manitoba*, a book that is in many ways a "prequel" to *Second Thoughts*. *Métis Lands* is a case study that ends with the sentence "To explain why I believe paternalism was and is not appropriate would require another publication." *Second Thoughts* is that publication, although Flanagan says in the first chapter of *Second Thoughts* that he decided to write the book in response to his perception of an "aboriginal orthodoxy," presumably that contained in the Report of the Royal Commission on Aboriginal Peoples (RCAP), which was both prevalent and, he believed, unuseful in discussing contemporary and historical Indigenous issues. Flanagan's statement of his own goals is characteristically modest and reasonable sounding: "I do not claim to say the last word on these difficult and controversial issues, only to offer some viewpoints that are seldom heard today. In particular, I do not present a plan for curing all the ills besetting aboriginal peoples. I do not believe in the validity of such plans."[20]

The power and the shortcomings of Flanagan's reasoning are illustrated clearly in *Métis Lands in Manitoba*. Here Flanagan argues that the procedures for assigning Métis land were fair and were fairly carried out for the benefit of the Métis, except in a few individual cases of fraud, which were almost all rectified in the end. Although he admits that, given the nature of the Métis economy and the way lands were parceled out, they were of virtually no use to most of the participants, he does not consider the possibility of administering the claims so as to produce the large block settlements that the Métis actually wanted in order to create their own land base. By damning anything other than a strictly market solution as "paternalism," Flanagan makes it impossible for the reader to engage in anything other than Flanagan's examination of procedures. The zebras, one might say, were given their fair share of lion meat, and they could even exchange it for grass under very fair terms. An excruciatingly careful study of the parceling out of the lion meat and of the rules for exchanging it for grass would, however, probably seem odd if it never actually mentioned that zebras do not, under most circumstances, eat lions.

Some of the *Second Thoughts* premises are not particularly significant, except from a rhetorical point of view. In the first of eight statements that Flanagan claims encapsulate "The Aboriginal Orthodoxy," he writes, *"Aboriginals differ from other Canadians because they were here first. As 'First Nations,' they have unique rights, including the inherent right of self-government"* (6). Flanagan seriously argues that First Nations and Inuit peoples were not really in Canada "first"; they moved around a lot and in some places, such as the prairies, their present locations date only from post-1492. This is simply irrelevant. A homesteader who arrived on a claim only ten minutes after another homesteader had staked it was out of luck. The movement of different groups of peoples across the continent and within regions certainly complicates the assignment of land and other rights among Native groups but has no relevance to the relation between the rights of Natives and newcomers. The current constitutions of both the United States and Canada clearly distinguish certain rights retained by Indigenous people because of their prior occupancy and sovereignty. Clearly, despite the deficiency models that the early explorers and settlers carried with them, they realized that Native people had a presence on the land and distinctive ways of land use that were systematic and understandable. As Flanagan would tell us—for word origins are an important part of his rhetoric—*prior* is from Old Latin and relates to such words as *prime* and *primitive,* derived from the prefix *pri-,* before. *Prior* means preceding in time. Native North American peoples possess prior rights to newcomer North American peoples. What, exactly, these rights may be is not clear from simple priority—but the priority exists. Nor are these rights, as Flanagan suggests, in some way "racially" based. Yes, they depend on inheritance within family lines, but so do most kinds of intergenerational property transfer. One could as well argue that all inheritance rights premised on passage from parent to child are racially based, and thus junior has no claim to inherit the family fortune. Collective rather than personal inheritance rights, however, do suggest alternative ways of understanding land ownership than the favoured fee simple of Amer-Europeans.

Other of Flanagan's summaries of pro-Aboriginal arguments are oversimplifications of extremely complex issues, as in his point number five: *"Aboriginal peoples can successfully exercise their inherent right of self-government on Indian reserves"* (7). In a world of multinational corporations, GATT, and a US president who claims the right to act unilaterally to protect US rights, all polities are too small for self-government. Reserve sovereignty is hampered by diseconomies of scale and by the lack of an economic base that does not depend on the outside for both money and goods. One can make exactly the same statement about Canada. Even the United States is not fully "sovereign," as the many protestors against having American military personnel serve under non-American United Nations commanders fervently point out. Determining both the limits and the expanses of First Nations sovereignty, like Canadian sovereignty, will require negotiation, accommodation, and change, and it will rarely if ever be defined only in terms of one reserve. Furthermore, as John Borrows argues, since most discussions involving Aboriginal rights take place at the federal level, Indigenous peoples and Indigenous principles of law must become part of general Canadian constitutional and common law, or the system will fail Indigenous and non-Indigenous persons alike.[21]

The moral heart of Flanagan's argument and the most misleading of his premises comes in his opposition to statement number two: "*Aboriginal cultures were on the same level as those of the European colonists. The distinction between civilized and uncivilized is a racist instrument of oppression*" (6). That Flanagan's wording of his oversimplification here is more grating than that of his other statements perhaps attests to his awareness that his elaboration of the argument will be untenable. Flanagan proposes the familiar Enlightenment theory that extensive agricultural societies are superior to hunter-gatherer societies because they can support more people. Since European technologies allowed two (or twenty) people to survive where only one had lived before, Europeans were justified in taking the land. Even if the land were not useful for agriculture, as in the North, other "beneficial" uses, such as eventual uranium mining, justified the taking of the land. The worldwide spread of agriculture and the organized states that it allowed to form were essentially processes like childbirth or death, floods or drought. They happen, and there is no way to even formulate the question of right or wrong. More's *Utopia* allows Utopians to set up their colonies and to assimilate the original inhabitants because all will be better off under Utopian rule. The colonists may fight and kill those of the invaded who do not wish to join the co-operative. (Flanagan acknowledges that More was writing at the beginning of English hegemony in North America but does not seem to recognize that More was speaking *for* colonization, not just coincidentally at the same time.) Subsequent philosophers have agreed that one overpopulated group may take land from another group with less technology and more land, as long as they allow the landed group to share their technology. Flanagan generously allows that the landholding group would suffer real losses in having to give up their cherished hunting-and-gathering way of life, but adds, "On the other hand, I cannot see a moral justification for telling the agriculturalists that they cannot make use of land that, from their point of view, is not being used" (44).

In his article "Civilization, Self-Determination, and Reconciliation," Michael Murphy examines the ideas of the nineteenth-century thinkers who provided the underpinnings for Flanagan's definitions of civilization: John Stuart Mill, John Locke, Immanuel Kant, and Karl Marx. He points out that, to some extent, all four asserted the inferiority of Indigenous North American society and thought to European society and thought. Thus, Flanagan can adopt their theories to proclaim that Indigenous North Americans could not have a civilization worthy of the name or with the moral weight to establish sovereignty among nations. Murphy points out that, in fact, agriculturalists were never told that they could not make use of the land: "The central moral failing of Flanagan's civilizationist paradigm of reconciliation is its unsatisfactory engagement with the question of consent."[22] Like his nineteenth-century predecessors, Flanagan is so sure that his civilization is better than the Indigenous alternatives that he does not consider the possibility that not everyone will agree. And certainly there are Indigenous people from many different tribes and walks of life who agree with him—see, for instance, William Wuttunee's 1971 book *Ruffled Feathers: Indians in Canadian Society*, an argument against Harold Cardinal's influential *Unjust Society*. Wuttunee accepts the White Paper's call for the abolition of the *Indian Act* and treaty relationships, and the levelling of Indian identity into simple Canadian citizenship along the same lines that Flanagan lays out. Flanagan also refuses to look at the loss of life, the cultural destruction, and the loss of personal freedom and autonomy that followed whitestream domination of North America.

I would add even more qualifications to Murphy's. The first would be to ask whether Indigenous hunter-gatherers could actually be assimilated into European agricultural society, as Flanagan suggests. In the case of Europeans coming to what they called the "New World," one would have to answer that in practice they did not incorporate the people they found. Something like 98 percent of the population of the Americas did not survive colonization.[23] For Canada, the population decline may not have been so precipitous—perhaps 75 to 80 percent of the Indigenous population perished. Flanagan suggests that because of the large-scale swapping of micro-organisms from animals to humans and from Asia to Africa to Europe, "civilized" people developed immunities, while "Indian cultures" were "inexperienced" and therefore died in huge numbers when exposed to these new diseases. Most such disease was not deliberate "germ warfare," but nonetheless, dead people do not have the choice to assimilate. Having to give up one's attachment to hunting and fishing is quite different than surviving (provided one did survive) the deaths from epidemics and the resultant social upheaval and starvation of 75 percent of one's community.

The assimilation assumption also requires that the colonists should welcome the hunter-gatherers, something that would not have been obvious to, among others, the Beothucks, the Plains Crees who were denied the use of agricultural equipment that alone allowed any chance of taking off a crop in the short Saskatchewan growing season, or the Blackfoot, Northern Cheyenne, and Crow ranchers who found their land leased away from them, who were prohibited from buying tractors, and who were generally harassed by the US government Indian service despite their demonstrated success at cattle raising. Certainly, it does not consider the children who were abused and died in great numbers at residential schools and continue to be abused and to die in foster care. Third, it requires that the Indigenous peoples should accept not only intensive agriculture but also a particular Protestant, European version of free market agriculture and economics in general. Variant practices that can and have worked, as shown by both economic development theory and the experiences of the "Five Civilized Tribes" in Oklahoma before allotment, are completely cancelled out of Flanagan's account.[24] In fact, the argument melts fairly dismally into an unaffected "might makes right" plaint.

Flanagan's defence of "civilization" is not actually any more useful for the "civilized" than it is for Indigenous peoples. Although he states his belief in free market economics as at least the least worst system yet devised, he admits that something better might come along.[25] Where will it come from if all competitors are ruthlessly and needlessly suppressed? Flanagan also makes occasional concessions to conservation, but if supporting more people is the highest form of land use, then all public and private parks and green spaces would appear to be unjustified. Furthermore, it would seem that if expansion is a virtually natural process, all immigration restrictions are unjustified and only partially enforceable. The idea of rights, whether derived from Locke or elsewhere, is to set up a framework other than "might makes right" so that each of us may be protected against whoever is stronger than one of us is today. Again, Flanagan negates the possibility of choice, the mainstay of democracy.

Furthermore, European humid agriculture is not always the highest use for agricultural land. The Great Plains, as it turns out, is not particularly conducive to technological agriculture. In many years, it appears that not only was Palliser right, but groups who could move with the buffalo herds and utilize different environments of the Great Plains, including its rivers and its nearby mountains, modelled a more sustainable form of agriculture than did the sedentary farmers who moved in and began ploughing at the end of the nineteenth century. If we look at the ways that monocropping, especially on the Great Plains, damages the fertility of the land, uses excessive amounts of water and unsustainable inputs of fossil fuels and fertilizers, and produces an unsaleable (Trudeau's wheat) or unhealthy (high fructose corn syrup or marbled grain-fed beef) product, we may conclude that the hunter-gatherers were right: the Great Plains produces more usable human food as a grazing, gathering, and horticultural area than as an intense monoculture. Obviously, the colonizers from the intensive agriculture and centralized state cultures with their belief in the free market did not choose to find ways to share the land of the hunter-gatherers with due respect for the integrity of those host societies, but that does not necessarily mean that such sharing was either impossible or undesirable. Nor is it impossible or undesirable to work back toward such a sharing today.

Although I find Flanagan's major moral premise untrue, his outlook is useful for understanding—and thus for countering—a number of neo-conservative positions on Indigenous issues and particular red flags to neo-conservatives and to a Liberal party that strives to dominate the centre of Canadian politics by absorbing neo-conservative ideas that seem to be gaining some public acceptance. Nepotism in band councils and reserve politics in general, for instance, is a tempting target, and Flanagan argues persuasively that small communities organized largely by family ties are particularly vulnerable to abuses of those family ties. But instead of dismantling the small communities or insisting that they must work on civil service lines designed for larger, more heterogeneous communities, how might public policy enable reserve communities to put into practice the checks and balances that the cultural heritage of the communities might suggest? We have seen cultural traditions used very effectively in place of some of the European-derived justice systems.[26] How might similarly imaginative groups address nepotism? Some of my colleagues who come from strongly clan-based cultures suggest that the clans, the men's and women's societies, and differentiated roles for women and men traditionally provided for checks and balances to nepotism and could work so again. In *Akat'stiman: A Blackfoot Framework for Decision-Making and Mediation Processes*, Reg Crowshoe and Sybille Manneschmidt painstakingly explain how the Blackfoot have adapted traditional bundle transfer ceremonies to use for decision making and mediation in such diverse fields as child welfare and business deals with the oil and gas industry.[27] Which other groups are finding successful ways to open up band decision making that do not fall back on European-style elections? What can we find in the focussed and pragmatic arguments that Deloria and Cardinal made forty years ago? How can these processes become more visible as counters to Flanagan's essentially ignorant argument that a paternalistic and sentimental government has prohibited Indigenous communities from complete immersion in "civilization," which offers the only real alternative to nepotism, cronyism, and the continuing degradation of "aboriginal people"? One could ask similar questions about Flanagan's prescriptions for economic development, resource management, and a host of other issues that are important to Indigenous communities. Although none of this may be of explicit use to our understanding of the Great Plains, it does provide us with practice in re-understanding the old deficiency arguments and moving them to ones of sufficiency.

Two books more or less contemporaneous with Flanagan's writings and considerably more imaginative and optimistic about the strengths of Indigenous North American philosophy and practice, particularly in Canada, provide a useful overview for understanding how these might counteract dysfunctional whitestream practices. Although both refer specifically to issues of criminal, common, and constitutional law—as do the news stories and artists we have already discussed—their counter to deficiency theories such as Flanagan's gives us another kind of analogy for understanding the Great Plains. Rupert Ross's *Dancing with a Ghost*, published in 1992, both the year after Flanagan's *Métis Lands* and the five hundredth anniversary of Columbus, is a pragmatic study of "Indian Reality" by a Crown attorney from Kenora who wanted to figure out why the justice system he was bringing to isolated Cree and Ojibway communities in northern Ontario was not working. Ross argues persuasively that the system is based on a "reality" so different from the traditional and formative world view of its Indigenous clients that it is literally senseless, and therefore lacking in basic human courtesy. The Western legal system, he points out, operates on a theory of "original sin," in which humans must be deterred, by fear and the threat of punishment, from doing the evil deeds prompted by base human nature. Indigenous people, he observes, work rather from a "doctrine of original sanctity," in which erring humans must be nurtured, through patient listening and counselling, to regain their natural balance in the universe. The proper response to crime, then, is not punishment and exclusion, but comforting and inclusion. Ross argues that this, as well as other aspects of Indigenous philosophy and practice, arise from a subsistence lifestyle but offer necessary corrections to current whitestream philosophy and practice deriving from a technological and highly individuated way of life.

John Borrows's *Recovering Canada: The Resurgence of Indigenous Law*, published two years after Flanagan's *Second Thoughts*, makes a similar case for common and constitutional law. Working from the premise that "one should not found a just country on stolen land and repressive government," Borrows argues that Canada cannot respect itself without living up to the responsibilities guaranteed in the treaties made when Indigenous people were the majority in the land. He expands the Supreme Court's holding in *Delgamuukw* that oral traditions be permissible in court by suggesting that oral tradition has functioned in the same way as common law to shape society and belief, and that it ought to be given the same weight in court. Thus he presents Nanabush (Anishinaabe Trickster) stories that can be analyzed in the same way as other legal precedents for understanding and putting into action Native law regarding such things as resource utilization. He also points out, as we mentioned in our discussion of Flanagan's ideas about sovereignty, that since all meaningful decisions about land use—and about recognizing land as a citizen—*are* made at the federal level, any meaningful Aboriginal sovereignty must include joint federal sovereignty.[28] Thus, while neo-conservative whitestream political philosophy attempts to pin us to nineteenth-century theory, reading Native news, art, law, and philosophy within an Indigenous context provides ethical, intellectual, and even spiritual and emotional alternatives to what we have. As Ross shows, Indigenous philosophy does not focus on the ills of the past but rather on the rebalancing necessary for the future.

How all this connects to the Great Plains, however, may not be intuitively obvious. Ursula Le Guin wrote a famous short story called "The Ones Who Walk Away from Omelas" (1974). It is about an isolated utopian city where peace and plenty abound and all is fair and beautiful—except that somewhere at the centre, there is a broken and deprived child whose existence is the antithesis of all the beauty. Yet upon her continuing deprivation, everyone's happiness depends. Those who walk away refuse to benefit from her destitution. A just society cannot be based on a fundamental and arbitrary injustice, yet as we have seen, Great Plains society (like all Western societies) rests not on historic dispossession of Indigenous peoples but on a present and continuing dispossession that has been dealt with by isolating and ignoring the people in the hope that they will simply vanish, by forced assimilation and marginalization in whitestream society, and by paternalistic and ineffective "welfare" interventions. None of these work. In most Indigenous philosophies, the people *are* the land. Whitestream society, especially on the Great Plains, stands to benefit from walking away from a concept of the land and its people that is based in deficiency and a punitive notion of restoration. We need paradigms and responsibilities that stem from the land and not just from the theories of Western Enlightenment that are engendered by another environment.

Conclusion

The Great Plains is my home. It is where my son was born and where my grandparents are buried. I have spent my career living on and teaching about the Great Plains. I own homes in Nebraska and Alberta, and perform my own annual migration north and west to a higher elevation every summer and south and east and down every fall. I want to live out my life in this region and to see it provide homes and lifework for my son and his children to be. Although issues relating to climate change may have a more disastrous effect on parts of the globe other than the Great Plains, I believe that at present, my region has neither a sustainable economy nor an aesthetic that will produce either a sustainable economy or a humanly satisfying way of living. This book has been about the choices we have made in the past and the implicit and explicit arguments behind those choices. Now, I believe, it is time to look at how we might think about constructing a plausible and positive future. Certainly, groups like the Parkland Institute, the Pembina Institute, the Center for Rural Affairs, the Land Institute, the Quivira Foundation, and others have done excellent work in examining problems, testing solutions, and planning for positive change. I have enormous respect for them and for the education I have received from their publications and practices.

Trying to frame a satisfactory conclusion to this study, however, I have found myself drawn less directly to their work than to analogies based on studies of the failures in the provision of justice to Native persons. Looking at both the land of the Great Plains and the Indigenous people who lived there, European and Euro–North American observers, administrators, and settlers perceived deficiency where there was actual functioning sufficiency, and in both cases, the outside invaders overlooked and instrumentally suppressed both the existing systems and the innovations put forth by Indigenous societies. We have seen the deficiency theories of Thomas Flanagan, the events dealing with Native justice issues around the time of the Columbus quincentenary, and the kinds of solutions posed by *Windspeaker* authors and featured artists. Now let us use this background to try to understand what our lagging knowledge of Native justice issues might mean for this place, the Great Plains.

The events around Oka and the Columbus quincentenary led to a number of inquiries, both the artistic ones discussed in the last chapter and more formal ones that we have only mentioned in passing. What all the inquiries agreed upon was that the "justice" system was not providing justice for Aboriginal people in Canada; that from birth onward, Canadians of Aboriginal descent were more likely than other Canadians to be touched and badly served by everything from child protective services to employment services, and frequently by the police and court systems. Aboriginal people were more likely than other Canadians to be both the victims and the perpetrators of crimes, and more likely than other Canadians to be incarcerated. As the RCAP report documented, "In the Prairie region, Natives make up about 5 per cent of the total population but 32 per cent of the penitentiary population. . . . Even more disturbing, the disproportionality is growing. . . . Placed in a historical context, the prison has become for many young Native people the contemporary equivalent of what the Indian residential school represented for their parents." For the last fifteen or twenty years (I cannot remember exactly when I began), I have been volunteering with Aboriginal groups in prisons in Nebraska and with ex-cons who have served their time, so these questions are not only clearly in my consciousness and shaping my view of the world, but they also carry an emotional and moral imperative that is impossible to dismiss. Rupert Ross, a Crown attorney who was seconded to study Aboriginal justice in northern Ontario—a study he extended to the United States, Australia, and New Zealand—has published two books as well as various articles and position papers that serve as primers for understanding how an Aboriginal justice system can work, and in some places, *is* working. Ross's description of how Aboriginal science studies things in context and thus can often provide better and more complex solutions to problems than more linear and technological science coincides with James Scott's evaluations of the indispensability of informal, experiential land-based knowledge in any kind of development. Ross emphasizes the idea of wholeness in most Aboriginal societies, which means reconciliation, not punishment or retribution. Instead of dividing the "victim" and "victimizer" as opposing entities, Aboriginal justice sees both as parts of a wounded community. Neither can heal unless both are healed and balance is restored. Ross points out that an adversarial justice system intensifies anger rather than defusing it, and even the presumption of innocence, so basic to Western liberal democracies and enshrined in the United Nations Universal Declaration of Human Rights (#11), can lead to denial of guilt instead of one's taking personal responsibility for harmful actions. Holistic healing circles have been used successfully in some Aboriginal communities to foster responsibility and restoration by engaging victim and perpetrator in the context of an understanding but also demanding community.[1]

Despite all the studies, we are only beginning to identify what is broken in the rightly vaunted British justice tradition as applied to Aboriginal peoples worldwide. We have hardly begun to identify the problems, let alone offer possible solutions, for the increasingly unworkable Euro–North American perception of the Great Plains, where fewer and fewer people grow unmarketable crops —or crops that promise, a bit wishfully, to assuage energy dependence—at huge environmental cost. If we try to apply the restorative principles developed in the justice systems to the Great Plains, what might we see? We must acknowledge that the Great Plains is not so obviously broken. While some farmers and ranchers feel that something is wrong with the system, others are quite pleased with their own successes or are confident that they will continue to expand and to succeed. Others see problems but internalize them, feeling that they are to blame for not keeping up the prosperity of the farm—especially if it has been in the family for a number of generations. We have agreed that mad cow disease, the most recent rural bogeyman, is acceptable on a low level as long as the more obviously whacko cows do not get into the food or feed chains. While some American ranchers strive to halt live cattle and beef imports from Canada, meat packers and government animal health experts insist that North American beef is all equally safe—and they are probably even right. We have only begun to address the effects of energy production on the Plains and the potential effects of global warming.

Although rural populations continue to decline, Euro–North American families losing the farm and moving to town are not as visible as Native people who are incarcerated. Many farmers are content to sell the land and move to town, and even those who have mixed feelings or are reluctant feel that they still have agency and at least some control over the decision. North American farm families blend into the cities culturally and educationally, and usually do not face ethnic or racial job discrimination. The supermarkets and fast food joints are stuffed with things to eat, and stuffed North Americans grow fatter and fatter, rarely noticing that the foods available to them are grown and produced far away and that much of the cost of food is for excessive processing or for transportation from half-way around the globe, not a payment to the farmer. Many farm people love and respect the land and value a way of life that allows them to be working outside and relying on nature to ripen the crop or feed the animals. Yet if all land is sacred, abandoned city lots are nature, too, and invasive English sparrows chirp quite endearingly. Prairie cities often have beautiful parks to comfort homesick farm folk, including the linear groves of the rivers (if one ignores the homeless people living under the bridges, another sign of the failure of the regional, as well as the national, economy to sustain all of society). Even street lighting can be directed downward so one can see the stars from the middle of the city, and besides, rural skies are polluted by various kinds of security lighting, especially if there are extractive industries nearby. Yet like the Aboriginal justice system, the mode of living on the

Great Plains is broken because it is based on a model of deficiency instead of a model of strength. The Great Plains does not have to be transformed to be useful or acceptable. Nor does the Great Plains of today have to be transformed *back* to Buffalo Commons to be viable, any more than Indigenous people have to recapture a lost and nostalgic past. As John Borrows says, to relegate Native rights only to aspects of life that have remained the same from pre-contact days is to deny the resiliency and flexibility of Native traditions to deal with post-contact issues.[2]

Humans exist, a fact of great importance to the humans, if not necessarily to the universe or even the particular biosphere we might call Earth or Turtle Island. If humans are to continue to exist, they will have to depend on an intact biosphere with earth, air, and water. Most current agriculture on the Great Plains is extractive and industrial, heavily dependent on petrochemical fuels, fertilizers, and pesticides. It relies on monocropping, which implies the extermination of biodiversity. Ironically, but not entirely coincidentally, prisons are also monocrops, requiring uniforms to designate inmates and to distinguish them from staff and visitors. As industries, prisons are very highly sought after by small Great Plains cities for the employment base that they provide. What are the models that all the studies have provided for Aboriginal justice, and how might we understand them in terms of the Great Plains? Let us list some qualities common to these models: (1) land-based; (2) restorative; (3) community-centred; (4) decentralized; (5) holistic. Obviously these are interlocking rather than separate, but let us look at them one at a time.

Except in science fiction, human communities have never existed without a particular *land* base. Most proposals for Aboriginal justice systems require community sovereignty of some sort, which implies a regional association, based on people living not only *on* a specific plot of land but *with* the land as a meaningful aspect of community. European systems of land use are not problem free, as one can see by problems of pollution in both Western and especially Eastern Europe, and by European rural depopulation. European Union agricultural policy has for the most part protected small (by Great Plains standards) farms and farmers, and has accepted agricultural surpluses to enable a cheap food policy. European animal rights groups have been more successful than those in the United

States and Canada in requiring adequate space and freedom of movement for food animals, and the bovine spongiform encephalopathy (BSE) and foot-and-mouth crises—and the subsequent widespread destruction of ruminant herds—have shocked Europeans even more than North Americans to move away from "unnatural" practices such as feeding sheep carcass renderings to cattle. Most important, despite bureaucratic attempts at control and uniformity, modern European agriculture developed *in Europe* in response to European land and climate, and was specialized by country and even by region, as can be seen in the European Union's rather draconian product labelling. Names derived from place names—Dijon, Champagne, Newcastle—cannot be used as generics.

Great Plains agriculture, as we have seen, is imposed and is as often defined by wishful thinking as by a sober estimation of the land and climate. We need to ask what the land does well, how to work with its strengths, and what we would like to see. Buffalo Commons is one possibility—and one would have to be emotionally dead not to stir at the image of the shaggy rivers flowing again over hundreds of miles—but it is only partial. Wes Jackson's experiments in re-establishing small communities in the Kansas Flint Hills and Nebraska's School at the Center project are other, still largely unfulfilled, possibilities intended to teach people, especially children, how to live productively and successfully on the land. Although the perennial grains that The Land Institute has been breeding would still be grown as partially diversified monocrops (since the actual variety of the tallgrass prairie is not attainable), they would provide for better cover for both wildlife and the land itself than crops that must be planted and tilled each year. Repurchase of lands from Saskatchewan farmers by Saskatchewan First Nations bands attempting to re-establish a land base is successfully refloating some regional economies for the time being and represents another possibility, as do the various successful enterprises of Ho-Chunk Inc. in northeastern Nebraska. The Ho-Chunk or Winnebago people have used their casino earnings to invest in regional businesses, such as gas stations and motels on the nearby interstate highways, at the same time as they are building up their buffalo herds to offer employment and cultural inspiration to young people and to provide nutritious, low-fat meat to Winnebago people at risk for diabetes.[3] Unlike Buffalo Commons, these latter solutions envisage twenty-first-century humans living in a conversation with the land, neither leaving the area nor becoming solely guides for ecotourism— not that ecotourism should not be a part of the economic mix. It is easy to satirize all these movements as utopian anachronisms that merely seek to invert nineteenth-century ideas of "progressives" and "traditionals," but none of these ideas is any more anachronistic than the twenty-first-century use of wind turbines to generate electricity.[4] Because the Great Plains is not like Europe in either climate or soil, and because it has not co-evolved with European people, animals, or crops, a future land-use system has to be based on a close study of what this land does well, not on how it can be made to behave more like the well-watered eastern regions of North America or like Europe.

That brings us to our second principle, the *restorative* nature of the future of the Great Plains. Restoration is the major principle of all Aboriginal justice systems. The focus is not on accusation or retribution or even "justice." Rather, it is on the restoration of balance to the community, of safety to the victim, of responsibility to the perpetrator, and of the strength to intercede in the community. As we learned from James Malin many years ago, restoration of the Great Plains or any other ecosystem to some past utopia or climax vegetation is not possible; it is not, in any particulars, even imaginable. Restoration here means, as it does in all the plans for justice, getting everyone to the point of working together for the future. How might federal and state/provincial tax and land-use policies promote population on the land? What kinds of plants have co-evolved with the land and how can they and their values be enhanced? What would happen if the grazing of domestic ungulates or captive buffalo were regulated to more closely resemble the grazing patterns of wild buffalo? What is the value of grass-fed cattle in preventing outbreaks of *E. coli* in beef? Can grass-fed free-ranging cattle avoid the pollution of feedlots and cut down on the ploughing, irrigation, pesticides, and possibly genetically modified seeds needed for feed crops? To what extent have grazing operations that can enhance grasslands become captives to the feedlots that are dumping grounds for the excess grain production that degrades grasslands? Does range feeding cattle enhance animal welfare? Range management that mimics the relationship of buffalo to the pastures is also labour intensive; could such restorative ranching stimulate sustainable population growth on the Great Plains and provide the basis for population elasticity in the creation of regional business and communication networks? Could increased labour costs be recouped by cutting the feedlot stage out of the meat-producing process? Hay could replace feed grains and relieve some excess production. Food grains, oil seed, and pulses such as wheat, canola, and dried beans could fill horticultural niches. Petroleum extraction could continue with safeguards for land, water, and air. Buffalo, elk, and other animals could begin to re-establish parts of their historic ranges, as well as their predators: wolves, cougars, and grizzly bears. Ecotourism would become a feasible part of the mix, especially if its proceeds indemnified ranchers who lost domesticated animals to the predators.

Humans are a large part of this mix, whether they be Native, non-Native, or in the process of developing an ethic of place. This brings us to the idea of *community*. Looking back to our models in the reinvention of Aboriginal justice systems, we see that no one can be "cured" unless everyone is cured. This is exactly why we see Aboriginal justice form healing "circles," where everyone is vitally engaged in working out a problem. To some extent, of course, our meaningful community is the entire globe. As we well know, social injustice or bombings in Afghanistan or Iraq affect the whole world, including the Great Plains. Depressed young people—whether reluctantly signing up for the army in sparsely populated South Dakota farm or reservation communities, or huffing gasoline on northern reserves, or joining Asian drug gangs in Calgary, or exploding themselves on London subways and buses, or simply feeling themselves unable to craft meaningful lives within an engaged community—are not only a danger to themselves and others, but also signal that something is terribly wrong, that disengagement from land, community, family, and self call for a systemic healing, not punishment or even rehabilitation that focusses only on the individuals who are alienated.

Contemporary rhetoric extolling the "family values" of small rural Great Plains towns seems to call on the idea of community, but as Thomas Frank has shown, it tends to lead to political behaviour that actually destroys community. A recent CBC radio exploration of small towns in Nebraska clearly shows this dichotomy. A farmer points out that the economics of farming have become untenable, with only one crop a year making a profit while three or four others offer a loss, and with farm prices not having kept up with inflation, particularly in inputs like energy. The farmer says that price supports only help out the largest farmers, while people like him are squeezed out of the business. Meanwhile, boosters in the small town of Superior put their hopes on their small-town moral values and their rock-ribbed Republicanism, not noticing that the leaders they elect are the ones forcing the family farms out of business and driving their potential market out of the local county. They tell the reporter that it is international markets and global progress that are putting the squeeze on the town, not recogniz-ing that their elected officials, particularly on the national level, are the ones determining the rules of the markets and of international progress. This is exactly the political de-skilling that Roger Epp discusses. The enormously energetic, hopeful, and hard-working boosters focus on "values" that have little effect on their lives—gay marriage and even abortion are not likely to change Superior any time soon—and that are to some extent mythical, as the discussion of the rise of crystal methamphetamine production and addiction in the county, raised by the reporter, suggests.[5] Planning based on doubtful premises and completely ignoring the mechanics of the economic squeeze is simply not going to work.

While it is clear that meaningful community planning must be an ongoing grassroots process, Prairie populism has never been particularly successful—with the partial exception of the CCF in Saskatchewan—usually because it has arisen from a single-minded ideology, whether it be the fet-tering of the railroads, grain elevators, and land speculators advocated by the Populists of the 1890s or the unfettering of private enterprise and gov-ernment capitalism advocated by the Reform Party in the 1990s. The chal-lenge of engaging a community that includes rural and urban areas, and Native, long-resident, and newly arrived populations, and that addresses issues from agriculture to child rearing is not only daunting but unheard of. The Royal Commission on Aboriginal Peoples (RCAP) spent five years and millions of dollars on a brilliant, if not perfect, study of one relatively small population in one country in response to fairly clear and definable stim-uli such as Oka, the Aboriginal justice inquiries, and high rates of youth suicide. Even then the report has languished with no sustained attempt to meet the ambitious schedule for innovation put forward. There is no evidence of a widespread will to attempt a study and redefinition of the Great Plains, though individual outfits such as the Center for Rural Affairs, the Parkland Institute, the Grassland Foundation, and Ho-Chunk Inc. are all supporting ongoing study and innovation.

My own sense, judging only from what has been published and from the people who pass through my classrooms or with whom I otherwise interact in Lincoln and Calgary, is that Aboriginal communities are more hopeful and innovative than non-Native rural communities, which are less likely to be propelled by a sense of absolute necessity and more likely to embrace whitestream norms of progress and the rational depopulation of the Great Plains. On the other hand, the rates of despair, substance abuse, violence, incarceration, and unemployment on many reserves and reservations are so high as to be almost life denying. Still, as we saw in the chapter on planning, innovation is most likely to come from those whose struggle to survive is precarious, not from those who are comfortable. And Native communities are definitely the most precarious. The small rural towns and farming populations are certainly willing to envision both smaller and larger versions of Great Plains community as the central focus for their own lives and for encountering the world.

While there is no particular point in large numbers of communities working separately to reinvent the wheel, the Aboriginal justice models we have been trying to follow do depend on a large degree of community autonomy and on decentralized planning and structure that includes centralized support but not governance. Again, as Scott and Ross point out, specific, contextualized knowledge that depends on gut feeling rather than on clear, articulable designs is crucial to positive change. In the justice studies, writers note that not only are urban and reserve communities very different, but there are different traditions of healing in different Aboriginal cultures. What is appropriate for a Cree may not be particularly appropriate for a Kiowa. A great deal of Plains intellectual history has focussed on rebellion against governments, whether they be in Ottawa or Washington, Edmonton, or Lincoln, or Bismarck or . . . Often, as Lorelei Hanson points out, these histories of rebellion are themselves romanticized. Albertans are delighted when author Aritha Van Herk calls them

Mavericks (and the Glenbow Museum develops a whole exhibition on the theme) and not eager to acknowledge to what extent "rebellion" is only a form of political "de-skilling."[6] As we have seen, Alberta's furious dissent from the National Energy Policy and its long-cherished grudge against Ottawa and the Liberal Party primarily benefited—and benefits—American oil companies. Similarly, current opposition to gay marriage and to abortion, and support for the death penalty in the "red" states of the US Great Plains and among the supporters of Reform/Alliance/Conservative politics in Canada, as Thomas Frank pointed out in *What's the Matter with Kansas*, however honestly intentioned, does serve to distract attention away from failures of economy, ecology, and social justice, a particularly pernicious form of de-skilling. Centralized agendas of dissent are as distracting from regional, place-based problem solving as are centralized agendas of assent.

Yet at the same time that decentralized, community-based formulations of solutions are necessary, our guiding principle is still the interconnectedness of all things and thus the insufficiency of any but *holistic* solutions. Inability to perform one task at a time is, of course, a recipe for dithering. Successful problem solving usually begins with defining what issue is the most bothersome, and then moving wider out, like ripples, to find the connections and to explore them. Aboriginal justice programs always begin with some limited jurisdiction, be it domestic and family court issues, juvenile justice, or the equivalents of municipal courts. Starting with communities, then, means that there will be many different "first problems," including those usually denominated "personal morality," "social justice," "economic," or "ecological." The more specifically and passionately each can be articulated—traced backward and forward from origins to desired outcomes—the more apparent nodes of interconnection will become, just as the justice inquiries found linkages by looking closely at individual cases. Only then did underlying assumptions about what justice systems were supposed to do come into obvious conflict with both physical conditions and Aboriginal philosophy. The assumption that a child caught in vandalism should be remanded and charged, for instance, was simply impractical in northern communities remote from remand centres, and it contradicted Aboriginal emphasis on the individual's taking responsibility for his or her own actions. Although the English justice system seems to set the greatest value on personal responsibility, it actually negates that responsibility from the point of view of an Aboriginal system based on connectedness. Taking personal responsibility for one's actions and working toward the mitigation of the harm one may have caused is diametrically opposed to being adjudged guilty by an outside source and punished for the harm one may have caused. Connections are hard to come by in an adversarial system. Think of the little warning printed on your insurance policy or on the proof of insurance card you carry in your car. It tells you to deny responsibility for a collision, even when you know you are in the wrong.

Great Plains farmers continue to leave the land. Those who stay manage larger and larger spreads dependent on government support that encourages consolidation and monocropping and on chemical fertilizers, pesticides, and genetically modified organisms. Or the farmers' main income comes from leases and easements from petroleum companies—bringing risks of environmental degradation, sour gas wells, and the dangers of sharing small gravel roads with heavy drilling and exploration equipment. Rural Great Plains communities are losing their ability to organize for their own economic benefit and are instead railing against elites and framing their arguments in extremely black and white "moral" terms. Tellingly, these "moral" terms never include issues such as poverty or social justice. The communities that once passionately supported leaders like Tommy Douglas and George Norris seem estranged from their own roots, despite the research and leadership of organizations such as the Parkland Institute or the Center for Rural Affairs. Similarly, reserve and reservation communities try to establish workable sovereignty in the context of a larger political system that requires a different kind of "democracy" from that of Aboriginal tradition, while urban Native people are disproportionately alienated and stigmatized in a vicious circle that keeps turning upon itself. Sovereignty cannot be confined to reserves and reservations when almost all higher level political decisions are made at the state, provincial, or federal level.[7]

Most solutions proposed by politicians who perceive that something is wrong on the Great Plains maintain the point of view of nineteenth- and early twentieth-century mainstream politicians—to some extent replicated in the "Second Thoughts" of new right politicians of the early twenty-first century—that Christian, Amer-European principles of economy, society, and culture are inherently superior to Indigenous principles of economy, society, and culture. James Malin maintained that the contriving brain of the human would always find ways of recognizing new layers of usefulness in any environment. Democracy and free market economics are not automatic utopias, as government scandals and market crises reliably remind us. Even their most ardent defenders can only claim that they are the least worst systems that humans have as yet devised. But if all alternatives are ruthlessly repressed, as they have been in the recent past of the Great Plains, how can new and better systems develop?

For most of the thousands of years of human home making on the Great Plains, human groups could move, like the buffalo herds, to utilize different environments, including riverine valleys and nearby mountains. They modelled a more sustainable form of agriculture than did the sedentary farmers who moved in and began ploughing and irrigating at the end of the nineteenth century. Obviously, the newcomers from the intensive monocultural agriculture and centralized states with their belief in the free market and their acceptance of fee simple ownership of square surveyed plots of land did not choose to find ways to share the land of the hunter/gatherer/horticulturalists with due respect for the integrity of those host societies. That does not necessarily mean that such sharing was either impossible or undesirable. We can continue to produce more and more surplus grain on bigger and bigger farms with fewer and fewer people and more and more water, herbicides, insecticides, and petroleum-sourced fertilizers and energy. And with federal subsidies in the United States and disaster payments in both countries. We can continue to burn corn as ethanol, to feed it to pigs in confinement sheds that create whole cities' worth of excrement, to feed grain to cattle in feedlots knee deep in muck, or to demand that our federal governments sell our grain abroad. We can continue to depopulate our rural areas and eventually our regional towns and cities. We can lose the last vestiges of native grass prairie, even as we abandon human habitations for vast ecotourist theme parks.

Or we can do something else.

Notes

Introduction

1. See Stan Rowe, *Home Place: Essays on Ecology*, rev. ed. (Edmonton: NeWest Press, 2002), 178.
2. See Robert Diffendahl, "Plate Tectonics, Space, Geologic Time, and the Great Plains: A Primer for Non-Geologists," *Great Plains Quarterly* 11, no. 2 (Spring 1991): 81–102; Norman Rosenberg, "Climate of the Great Plains Region of the United States," *Great Plains Quarterly* 7, no. 1 (Winter 1987): 22–33; Kathy Keeler, "Influence of Past Interactions on the Prairie Today: A Hypothesis," *Great Plains Research* 10, no. 1 (Spring 2000): 107–26; Don Gayton, *The Wheatgrass Mechanism: Science and Imagination in the Western Canadian Landscape* (Saskatoon: Fifth House, 1990), 125–33.
3. Mary K. Stillwell coined the word "grassful" as an antonym to "treeless," so often used for the Great Plains. Mary K. Stillwell, personal communication, September 2002.
4. See Linea Sundstrom,"The Sacred Black Hills: An Ethnohistorical Review," *Great Plains Quarterly* 17, no. 3/4 (Summer/Fall 1997): 185–212; N. Scott Momaday, *The Way to Rainy Mountain* (New York: Ballantine Books, 1970), 6–7; Gilbert L. Wilson, *Buffalo Bird Woman's Garden: Agriculture of the Hidatsa Indians* (1917; rept., St. Paul: Minnesota Historical Society Press, 1987).
5. See Theodore Binnema, *Common and Contested Ground: A Human and Environmental History of the Northwestern Plains* (Toronto: University of Toronto Press, 2004), 18–61; Trevor R. Peck and J. Rod Vickers, "Buffalo and Dogs: The Prehistoric Lifeways of Aboriginal People on the Alberta Plains, 1004–1005," in *Alberta Formed, Alberta Transformed*, ed. Michael Payne, Donald Wetherell, and Catherine Cavanaugh (Edmonton: University of Alberta Press, and Calgary: University of Calgary Press, 2005), 1:55–86. See also Ella Deloria, *Waterlily* (Lincoln: University of Nebraska Press, 1988).
6. See Waldo Wedel, "Coronado, Quivira, and Kansas: An Archeologist's View," *Great Plains Quarterly* 10, no. 3 (Summer 1990): 139–51; Germaine Warkentin, ed., *Canadian Exploration Literature: An Anthology* (Toronto: Oxford University Press,

1993), 194–242; Bernard De Voto, ed., *The Journals of Lewis and Clark* (Boston: Macmillan, 1953), 18–22; Henry Nash Smith, *Virgin Land: The American West as Symbol and Myth* (1950; rept., Cambridge: Harvard University Press, 2007); Patricia Nelson Limerick, *Legacy of Conquest: The Unbroken Past of the American West* (New York: W.W. Norton, 1987); Douglas Owram, *Promise of Eden: The Canadian Expansionist Movement and the Idea of the West, 1856-1900* (Toronto: University of Toronto Press, 1992).

7. See Paul Sharp, *Whoop-Up Country: The Canadian-American West, 1865-1885* (St. Paul: University of Minnesota Press, 1955); Seymour Martin Lipset, *The First New Nation: The United States in Historical Comparative Perspective* (New York: W.W. Norton, 1979); White, *"It's Your Misfortune and None of My Own": A New History of the American West* (Norman: University of Oklahoma Press, 1993); Limerick, *Legacy of Conquest*; Gerald Friesen, *The Canadian Prairies: A History* (Toronto: University of Toronto Press, 1984); Roger Epp, "The Political De-skilling of Rural Communities," in *Writing Off the Rural West: Globalization, Governments, and the Transformation of Rural Communities*, ed. Roger Epp and Dave Whitson (Edmonton: University of Alberta Press, 2001), 301–24; Sarah Carter, *Lost Harvests: Prairie Indian Reserve Farmers and Government Policy* (Montreal and Kingston: McGill-Queen's University Press, 1990); Sarah Carter, *The Importance of Being Monogamous: Marriage and Nation Building in Western Canada to 1915* (Edmonton: University of Alberta Press and Athabasca University Press, 2008); Barbara Belyea, *Dark Storm Moving West* (Calgary: University of Calgary Press, 2007); Paul Voisey, *Vulcan: The Making of a Prairie Community* (Toronto: University of Toronto Press, 1988); James M. Pitsula and Ken Rasmussen, *Privatizing a Province: The New Right in Saskatchewan* (Vancouver: New Star Books, 1990); James M. Pitsula, "The Thatcher Government in Saskatchewan and the Revival of Métis Nationalism, 1964–71," *Great Plains Quarterly* 17, no. 3/4 (Fall/Summer 1997): 213–36; Angie Debo, *And Still the Waters Run: The Betrayal of the Five Civilized Tribes* (Princeton: Princeton University Press, 1940); James C. Malin, *History and Ecology: Studies of the Grassland*, ed. Robert Swierenga (Lincoln: University of Nebraska Press, 1984); John Joseph Mathews, *Wah'Kon-Tah: The Osage and the White Man's Road* (Norman: University of Oklahoma Press, 1932); John Joseph Mathews, *The Osages: Children of the Middle Waters* (Norman: University of Oklahoma Press, 1961); Hamlin Garland, *A Son of the Middle Border* (New York: Macmillan, 1936).

8. "Treaty Land Entitlement," Peguis First Nation website, www.peguisfirstnation.ca/tle.html, accessed 23 June 2009.

9. "'Kill the Indian, and Save the Man': Capt. Richard H. Pratt on the Education of Native Americans," *History Matters: The U.S. Survey Course on the Web*, www.historymatters.gmu.edu/d/4929/, accessed 18 February 2011.

10. John G. Neihardt, *Black Elk Speaks* (1932; rept., Lincoln: University of Nebraska Press, 1961), 276. But cf. Raymond Demallie, *The Sixth Grandfather: Black Elk's Teachings Given to John G. Neihardt* (Lincoln: University of Nebraska Press, 1985), which shows that the "death of a dream" speech did not originate from anything Black Elk said.

1 A Unified Field Theory of the Great Plains

1. Janine Brodie, "The Concept of Region in Canadian Politics," in *Federalism and Political Community: Essays in Honour of Donald Smiley*, ed. David P. Shugarman and Reg Whitaker (Peterborough, ON: Broadview Press, 1989), 33–53, quotation on 36 (emphasis in the original).

2. See Robert Diffendahl, "Plate Tectonics, Space, Geologic Time, and the Great Plains: A Primer for Non-Geologists," *Great Plains Quarterly* 11, no. 2 (Spring 1991): 81–102; Norman Rosenberg, "Climate of the Great Plains Region of the United States," *Great Plains Quarterly* 7, no. 1 (Winter 1987): 22–33.

3. Theodore Binnema, *Common and Contested Ground: A Human and Environmental History of the Northwestern Plains* (Toronto: University of Toronto Press, 2004), 38–39, 49.

4. See Harold Innis, *The Fur Trade in Canada: An Introduction to Canadian Economic History*, rev. ed. (Toronto: University of Toronto Press, 1956; orig. ed. 1930); J.M.S. Careless, *Frontier and Metropolis: Regions, Cities, and Identities in Canada Before 1914* (Toronto: University of Toronto Press, 1989); Paul Voisey, *Vulcan: The Making of a Prairie Community* (Toronto: University of Toronto Press, 1988); William Cronon, *Nature's Metropolis: Chicago and the Great West* (New York: W.W. Norton, 1991).

5. Linea Sundstrom, "The Sacred Black Hills: An Ethnohistorical Review," *Great Plains Quarterly* 17, no. 3/4 (Summer/Fall 1997): 185–212; David Wishart, ed., *Encyclopedia of the Great Plains* (Lincoln: University of Nebraska Press, 2004), 555, 557–58; Paul Gates, *Fifty Million Acres: Conflicts over Kansas Land Policy, 1854-1890* (Ithaca, NY: Cornell University Press, 1954).

6. Gerald Friesen, *The Canadian Prairies: A History* (Toronto: University of Toronto Press, 1984), 66–128; Wishart, *Encyclopedia of the Great Plains*, 346–47; N. Scott Momaday, *The Way to Rainy Mountain* (New York: Ballantine Books, 1970); John R. Wunder, *The Kiowa* (New York: Chelsea House, 1989).

7. See Dan Flores, "Bison Ecology and Bison History Redux," chap. 3 in *The Natural West: Environmental History in the Great Plains and Rocky Mountains* (Norman: University of Oklahoma Press, 2003), 50–70; Andrew C. Isenberg, "Toward a Policy of Destruction: Buffaloes, Law, and the Market, 1803–83," *Great Plains Quarterly* 12, no. 4 (Fall 1992): 227–41; Andrew C. Isenberg, *The Destruction of the Bison: An Environmental History, 1750-1920* (Cambridge: Cambridge University Press, 2001); Ken Zontek, "Hunt, Capture, Raise, Increase: The People Who Saved the Bison," *Great Plains Quarterly* 15, no. 2 (Spring 1995): 133–49; Ken Zontek, *Buffalo Nation: American Indian Efforts to Restore the Bison* (Lincoln: University of Nebraska Press, 2007); James C. Malin, *History and Ecology: Studies of the Grassland*, ed. Robert Swierenga (Lincoln: University of Nebraska Press, 1984); Don Gayton, *The Wheatgrass Mechanism: Science and Imagination in the Western Canadian Landscape* (Saskatoon: Fifth House, 1990); John E. Weaver, *Grasslands of the Great Plains* (Lincoln: Johnsen Pub., 1956). See also Geoff Cunfer, *On the Great Plains: Agriculture and Environment* (College Station: Texas A&M University Press, 2005), 43–45.

8. Momaday, *Way to Rainy Mountain*, 10.

9. Warren Elofson, *Cowboys, Gentlemen and Cattle Thieves: Ranching on the Western Frontier* (Montreal and Kingston: McGill-Queen's University Press, 2000), 15, 20.

10. Howard Lamar, ed., *The Reader's Encyclopedia of the American West* (New York: Harper and Rowe, 1977), 290.

11. See Sarah Carter, *Lost Harvests: Prairie Indian Reserve Farmers and Government Policy* (Montreal and Kingston: McGill-Queen's University Press, 1990), and *Aboriginal People and Colonizers of Western Canada to 1900* (Toronto: University of Toronto Press, 1999), 169–70; Gerhart Ens, *Homeland to Hinterland: The Changing Worlds of the Red River Métis in the Nineteenth Century* (Toronto: University of Toronto Press, 1996), 9–28; James C. Scott, *Seeing Like a State: How Certain Schemes to Improve the Human Condition Have Failed* (New Haven: Yale University Press, 1998), 33–45; Gerald Friesen, *River Road: Essays on Manitoba and Prairie History* (Winnipeg: University of Manitoba Press, 1996), 64; Hannah Samek, *The Blackfoot Confederacy, 1880–1920: A Comparative Study in Canadian and U.S. Indian Policy* (Albuquerque: University of New Mexico Press, 1987), 119; Peter Elias, *The Dakota of the Canadian Northwest: Lessons for Survival* (Regina: Canadian Plains Research Center, 2002), 118–22.

12. Robert Fogel, *The Union Pacific Railroad: A Case in Premature Enterprise* (Baltimore: Johns Hopkins Press, 1960); Don McLean, *1885: Métis Rebellion or Government Conspiracy?* (Winnipeg: Pemmican Publications, 1985).

13. Paul Gates, *The Illinois Central Railroad and Its Colonization Work* (Charlottesville: University of Virginia Press, 1934).

14. See Irene M. Spry, "The Tragedy of the Loss of the Commons in Western Canada," in *As Long as the Sun Shines and the Water Flows: A Reader in Canadian Native Studies*, ed. Ian A.L. Getty and Antoine S. Lussier (Vancouver: UBC Press, 1983), 203–28; Charles Gore, *Regions in Question: Space, Development Theory and Regional Policy* (London: Methuen, 1984), 105; Voisey, *Vulcan*, 135–40, 214–15.

15. Patricia Nelson Limerick, *Legacy of Conquest: The Unbroken Past of the American West* (New York: W.W. Norton, 1987), esp. 35–54.

16. Cronon, *Nature's Metropolis*, 97–147.

17. Dan Morgan, *Merchants of Grain: The Power and Profits of the Five Giant Companies at the Center of the World's Food Supply* (New York: Viking, 1979), 150–51, 182–83.

18. The image of "straight, dark rows" is from the Stan Rogers's song "The Field Behind the Plow," on the *Northwest Passage* compact disc (Dundas, ON: Fogarty's Cove Music, 1981).

19. See Walter Stewart, *Shrug: Trudeau in Power* (Toronto: New Press, 1971), 41–42.

20. Stan Rowe, *Home Place: Essays on Ecology*, rev. ed. (Edmonton: NeWest Press, 2002), 189; Raj Patel, *Stuffed and Starved: Markets, Power and the Hidden Battle for the World's Food System* (Toronto: HarperCollins, 2007), 90–92, 49–51.

21. David Pimentel and Tad W. Patzek, "Ethanol Production Using Corn, Switchgrass, and Wood; Biodiesel Production Using Soybean and Sunflower," *Natural Resources Research* 14, no. 1 (March 2005): 65–76.

22. David Jones, *Empire of Dust: Settling and Abandoning the Prairie Dry Belt* (Edmonton: University of Alberta Press, 1987); Vernon Carstensen, "*The Plow That Broke the Plains*: Film Legacy of the Great Depression," in *Americans View Their Dust Bowl Experience*, ed. Vernon Carstensen, John Wunder, and Frances W. Kaye (Boulder: University Press of Colorado, 1999), 303–20.

23. Friesen, *Canadian Prairies*, 188–89; Richard White, "*It's Your Misfortune and None of My Own": A New History of the American West* (Norman: University of Oklahoma Press, 1993), 263, 370–77.

24. Frank Popper, *The Politics of Land Use Reform* (Madison: University of Wisconsin Press, 1981); Deborah and Frank Popper, "The Great Plains: From Dust to Dust," *Planning* 53 (1987): 12–18.

25. Rowe, *Home Place*, esp. 20–21, 200; Cunfer, *On the Great Plains*, 6–7, 14, 234; Robert Fletcher, *Free Grass to Fences: The Montana Cattle Range Story* (New York: University Publishers, 1960), 149–50.

26. Rowe, *Home Place*, ix; Wes Jackson, *Becoming Native to This Place* (New York: Counterpoint, 1996), 12–13, x.

27. Blair Stonechild and Bill Waiser, *Loyal till Death: Indians and the North-West Rebellion* (Calgary: Fifth House, 1997), 55–59; Angie Debo, *And Still the Waters Run: The Betrayal of the Five Civilized Tribes* (1940; rept., Princeton: Princeton University Press, 1968), 21, 22, 132; see also Angie Debo, *The Road to Disappearance: A History of the Creek Indians* (Norman: University of Oklahoma Press, 1941).

28. Michael Lawson, *Dammed Indians: The Pick-Sloan Plan and the Missouri River Sioux, 1944–1980* (Norman: University of Oklahoma Press, 1982); Paul VanDevelder, *Coyote Warrior: One Man, Three Tribes, and the Trial That Forged a Nation* (Lincoln: University of Nebraska Press, 2005).

29. White, "*It's Your Misfortune*," 507, 488–89; "The Smithsonian and the Enola Gay," an Airforce Association Special Report (Arlington, VA: Aerospace Education Foundation, 2004), available at www.airforce-magazine.com/, Article Collections, Enola Gay Controversy.

30. Preston Manning, *The New Canada* (Toronto: Macmillan, 1992), 126.

31. Peter Lougheed, "The People of Alberta Are the Owner of the Reserve," interview with Adam Radwanski, *The Globe and Mail*, 19 June 2009, www.theglobeandmail.com; David G. Wood, *The Lougheed Legacy* (Toronto: Key Porter, 1985), 122; Milton Friedman, quoted in Robert Sherrill, *The Oil Follies of 1970–1980: How the Petroleum Industry Stole the Show (and Much More Besides)* (Garden City, NY: Anchor/Doubleday, 1983), 33.

32. James M. Pitsula and Ken Rasmussen, *Privatizing a Province: The New Right in Saskatchewan* (Vancouver: New Star Books, 1990); Margaret Laurence, *The Diviners* (Toronto: McClelland and Stewart, 1974); Francis Fukuyama, *The End of History and the Last Man* (New York: Free Press, 2006).

33. John Richards and Larry Pratt, *Prairie Capitalism: Power and Influence in the New West*, Canada in Transition Series (Toronto: McClelland and Stewart, 1979), 72, 156; cf. Sherrill, *Oil Follies*, 90; Roger Epp, "The Political De-skilling of Rural

Communities," in *Writing Off the Rural West: Globalization, Governments and the Transformation of Rural Communities*, ed. Roger Epp and Dave Whitson (Edmonton: University of Alberta Press, 2001), 304; Thomas Frank, *What's the Matter with Kansas? How Conservatives Won the Heart of America* (New York: Henry Holt, 2005).

2   Exploring the Explorers

1. William H. Goetzmann, *Exploration and Empire: The Explorer and the Scientist in the Winning of the American West* (New York: Alfred A. Knopf, 1966), 199.

2. Mary Louise Pratt, *Imperial Eyes: Travel Writing and Transculturation* (London: Routledge, 1992), 32–39, 51. See also Dean Neu and Richard Therrien, *Accounting for Genocide: Canada's Bureaucratic Assault on Aboriginal People* (Winnipeg: Fernwood Books, 1993), 96.

3. Goetzmann, *Exploration and Empire*, 53, 107.

4. See, for instance, Sylvia Van Kirk, *Many Tender Ties: Women in Fur-Trade Society, 1670–1870* (Winnipeg: Watson and Dwyer, 1980), 4, 8, 28–29, 53.

5. Goetzmann, *Exploration and Empire*, 328, 169, 177, 165.

6. Ibid., 328.

7. Douglas Owram, *Promise of Eden: The Canadian Expansionist Movement and the Idea of the West, 1856–1900* (Toronto: University of Toronto Press, 1980), 117.

8. Germaine Warkentin, ed., *Canadian Exploration Literature* (Toronto: Oxford University Press, 1993), 152. See also, in the same volume, Matthew Cocking, "An Adventurer from Hudson Bay: Journal of Matthew Cocking, from York Factory to the Blackfoot Country, 1772–1773," ed. Lawrence J. Burpee (orig. publ. *Transactions of the Royal Society of Canada*, 3rd ser., vol. 2, sec. 2 [1908]: 89–121), 206–19.

9. Barbara Belyea, *A Year Inland: The Journal of a Hudson's Bay Company Winterer* (Waterloo: Wilfrid Laurier University Press, 2000), 40.

10. Barbara Belyea, "Mapping the Marias: The Interface of Native and Scientific Cartographies," *Great Plains Quarterly* 17, no. 3/4 (Summer/Fall 1997): 165–84. See also Barbara Belyea, *Dark Storm Moving West* (Calgary: University of Calgary Press, 2007); Mark Warhus, *Another America: Native American Maps and the History of Our Land* (New York: St. Martin's Press, 1997); James P. Ronda, "'A Chart in His Way': Indian Cartography and the Lewis and Clark Expedition," *Great Plains Quarterly* 4, no. 2 (Spring 1984): 81–90; G. Malcolm Lewis, "Indian Maps: Their Place in the History of Plains Cartography," *Great Plains Quarterly* 4, no. 2 (Spring 1984): 91–108; G. Malcolm Lewis, ed., *Cartographic Encounters: Perspectives on Native American Mapmaking and Map Use* (Chicago: University of Chicago Press, 1998), esp. chap. 1, "Frontier Encounters in the Field: 1511–1925," by G. Malcolm Lewis (9–32), and chap. 6, "Inland Journeys, Native Maps," by Barbara Belyea (135–56).

11. Lewis, "Indian Maps."

12. Amos Bad Heart Bull and Helen Blish, *A Pictographic History of the Oglala Sioux* (Lincoln: University of Nebraska Press, 1967).

13. Goetzmann, *Exploration and Empire*, 10; Warkentin, *Canadian Exploration Literature*, 195.

14. David Thompson, *David Thompson's Narrative, 1784–1812*, ed. Richard Glover (orig. publ. Toronto: The Champlain Society, 1962), in Warkentin, ed., *Canadian Exploration Literature*, 217–19, 208.

15. Ibid., 194–95.

16. Nelson wrote extensively about the beliefs of the northern Anishinaabeg and Cree peoples and of the ways those beliefs appeared in the everyday life of the people. See Jennifer S.H. Brown and Robert Brightman, eds., *"The Orders of the Dreamed": George Nelson on Cree and Northern Ojibwa Religion and Myth* (Winnipeg: University of Manitoba Press, 1988).

17. Thompson, *David Thompson's Narrative*, in Warkentin, *Canadian Exploration Literature*, 290.

18. Ibid., 111. In the headnote to the entry on Samuel Hearne, which includes the attack at Bloody Falls, in Russell Brown, Donna Bennett, and Nathalie Cooke, eds., *An Anthology of Canadian Literature in English*, rev. and abr. ed. (Toronto: Oxford University Press, 1990), the editors write that "although not free from cultural bias, Hearne provides coolly dispassionate accounts of the native peoples he encountered, which have remained valuable to ethnographers and stand in sharp contrast to earlier portrayals of Indians that idealized them or saw them as irredeemable savages" (18). Ian MacLaren, however, noting that none of the particularly memorable and gory details appeared in Hearne's original fieldnotes and that the most gory event happens in explicit contradiction to the description in the fieldnotes, suggests that Hearne or his editor may have invented Gothic horrors that would definitely increase the narrative's reader appeal. The final, heart-wrenching account that was published—after Hearne's death—cannot be regarded as eyewitness ethnography. See Ian MacLaren, "Samuel Hearne's Accounts of the Massacre at Bloody Fall, 17 July 1771," *ARIEL: A Review of International English Literature* 22, no. 1 (1991): 25–51.

19. Owram, *Promise of Eden*, 12.

20. James Ronda, *Lewis and Clark Among the Indians* (Lincoln: University of Nebraska Press, 1984), and James Ronda, "Exploring the Explorers: Great Plains Peoples and the Lewis and Clark Expedition," *Great Plains Quarterly* 13, no. 2 (Spring 1993): 81–90.

21. Matt Jones, conversations with the author, Lincoln, NE, 2006–7.

22. Thompson, *David Thompson's Narrative*, in Warkentin, ed., *Canadian Exploration Literature*, 206, 211.

23. See, for instance, Linea Sundstrom, "The Sacred Black Hills: An Ethnohistorical Review," *Great Plains Quarterly* 17, no. 3/4 (Summer/Fall 1997): 185–212.

24. See, for instance, George F.G. Stanley, "As Long as the Sun Shines and Water Flows: An Historical Comment," in *As Long as the Sun Shines and Water Flows*, ed. Ian Getty and Antoine Lussier (Vancouver: UBC Press, 1983), 1–28.

25. Goetzmann, *Exploration and Empire*, 303.

26. Ibid., 314, 321, 322, 468, 498, 496.

27. Henry Youle Hind, *Narrative of the Canadian Red River Exploring Expedition of 1858* (orig. publ. London: Longman, Green, Longman and Roberts, 1860), in Warkentin, ed., *Canadian Exploration Literature*, 427, 428, 433, 426–27. Salt cedar or tamarisk is an invasive Eurasian plant that grows on riverbanks and in other moist areas in the Great Plains and US Southwest. It uses large amounts of water relative to native plants and leaves deposits of salt around it, both behaviours that inhibit the growth of native plants. See "Weed of the Week": Saltcedar, 10 March 2006, na.fs.fed.us.

28. See John Palliser, *The Papers of the Palliser Expedition*, ed. Irene M. Spry (orig. publ. Toronto: The Champlain Society, 1968), in Warkentin, ed., *Canadian Exploration Literature*, 441; Owram, *Promise of Eden*, 66.

29. Owram, *Promise of Eden*, 159–61.

30. Donald Creighton, *John A. Macdonald: The Young Politician* (1952; rept., Toronto: University of Toronto Press, 1998), 60–67.

31. Howard Lamar, ed., *The Reader's Encyclopedia of the American West* (New York: Harper and Rowe, 1977), 749–50.

32. James H. Howard, *The Canadian Sioux* (Lincoln: University of Nebraska Press, 1984), 34–38.

33. Alan Brinkley, *The Unfinished Nation: A Concise History of the American People*, vol. 1, 5th ed. (New York: McGraw-Hill, 2006; orig. ed. 1993), 423–24.

34. See, for instance, Maggie Siggins, *Riel: A Life of Revolution* (Toronto: HarperCollins, 1994), 204–5; J.M. Bumsted, *Louis Riel v. Canada: The Making of a Rebel* (Winnipeg: Great Plains Publications, 2001), 174–76.

35. Elizabeth Cady Stanton, Susan B. Anthony, and Matilda Joslyn Gage, *The History of Woman Suffrage* (1882; rept., New York: Arno Press, 1969), 2:545.

36. Owram, *Promise of Eden*, 162.

37. I am obliged to an anonymous reader for pointing out how this term, so common in Canadian geography, where longitudinal lines run ever closer together, presupposes deficiency.

3   Spiritual and Intellectual Resistance to Conquest, Part 1

1. Richard Slotkin, *Regeneration Through Violence: The Mythology of the American Frontier, 1600–1860* (1973; rept., Norman: University of Oklahoma Press, 2000), 57–93; Douglas Owram, *Promise of Eden: The Canadian Expansionist Movement and the Idea of the West, 1856–1900* (Toronto: University of Toronto Press, 1980), 130–35; Olive P. Dickason, *The Myth of the Savage and the Beginning of Colonialism in the Americas* (Edmonton: University of Alberta Press, 1984).

2. Gerald Vizenor, *Manifest Manners: Narratives on Postindian Survivance* (Lincoln: University of Nebraska Press, 1999); *Narrative Chance: Postmodern Discourse on Native American Indian Literature* (Norman: University of Oklahoma Press, 1993).

3. Evan Connell, *Son of the Morning Star: Custer and the Little Bighorn* (New York: North Point Press, 1984); Elizabeth B. Custer, *"Boots and Saddles": Or, Life in Dakota with General Custer* (1885; rept., Norman: University of Oklahoma Press, 1962).

4. Preston Manning, *The New Canada* (Toronto: Macmillan, 1992), 20.

5. Maggie Siggins, *Riel: A Life of Revolution* (Toronto: HarperCollins, 1994); J.M. Bumsted, *Louis Riel v. Canada: The Making of a Rebel* (Winnipeg: Great Plains Publications, 2001).

6. See, for instance, James Welch and Paul Stekler, *Killing Custer: The Battle of the Little Bighorn and the Fate of the Plains Indians* (New York: W.W. Norton, 1994); Theresa Gowanlock and Theresa Delaney, *Two Months in the Camp of Big Bear*, ed. Sarah Carter (Regina: Canadian Plains Research Center, 1999); George Woodcock, *Gabriel Dumont: The Métis Chief and His Lost World* (Edmonton: Hurtig, 1975); Brian Dippie, *Custer's Last Stand: The Anatomy of an American Myth* (Lincoln: University of Nebraska Press, 1976), esp. 34–37.

7. Rex Deverell, *Beyond Batoche*, in *Deverell of the Globe: Selected Plays*, ed. Don Perkins (Edmonton: NeWest Press, 1989).

8. For an additional perspective, see Owram, *Promise of Eden*, 92.

9. See Blair Stonechild and Bill Waiser, *Loyal till Death: Indians and the North-West Rebellion* (Calgary: Fifth House, 1997).

10. Frits Pannekoek, *A Snug Little Flock: The Social Origins of the Riel Resistance of 1869–70* (Winnipeg: Watson and Dwyer, 1991), 97–98; Owram, *Promise of Eden*, 82, 85.

11. Sylvia Van Kirk, *Many Tender Ties: Women in Fur-Trade Society, 1670–1870* (Norman: University of Oklahoma Press, 1980).

12. Louis Riel, *The Diaries of Louis Riel*, ed. and trans. Thomas Flanagan (Edmonton: Hurtig, 1976), 146.

13. Van Kirk, *Many Tender Ties.*

14. Maggie Siggins, *Marie-Anne: The Extraordinary Life of Louis Riel's Grandmother* (Toronto: McClelland and Stewart, 2008).

15. For the basic facts of Riel's life, see Siggins, *Riel*, and Bumsted, *Louis Riel v. Canada*, and also earlier studies of Riel, including George Stanley, *Louis Riel* (Toronto: Ryerson, 1963); Joseph Kinsey Howard, *Strange Empire: The Story of Louis Riel* (New York: William Morrow, 1952); Thomas Flanagan, *Louis "David" Riel: Prophet of the New World* (Toronto: University of Toronto Press, 1979).

16. See Thomas Flanagan, *Riel and the Rebellion: 1885 Reconsidered* (Saskatoon: Western Producer Prairie Books, 1983).

17. Heather Devine, "New Light on the Plains Métis: The Buffalo Hunters of Pembinah, 1870–71," in *The Long Journey of a Forgotten People: Métis Identities and Family Histories*, ed. David W. McNab and Ute Lischke (Waterloo: Wilfrid Laurier University Press, 2007), 207–11.

18. *Dictionary of Canadian Biography Online*, vol. 11, 1881–1890, s.v. "Norquay, John," by Gerald Friesen, www.biographi.ca.

19. Thomas Flanagan, *Métis Lands in Manitoba* (Calgary: University of Calgary Press, 1991); D.N. Sprague, *Canada and the Métis, 1869-1885* (Waterloo: Wilfrid Laurier University Press, 1988), 92–122.

20. Siggins, *Riel*, 121.

21. See, for instance, Jill St. Germaine, *Indian Treaty Making Policy in the United States and Canada, 1867-77* (Lincoln: University of Nebraska Press, 2001), 111, 121, 134–37; John L. Tobias, "Canada's Subjugation of the Plains Cree, 1879-1885," in *Sweet Promises: A Reader on Indian-White Relations in Canada*, ed. J.R. Miller (Toronto: University of Toronto Press, 1991), 212–42.

22. Owram, *Promise of Eden*, 130–35; Dick Harrison, "Introduction," in *Best Mounted Police Stories* (Edmonton: University of Alberta Press, 1978), 1–18; Andrew Graybill, *Policing the Great Plains: Rangers, Mounties, and the North American Frontier, 1875-1910* (Lincoln: University of Nebraska Press, 2008), esp. 15, 21, 38.

23. Frances W. Kaye, *Hiding the Audience: Viewing Arts and Arts Institutions on the Prairies* (Edmonton: University of Alberta Press, 2003), 205.

24. Howard Lamar, *The Reader's Encyclopedia of the American West* (New York: Harper and Row, 1977), 671–72, 1162–63.

25. Dippie, *Custer's Last Stand*, 10–11.

26. See St. Germaine, *Indian Treaty Making*, 139–52.

27. James H. Howard, *The Canadian Sioux* (Lincoln: University of Nebraska Press, 1984), 34.

28. Dippie, *Custer's Last Stand*, 7.

29. Douglas D. Scott, *Archaeological Perspectives on the Battle of the Little Bighorn* (Norman: University of Oklahoma Press, 2000).

30. Dippie, *Custer's Last Stand*, 90–91; Louis Warren, *Buffalo Bill's America: William Cody and the Wild West Show* (New York: Random House, 2005), 170–74.

31. Kingsley Bray, *Crazy Horse: A Lakota Life* (Norman: University of Oklahoma Press, 2006), 385–89.

4   Spiritual and Intellectual Resistance to Conquest, Part 2

1. Gilles Martel, "Les Indiens dans la pensée messianique de Louis Riel," in *Louis Riel and the Métis*, ed. A.S. Lussier (Winnipeg: Pemmican Publications, 1979).

2. Michael Hittman, *Wovoka and the Ghost Dance*, exp. ed. (Lincoln: University of Nebraska Press, 1990); Robert Utley, *The Last Days of the Sioux Nation* (New Haven: Yale University Press, 1963).

3. Alice Beck Kehoe, *The Ghost Dance: Ethnohistory and Revitalization* (Fort Worth: Holt, Rinehart and Winston, 1989), 114.

4. William Dean Howells, *The Leatherwood God* (1916; rept., Bloomington: Indiana University Press, 1976).

5. George Stanley, *Louis Riel* (1963; rept., Toronto: McGraw-Hill Ryerson, 1985), 214–16, 226; Maggie Siggins, *Riel: A Life of Revolution* (Toronto: HarperCollins,

1994), 246; *Catholic Encyclopedia*, s.v. "Bourget, Ignace," accessed 6 July 2009, www.newadvent.org/cathen/b-ce.htm.

6. James Mooney, *The Ghost-Dance Religion and the Sioux Outbreak of 1890* (1896; rept., Lincoln: University of Nebraska Press, 1991), 716–91; Kehoe, *Ghost Dance*, 32; Russell Thornton, *We Shall Live Again: The 1870 and 1890 Ghost Dance Movements as Demographic Revitalization* (Cambridge: Cambridge University Press, 1986), 5–6; Hittman, *Wovoka*, 63–106.

7. Utley, *Last Days*, 71–72; Gary C. Anderson, *Sitting Bull and the Paradox of Lakota Nationhood* (New York: HarperCollins, 1996), 135.

8. Thornton, *We Shall Live Again*, 17; Blair Stonechild and Bill Waiser, *Loyal till Death: Indians and the North-West Rebellion* (Calgary: Fifth House, 1997), 52–53; Anderson, *Sitting Bull*, 152; Kehoe, *Ghost Dance*, 17.

9. Kehoe, *Ghost Dance*, 130–32; Siggins, *Riel*, 286–87.

10. Stanley, *Louis Riel*, 233–51; J.M. Bumsted, *Louis Riel v. Canada: The Making of a Rebel* (Winnipeg: Great Plains Publications, 2001), 239.

11. Bumsted, *Louis Riel*, 248.

12. Ibid., 249.

13. Stanley, *Louis Riel*, 292–314; Louis Riel, *The Diaries of Louis Riel*, ed. and trans. Thomas Flanagan (Edmonton: Hurtig, 1976).

14. Stanley, *Louis Riel*, 319.

15. Anderson, *Sitting Bull*, 88–89.

16. Ibid., 150–51.

17. Stonechild and Waiser, *Loyal till Death*, 66, 134–36.

18. Utley, *Last Days*, 112, 103; Charles Alexander Eastman, *From Deep Woods to Civilization* (1916; rept., Lincoln: University of Nebraska Press, 1977), 97–98.

19. American Horse, quoted in Eastman, *From Deep Woods*, 94.

20. George Woodcock, *Gabriel Dumont* (Edmonton: Hurtig, 1976); Manfred Mossman, "The Charismatic Pattern: Canada's Riel Rebellion of 1885 as a Millenarian Protest Movement," in *Louis Riel: Selected Readings*, ed. Hartwell Bowsfield (Toronto: Copp Clark Pitman, 1988), 240.

21. Thomas Flanagan, *Riel and the 1885 Rebellion Reconsidered*, 2nd ed. (Toronto: University of Toronto Press, 2000), 22; Don McLean, *1885: Métis Rebellion or Government Conspiracy?* (Winnipeg: Pemmican Publications, 1985); Stanley, *Louis Riel*, 315–16.

22. Wilfrid Laurier, House of Commons, *Debates*, 16 March 1886, quoted in John Ralston Saul, *Reflections of a Siamese Twin: Canada at the End of the Twentieth Century* (Toronto: Penguin, 1997), 216.

23. Utley, *Last Days*, 112.

24. General Nelson A. Miles, quoted in Mooney, *Ghost-Dance Religion*, 816.

25. Stanley, *Louis Riel*, 222.

26. Ignace Bourget to Louis Riel, 14 July 1875, quoted in Stanley, *Louis Riel*, 222.

27. Siggins, *Riel*, 247.

28. Edmond Mallet, quoted in Stanley, *Louis Riel*, 222.

29. Thomas Flanagan, *Louis "David" Riel: Prophet of the New World* (Toronto: Goodread Biographies, 1979), 11; Hittman, *Wovoka*, 75–77.

30. Hittman, *Wovoka*, 51; Kehoe, *Ghost Dance*, 33.

31. Mooney, *Ghost-Dance Religion*, 1060.

32. Eugene Buechel, *Lakota Tales and Texts*, ed. Paul Manhart (Pine Ridge, SD: Red Cloud Lakota Language and Cultural Center, 1978), 277–313; Mooney, *Ghost-Dance Religion*, 953, 1065–66.

33. Riel, *Diaries*, 66.

34. Louis Riel, *Collected Writings of Louis Riel/Les écrits complets de Louis Riel*, ed. George G.F. Stanley et al. (Edmonton: University of Alberta Press, 1985), 2:376.

35. Sarah Carter, *Aboriginal People and Colonizers of Western Canada to 1900* (Toronto: University of Toronto Press, 1999), 123; Peter Douglas Elias, *The Dakota of the Canadian Northwest: Lessons for Survival* (Regina: Canadian Plains Research Center, 2002), esp. 72–73; Riel, *Diaries*, 165, 52.

36. Stanley, *Louis Riel*, 288; Mossman, "Charismatic Pattern," 234; Riel, *Diaries*, 163–65.

37. Bumsted, *Louis Riel*, 249; Albert Braz, *The False Traitor: Louis Riel in Canadian Culture* (Toronto: University of Toronto Press, 2003), 31; Stanley, *Louis Riel*, 230.

38. Mooney, *Ghost-Dance Religion*, 763; Hittman, *Wovoka*, 84–85, 58–60; Leslie Marmon Silko, *Gardens in the Dunes* (New York: Simon and Schuster, 2000); Elaine Pagels, *Beyond Belief: The Secret Gospel of Thomas* (New York: Random House, 2003), 32.

39. Birger A. Pearson, *Gnosticism, Judaism, and Egyptian Christianity* (Minneapolis: Fortress Press, 1990), 7–8.

40. Mooney, *Ghost-Dance Religion*, 786; cf. John Norton, *The Journal of Major John Norton, 1816*, ed. Carl Klinck and James Talman (Toronto: The Champlain Society, 1970); James Mooney, *Myths of the Cherokee* (1900; rept., www.sacred-texts.com/ nam/cher/motc/, scanned January–February 2001), 1; James Walker, *Lakota Myth*, ed. Elaine Jahner (Lincoln: University of Nebraska Press, 1980), 8.

41. Pagels, *Beyond Belief*, 123; Mooney, *Ghost-Dance Religion*, 1061-62, for instance.

42. Pagels, *Beyond Belief*, 139, 84, 17–18; George E. Tinker, speech at "Written in the Stars: Osages and Their Literature," Pawhuska, OK, 17 May 2003.

43. Pagels, *Beyond Belief*.

44. Mooney, *Ghost-Dance Religion*, 928–45.

45. Captain H.L. Scott to Indian Office, 10 February 1891, quoted in Mooney, *Ghost-Dance Religion*, 897.

46. James McLaughlin to Indian Office, 17 October 1890, quoted in ibid., 787.

47. Pagels, *Beyond Belief*, 133–35.

48. John G. Neihardt, *Black Elk Speaks: Being the Life Story of a Holy Man of the Oglala Sioux* (1932; rept., Lincoln: University of Nebraska Press, 1961), 276; cf. Raymond J. DeMallie, ed., *The Sixth Grandfather: Black Elk's Teachings Given to John G. Neihardt* (Lincoln: University of Nebraska Press, 1984), 282.

49. Stanley, *Louis Riel*, 370; Stonechild and Waiser, *Loyal till Death*, 235.

50. Siggins, *Riel*, 365.

51. Carter, *Aboriginal People*, 123, 139–40, 152–54; Stonechild and Waiser, *Loyal till Death*, 50–53.

52. Collingwood Schreiber to Charles Tupper, n.d. [June 1885], Tupper papers, vol. 6, National Archives of Canada, quoted in Pierre Berton, *The Last Spike: The Great Railway, 1881–1885* (Toronto: McClelland and Stewart, 1971), 401.

53. See John Tobias, cited in Carter, *Aboriginal People*, 150.

54. Letitia Hargrave to Mrs. Dugald Mactavish, 28 March 1849, *The Letters of Letitia Hargrave*, ed. Margaret Arnett McLeod (orig. publ. Toronto: The Champlain Society, 1947), in Germaine Warkentin, ed., *Canadian Exploration Literature* (Toronto: Oxford University Press, 1993), 420.

55. Helen Anne English, "The Journals of Helen Anne English, Field Matron on the Little Pine Reserve, 1913–1917," *Saskatchewan History* 45, no. 2 (1993): 39. In English's defence, however, one must mention that she did see her Cree neighbours suffering horribly from tuberculosis and other diseases and frequently noted that most families were going hungry.

56. Diane Paulette Payment, *"The Free People—Otipemisiwak": Batoche, Saskatchewan, 1870–1930* (Hull, QC: Parks Service, Environment Canada, 1990), 311.

57. DeMallie, *Sixth Grandfather*, 282; Thornton, *We Shall Live Again*, 17–19; Kehoe, *Ghost Dance*, 106, 39; Hittman, *Wovoka*, 136–42.

5   Spiritual and Intellectual Resistance to Conquest, Part 3

1. Most of the material about Mathews in this paragraph comes from Terry P. Wilson, "Osage Oxonian: The Heritage of John Joseph Mathews," *Chronicles of Oklahoma* 59 (Fall 1981): 264–93.

2. John Joseph Mathews, *Talking to the Moon* (Chicago: University of Chicago Press, 1945), 221.

3. John Joseph Mathews, *The Osages: Children of the Middle Waters* (Norman: University of Oklahoma Press, 1961), 698–99.

4. John Joseph Mathews, *Wah'Kon-Tah: The Osage and the White Man's Road* (Norman: University of Oklahoma Press, 1932), 19; hereafter cited parenthetically in the text.

5. John Joseph Mathews, *Life and Death of an Oilman: The Career of E.W. Marland* (Norman: University of Oklahoma Press, 1951), 165–93.

6. Wilson, "Osage Oxonian," 265.

7. Julian Rice, *Black Elk's Story: Distinguishing Its Lakota Purpose* (Albuquerque: University of New Mexico Press, 1991); David C. Young, "Crazy Horse on the Trojan Plain: A Comment on the Classicism of John G. Neihardt," *Classical and Modern Literature* 3 (1982): 45–53.

8. Garrick A. Bailey, *The Osage and the Invisible World: From the Works of Francis La Flesche* (Norman: University of Oklahoma Press, 1995), 76–221.

9. Ibid., 18–21.

10. Wilson, "Osage Oxonian," 280–83.

11. Bailey, *Osage and the Invisible World,* 5.

12. Robert Allen Warrior, *Tribal Secrets: Recovering American Indian Intellectual Traditions* (Minneapolis: University of Minnesota Press, 1995), 103.

6   Intellectual Justification for Conquest

1. Carol Higham and Robert Thacker, eds., *One West, Two Myths,* vol. 1, *A Comparative Reader* (Calgary: University of Calgary Press, 2004); *One West, Two Myths,* vol. 2, *Essays on Comparison* (Calgary: University of Calgary Press, 2007); Beth Ladow, *The Medicine Line: Life and Death on a North American Frontier* (New York: Routledge, 2001); Sheila McManus, *The Line Which Separates: Race, Gender, and the Making of the Alberta Borderlands* (Edmonton: University of Alberta Press, and Lincoln: University of Nebraska Press, 2005); Andrew Graybill, *Policing the Great Plains: Rangers, Mounties, and the Frontier, 1875-1910* (Lincoln: University of Nebraska Press, 2007).

2. Harold A. Innis, *The Fur Trade in Canada: An Introduction to Canadian Economic History,* rev. ed. (1930; Toronto: University of Toronto Press, 1956); J.M.S. Careless, "Frontierism, Metropolitanism, and Canadian History," *Canadian Historical Review* 35, no. 1 (March 1954): 1–21; Donald Creighton, *The Commercial Empire of the St. Lawrence* (Toronto: Ryerson Press, 1937); William Cronon, *Nature's Metropolis: Chicago and the Great West* (New York: W.W. Norton, 1991).

3. Frederick Jackson Turner, "The Significance of the Frontier in American History," in *The Frontier in American History* (New York: H. Holt, 1920), 1.

4. W.J. Eccles, *The Canadian Frontier, 1534-1760* (Albuquerque: University of New Mexico Press, 1974).

5. Turner, "Significance of the Frontier," 12.

6. Ibid., 4.

7. Innis, *Fur Trade,* 392.

8. Ibid., 262; Douglas Owram, *Promise of Eden: The Canadian Expansionist Movement and the Idea of the West, 1856-1900* (Toronto: University of Toronto Press, 1980).

9. Owram, *Promise of Eden;* Henry Nash Smith, *Virgin Land: The American West as Symbol and Myth* (Cambridge, MA: Harvard University Press, 1950).

10. John Herd Thompson, *Forging the Prairie West* (Toronto: Oxford University Press, 1998), 21; *Dictionary of Canadian Biography Online,* vol. 5., 1801–1820, s.v. "Semple, Robert," by Hartwell Bowsfield, www.biographi.ca; George Bryce, *The Romantic Settlement of Lord Selkirk's Colonists: The Pioneers of Manitoba* (Winnipeg: Clark Bros., 1909; orig. ed. Toronto: Musson Book Co., 1909; also available at www.gutenberg.org/ebooks/17358), 133.

11. Richard Slotkin, *Regeneration Through Violence: The Mythology of the American Frontier, 1600-1860* (Middletown, CT: Wesleyan University Press, 1973).

12. Walter Prescott Webb, *The Great Plains* (New York: Ginn and Co., 1931); Ray Allen Billington, *America's Frontier Heritage* (New York: Holt, Rinehart and Winston, 1966); Arthur S. Morton, *A History of the Canadian West to 1870-71*, rev. ed. (1939; Toronto: University of Toronto Press, 1967).

13. Patricia Limerick, *The Legacy of Conquest: The Unbroken Past of the American West* (New York: W.W. Norton, 1987); Richard White, *"It's Your Misfortune and None of My Own": A New History of the American West* (Norman: University of Oklahoma Press, 1991); Gerald Friesen, *The Canadian Prairies: A History* (Toronto: University of Toronto Press, 1984).

14. Limerick, *Legacy of Conquest*, 292.

15. See White, *"It's Your Misfortune,"* 518, 561–72.

16. Angie Debo, *And Still the Waters Run: The Betrayal of the Five Civilized Tribes* (Princeton: Princeton University Press, 1940); Michael Lawson, *Dammed Indians: The Pick-Sloan Plan and the Missouri River Sioux, 1944–1980* (Norman: University of Oklahoma Press, 1982); Linda Hogan, *Mean Spirit* (New York: Ivy Books, 1992); Thomas King, *Green Grass, Running Water* (Boston: Houghton Mifflin, 1993).

17. Friesen, *Canadian Prairies*, 47.

7 Homesteading as Capital Formation on the Great Plains

1. Mark Twain (Samuel L. Clemens) and Charles Dudley Warner, *The Gilded Age: A Tale of Today* (1873; rept., New York: New American Library, 1980; also available at www.gutenberg.org/ebooks/3178); "An Indian Swindle: History of the Ottawa (Kansas) University Project—How a Tribe Was Cheated Out of Their Lands," *New York Times*, 22 May 1871.

2. Robert Fogel, *The Union Pacific Railroad: A Case in Premature Enterprise* (Baltimore: Johns Hopkins Press, 1960); John W. DeForrest, *Honest John Vane* (New Haven, CT: Richmond and Patten, 1875); "The Pacific Scandal" (last modified 20 September 2005), www.collectionscanada.gc.ca/decret-executif/023004-3052-e.html.

3. Paul Wallace Gates, *The Jeffersonian Dream: Studies in the History of American Land Policy and Development*, ed. Allan G. Bogue and Margaret Beattie Bogue (Albuquerque: University of New Mexico Press, 1996), 52; Robert Stanfield, *Hearings Before the Senate Committee on Public Lands and Surveys on Grazing Facilities on Public Lands*, 1926, p. 101, quoted in Paul Wallace Gates, *History of Public Land Law Development* (Washington: USGPO, 1968), 479.

4. "An Act Respecting the Public Lands of the Dominion of Canada, 1872," in *Statutes of Canada (Acts of the Parliament of the United Kingdom of Great Britain and Ireland . . . Being the First, Second, Third, and Fourth Sessions of the Twentieth Parliament of the United Kingdom)* (Ottawa: Brown Chamberlin, 1872), 56–92 (chap. 23); Wilcomb Washburn, *The Assault on Indian Tribalism: The General Allotment Law (Dawes Act) of 1887* (Philadelphia: Lippincott, 1975).

5. Dan Jaffe, *Dan Freeman* (Lincoln: University of Nebraska Press, 1967), 13, 14, 61, 63.

6. Gates, *History of Public Land*, 402; Clarence Danhof, "Farm-Making Costs and the 'Safety Valve,' 1850-1860," *Journal of Political Economy* 49 (June 1941): 317-59.

7. Quoted in Benjamin Hibbard, *A History of the Public Land Policies* (New York: Peter Smith, 1924), 363.

8. Thomas Flanagan, *Métis Lands in Manitoba* (Calgary: University of Calgary Press, 1991), 123.

9. Gates, *History of Public Land*, 454n38; Russell Errett (US Congress, House Committee on Indian Affairs, *Lands in Severalty to Indians: Report to Accompany H.R. 5038*, 46th Cong., 2d Sess., May 28, 1880, H. Rept. 1576, pp. 7-10), quoted in Washburn, *Assault on Indian Tribalism*, 39, 40.

10. Jaffe, *Dan Freeman*, 65-66.

11. Gates, *Jeffersonian Dream*, 45.

12. Ibid., 46; Hibbard, *History of the Public Land Policies*, 387.

13. Gates, *History of Public Land*, 479; Gates, *Jeffersonian Dream*, 47; David C. Jones, *Empire of Dust: Settling and Abandoning the Prairie Dry Belt* (Calgary: University of Calgary Press, 2002).

14. Gates, *Jeffersonian Dream*, 47, 52.

15. Irene M. Spry, "The Tragedy of the Loss of the Commons in Western Canada," in *As Long as the Sun Shines and Water Flows: A Reader in Canadian Native Studies*, ed. Ian Getty and Antoine Lussier (Vancouver: UBC Press, 1983), 203-38.

16. Kenneth Norrie and Douglas Owram, *A History of the Canadian Economy* (Toronto: HBJ Canada, 1991), 324-25, 322, 329, 331, 336, 360, 332.

17. Roger Epp, "1996—Two Albertas: Rural and Urban Trajectories," in *Alberta Formed, Alberta Transformed*, ed. Michael Payne, Donald Wetherell, and Catherine Cavanaugh (Edmonton: University of Alberta Press, and Calgary: University of Calgary Press, 2005), 2:725-46; Pierre Elliott Trudeau, quoted in Walter Stewart, *Shrug: Trudeau in Power* (Toronto: New Press, 1971), 39-40; Jackie Skelton on *Ideas*, CBC Radio, 27 September 2005; Raj Patel, *Stuffed and Starved: Markets, Power, and the Hidden Battle for the World's Food System* (Toronto: HarperCollins, 2007); James Scott, *Seeing Like a State: How Certain Schemes to Improve the Human Condition Have Failed* (New Haven: Yale University Press, 1998).

18. Hamlin Garland, *A Son of the Middle Border* (1917; rept., St. Paul: Minnesota State Historical Society Press, 2007), 317.

19. Paul Voisey, *Vulcan: The Making of a Prairie Community* (Toronto: University of Toronto Press, 1988), 33, 33-34, 37.

20. Paula Nelson, *After the West Was Won: Homesteaders and Town-Builders in Western South Dakota, 1900-1917* (Iowa City: University of Iowa Press, 1986).

21. Voisey, *Vulcan*, 41.

22. John Ise, *Sod and Stubble* (1936; rept., Lawrence: University Press of Kansas, 1997); Deborah Fink, *Agrarian Women: Wives and Mothers in Rural Nebraska, 1880-1940* (Chapel Hill: University of North Carolina Press, 1992), esp. 58-95.

23. Roger Epp, "The Political De-skilling of Rural Communities," in *Writing Off the Rural West: Globalization, Governments, and the Transformation of Rural*

Communities, ed. Roger Epp and Dave Whitson (Edmonton: University of Alberta Press, 2001), 301–24.

24. Scott, *Seeing Like a State*, 33–47.

25. Garland, *Son of the Middle Border*, 303.

8 The Women's West

1. Sarah Carter, *The Importance of Being Monogamous: Marriage and Nation Building in Western Canada to 1915* (Edmonton: University of Alberta Press and Athabasca University Press, 2008), 5.

2. Ibid., 59, 78; see also 74–76.

3. Ibid., 188, 203.

4. Terry Wilson, *The Underground Reservation: Osage Oil* (Lincoln: University of Nebraska Press, 1985), esp. 99–147.

5. See, for example, Lisa Emmerich, "Marguerite LaFlesche Diddock: Office of Indian Affairs Matron," *Great Plains Quarterly* 13, no. 3 (Summer 1993): 162–71; Peggy J. Blair, "Fact Sheet, Rights of Aboriginal Women On- and Off-Reserve," Vancouver, Scow Institute, 2005, www.scowinstitute.ca/library.html.

6. David Stannard, "Disease and Infertility: A New Look at the Demographic Collapse of Native Populations in the Wake of Western Contact," *Journal of American Studies* 24 (1990): 325–50; Theodore Binnema, *Common and Contested Ground: A Human and Environmental History of the Northwestern Plains* (Toronto: University of Toronto Press, 2004), 107–28; Richard White, "The Cultural Landscape of the Pawnees," *Great Plains Quarterly* 2, no. 1 (Winter 1982): 31–40; John Ewers, *The Blackfeet: Raiders on the Northwestern Plains* (Norman: University of Oklahoma Press, 1958), 99–100.

7. See Sylvia Van Kirk, *Many Tender Ties: Women in Fur-Trade Society, 1670–1870* (Norman: University of Oklahoma Press, 1983), esp. 181–230; Carter, *Importance of Being Monogamous*, 70; Daniel Heath Justice, *Our Fire Survives the Storm: A Cherokee Literary History* (Minneapolis: University of Minnesota Press, 2006), 39.

8. Anne M. Butler, *Daughters of Joy, Sisters of Misery: Prostitution in the American West, 1865–90* (Urbana: University of Illinois Press, 1985); Carter, *Importance of Being Monogamous*, 273; James H. Gray, *Red Lights on the Prairie* (Toronto: Macmillan, 1971); see also Charlene Porsild, *Gamblers and Dreamers: Women, Men, and Community in the Klondike* (Vancouver: UBC Press, 1998), 11–29; Van Kirk, *Many Tender Ties*, 160–63, 240.

9. Carroll Smith-Rosenberg, *Disorderly Conduct: Visions of Gender in Victorian America* (New York: Oxford University Press, 1985), 53–76, 254; Virginia Scharff, *Twenty Thousand Roads: Women, Movement, and the West* (Berkeley: University of California Press, 2002), 101–3.

10. See, for instance, Elinore Pruitt Stewart, *Letters of a Woman Homesteader* (1914; rept., Lincoln: University of Nebraska Press, 1989), 17–18. As Susanne George

(Bloomfield) has shown, Stewart fictionalized many of her friends and social connections, but her depictions of everyday life are extremely accurate. See Susanne K. George, *The Adventures of the Woman Homesteader: The Life and Letters of Elinore Pruitt Stewart* (Lincoln: University of Nebraska Press, 1993).

11. Joanna L. Stratton, *Pioneer Women: Voices from the Kansas Frontier* (New York: Touchstone, 1982), 258–62; Susan C. Peterson, "A Widening Horizon: Catholic Sisterhoods on the Northern Plains, 1874–1910," *Great Plains Quarterly* 5, no. 2 (Spring 1985): 125–32; Paul Voisey, *Vulcan: The Making of a Prairie Community* (Toronto: University of Toronto Press, 1988).

12. Carter, *Importance of Being Monogamous*, 75; photograph reproduced in John Herd Thompson, *Forging the Prairie West* (Toronto: Oxford University Press, 1998), 68.

13. "Clara Dorothy Bewick Colby," in *American National Biography*, ed. John A. Garraty and Mark C. Carnes (New York: Oxford University Press, 1999), 5:194–96.

14. Carol Lee Bacchi, *Liberation Deferred? The Ideas of the English-Canadian Suffragists, 1877–1918* (Toronto: University of Toronto Press, 1983), 40–57.

15. Nancy Millar, *The Famous Five: Emily Murphy and the Case of the Missing Persons* (Cochrane, AB: Western Heritage Centre, 1999), 72–82.

16. Ibid., 9–15, 66–71, 83–97; Patricia Roome, "'From One Whose Home Is Among the Indians': Henrietta Muir Edwards and Aboriginal Peoples," in *Unsettled Pasts: Reconceiving the West through Women's History*, ed. Sarah Carter et al. (Calgary: University of Calgary Press, 2005), 47–78.

17. Roome, "'From One Whose Home,'" 48–50.

18. "Eugenics in Alberta," Albertasource.ca, accessed 14 July 2009, www.abheritage.ca/abpolitics/people/influ_eugenics.html.

19. Patricia G. Holland, "George Francis Train and the Women's Suffrage Movement, 1867–70," University of Iowa Special Collections, April 1987, www.lib.uiowa.edu/spec-coll/bai/holland.htm.

20. Alan P. Grimes, *The Puritan Ethic and Woman Suffrage* (New York: Oxford University Press, 1967), 47–71, esp. 59–60.

21. Frances W. Kaye, "The Ladies Department of the Ohio Cultivator, 1845–1855: Feminist Forum," *Agricultural History* 50 (April 1976): 414–23; Frances W. Kaye, "Francis Marion Beynon and the *Grain Growers' Guide*" (conference paper, Midwest Association for Canadian Studies, Minneapolis, MN, September 1986).

22. Carrie Chapman Catt Girlhood Home, "Carrie Chapman Catt: A Biography," accessed 20 July 2007, www.catt.org/ccabout.html.

23. See Scott G. McNall, *The Road to Rebellion: Class Formation and Kansas Populism, 1865–1900* (Chicago: University of Chicago Press, 1988).

24. "Clara Dorothy Bewick Colby."

25. "Lost Bird of Wounded Knee," South Dakota Public Broadcasting, accessed 12 August 2010, www.sdpb.org/lostbird/summary.asp.

26. Veronica Strong-Boag and Michelle Lynn Rosa, "Some Small Legacy of Truth," introduction to *Clearing in the West* and *The Stream Runs Fast: The Complete Autobiography*, by Nellie McClung (Peterborough, ON: Broadview Press, 2003), 16.

27.	Roome, "From One Whose Home.'"

28.	Nellie McClung, *Purple Springs* (1922; rept., n.p.: BiblioBazaar, 2007), 196–204; Randi Warne, *Literature as Pulpit: The Christian Social Activism of Nellie L. McClung* (Waterloo: Wilfrid Laurier University Press, 1993); Francis Marion Beynon, *Aleta Dey* (Great Britain [n.p.]: C.W. Daniel, 1919), accessed 14 July 2009, digital. library.upenn.edu/women/beynon/aleta/aleta.html.

29.	Millar, *Famous Five*, 79, 96–97.

30.	"Persons Case," Canada Online, accessed 14 July 2009, canadaonline.about.com/cs/ women/a/personscase.htm; Joe Starita, *"I Am a Man": Chief Standing Bear's Journey for Justice* (New York: St. Martin's Press, 2009).

31.	Walter Stewart, *MJ: The Life and Times of M.J. Coldwell* (Toronto: Stoddard, 2000), 105–9.

32.	Ron Walters, "The Great Plains Sit-In Movement, 1958–60," *Great Plains Quarterly* 16, no. 2 (Spring 1986): 85–96.

33.	Jeannette Rankin interviews, accessed 14 July 2009, http://content.cdlib.org/ ark:/13030/kt758005dx/.

34.	"The Sign: Murdoch vs. Murdoch," AlbertaSource.ca, accessed 14 July 2009 (www. albertasource.ca/lawcases/civil/murdoch/murdoch.htm) but no longer available; *Matrimonial Property Act*, Alberta, Sections 8a and 8b, Canadian Legal Information Institute website, accessed 14 July 2009, www.canlii.org/en/ab/laws/stat/rsa-2000-c-m-8/latest/rsa-2000-c-m-8.html; Wava G. Haney and Lorna Clancy Miller, "U.S. Farm Women, Politics and Policy," *Journal of Rural Studies* 7, no. 1/2 (1991): 115–21.

## 9	And Still the Waters

1.	Howard R. Lamar, *The Reader's Encyclopedia of the American West* (New York: Harper and Row, 1977), 290; E. Jethro Gaede, "Termination and Relocation Programs," *Encyclopedia of Oklahoma History and Culture*, Oklahoma Historical Society, accessed 16 July 2009, digital.library.okstate.edu/encyclopedia/entries/T/TE014.html; see also Donald Fixico, *Termination and Relocation, Federal Indian Policy, 1945–1960* (Albuquerque: University of New Mexico Press, 1986); Angie Debo, *A History of the Indians of the United States* (Norman: University of Oklahoma Press, 1970), 349–56.

2.	"The History of Treaty Land Entitlement in Saskatchewan," Government of Saskatchewan website, accessed 15 July 2009, www.fnmr.gov.sk.ca/lands/tle/history; Pequis First Nation, "Newsletter No. 8," May 2006, www.peguisfirstnation.ca/pdf/ newsletter8.pdf; see also J.R. Miller, *Skyscrapers Hide the Heavens: A History of Indian-White Relations*, 3rd ed. (Toronto: University of Toronto Press, 2000), 367–68, 384.

3.	Sarah Carter, *Lost Harvests: Prairie Indian Reserve Farmers and Government Policy* (Montreal and Kingston: McGill-Queen's University Press, 1990), 209–13; Russel Barsh, "The Substitution of Cattle for Bison on the Great Plains," in *The Struggle for the Land: Indigenous Insight and Industrial Empire in the Semiarid World*, ed. Paul A. Olson (Lincoln: University of Nebraska Press, 1990), 103–26.

4. D.W. Adams, *Education for Extinction: American Indians and the Boarding School Experience* (Lawrence: University Press of Kansas, 1995); Robert Bensen, ed., *Children of the Dragonfly: Native American Voices on Child Custody and Education* (Tucson: University of Arizona Press, 2001).

5. But cf. Bruce Trigger, "The Historians' Indian: Native Americans in Canadian Historical Writing from Charlevoix to the Present," *Canadian Historical Review* 67, no. 3 (1986): 315–42.

6. Isaiah Berlin, "On Political Judgement," in *The Sense of Reality: Studies in Ideas and Their History* (New York: Farrar, Straus and Giroux, 1996), 40–53, quoted in James C. Scott, *Seeing Like a State: How Certain Schemes to Improve the Human Condition Have Failed* (New Haven: Yale University Press, 1998), 347.

7. Shirley A. Leckie, *Angie Debo: Pioneering Historian* (Norman: University of Oklahoma Press, 2000), 76–78, 80, 84.

8. Jill St. Germaine, *Indian Treaty-Making Policy in the United States and Canada, 1867-1877* (Toronto: University of Toronto Press, 2000), 165.

9. Joan Marks, *A Stranger in Her Native Land: Alice Fletcher and the American Indians* (Lincoln: University of Nebraska Press, 1989), 82–134.

10. Carter, *Lost Harvests*, 209–12; Paul Voisey, *Vulcan: The Making of a Prairie Community* (Toronto: University of Toronto Press, 1988), 90; Hana Samek, *The Blackfoot Confederacy, 1880-1920: A Comparative Study of Canadian and U.S. Indian Policy* (Albuquerque: University of New Mexico Press, 1987), 53–54, 75, 80.

11. James Axtell, *The Invasion Within: The Context of Cultures in Colonial North America* (New York: Oxford University Press, 1986); "Cherokee Removal," *The New Georgia Encyclopedia* (updated 5 November 2004), www.georgiaencyclopedia.org/nge/article.jsp?id=h-2722&sug=y; Charles Neider, ed., *Autobiography of Mark Twain* (New York: Harper and Row, 1959), 346.

12. See Angie Debo, *The Rise and Fall of the Choctaw Republic* (Norman: University of Oklahoma Press, 1934); Angie Debo, *And Still the Waters Run: The Betrayal of the Five Civilized Tribes* (Princeton: Princeton University Press, 1940).

13. "Sequoyah," *The New Georgia Encyclopedia*, published 3 September 2002, www.georgiaencyclopedia.org/nge/article.jsp?id=h-618&sug=y; Murray R. Wickett, *Contested Territory: Whites, Native Americans, and African Americans in Oklahoma, 1865-1907* (Baton Rouge: Louisiana State University Press, 2000), 163.

14. Marks, *Stranger in Her Native Land*, 80–134.

15. Ernest Renan, *The Life of Jesus*, accessed 20 July 2007, www.infidels.org/library/historical/ernest_renan/life_of_jesus.html; Chuck Neighbors, "The Story of 'In His Steps,'" accessed 20 July 2007, www.mastersimage.com/articles/ihs.htm.

16. Charles A. Eastman, *From Deep Woods to Civilization* (1916; rept., Lincoln, University of Nebraska Press, 1977), 143.

17. Wilcomb Washburn, *The Assault on Indian Tribalism: The General Allotment Law (Dawes Act) of 1887* (Philadelphia: Lippincott, 1975).

18. Janet A. McDonnell, *Dispossession of the American Indian, 1887-1934* (Bloomington: University of Indiana Press, 1991).

19. Debo, *And Still the Waters Run*, 91, 93.
20. Richard Maxwell Brown, "Violence," in *The Oxford History of the American West*, ed. Clyde A. Milner, Carol A. O'Connor, and Martha Sandweiss (New York: Oxford University Press, 1994), 393–426. See, for instance, John Mack Faragher et al., *Out of Many: A History of the American People*, one of the most inclusive American history textbooks, first published in 1993 and now in its seventh edition. Although its authors include two Western historians (Faragher and Susan H. Armitage), it said nothing about twentieth-century dispossession.
21. Debo, *And Still the Waters Run*, 21, 22, 132.
22. Dennis McAuliffe, *Bloodland: A Family Story of Oil, Greed, and Murder on the Osage Reservation* (Tulsa: Council Oak Books, 1999); Terry P. Wilson, *The Underground Reservation: Osage Oil* (Lincoln: University of Nebraska Press, 1985).
23. Wickett, *Contested Territory*, 196; Debo, *And Still the Waters Run*, 292, 200, 333.
24. Debo, *And Still the Waters Run*, 96, 97.
25. Ibid., 103, 111, 113, 309.
26. Ibid., 219–20.
27. Ibid., 184.
28. Wilson, *Underground Reservation*.
29. Debo, *And Still the Waters Run*, 171, 291–92, 327–30.

10   Dust Bowls

1. US Census, 1930. See also Jeffrey Walser and John Anderlik, "Rural Depopulation: What Does It Mean for the Future Economic Health of Rural Areas and the Community Banks That Support Them?" (last updated 11 February 2005), http://www.fdic.gov/bank/analytical/banking/2005jan/article2.html; Richard Rathge and Paula Highman, "Population Change in the Great Plains: A History of Prolonged Decline," *Rural Development Perspectives* 13, no. 1 (July 1998): 19–26.
2. James C. Malin, *History and Ecology: Studies of the Grassland*, ed. Robert P. Swierenga (Lincoln: University of Nebraska Press, 1985), 49–54, 116–23; Theodore Binnema, *Common and Contested Ground: A Human and Environmental History of the Northwestern Plains* (Toronto: University of Toronto Press, 2004), 69; Russel Barsh, "Bison" (paper presented at Center for Great Plains Studies symposium, Lincoln, NE, 2000); David Wishart, ed., *Encyclopedia of the Great Plains* (Lincoln: University of Nebraska Press, 2004), 47, 575, 579, 582–83, 586–87, 590.
3. Wishart, *Encyclopedia*, 555, 561, 567, 601.
4. Andrew Isenberg, "Toward a Policy of Destruction: Buffaloes, Law, and the Market, 1803–83," *Great Plains Quarterly* 12, no. 4 (Fall 1992): 227–41; Andrew Isenberg, *The Destruction of the Bison: An Environmental History, 1750–1920* (New York: Cambridge University Press, 2000); Ken Zontek, *Buffalo Nation: American Indian Efforts to Restore the Bison* (Lincoln: University of Nebraska Press, 2007), esp. 1–32.
5. Richard White, *"It's Your Misfortune and None of My Own": A New History of the*

American West (Norman: University of Oklahoma Press, 1991), 228–29, 263–64, 373–77; Jackie Crowley, "Parity! Still a Viable Word," American Agriculture Movement website, www.aaminc.org/newsletter/v8i6/v8i6-1.htm.

6.   Wishart, *Encyclopedia*, 240, 847.

7.   Duff Crerar, "1916 and the Great War," in *Alberta Formed, Alberta Transformed*, ed. Michael Payne, Donald Wetherell, and Catherine Cavanaugh (Calgary: University of Calgary Press, and Edmonton: University of Alberta Press, 2005), 2:393–94; Nancy Shoemaker, *American Indian Population Recovery in the Twentieth Century* (Albuquerque: University of New Mexico Press, 1999); Sarah Carter, *Lost Harvests: Prairie Indian Reserve Farmers and Government Policy* (Montreal and Kingston: McGill-Queen's University Press, 1990), 249–53.

8.   Paul Voisey, *Vulcan: The Making of a Prairie Community* (Toronto: University of Toronto Press, 1988), 90; Edward McCourt, *Music at the Close* (Toronto: Ryerson, 1947), 7.

9.   David C. Jones, *Empire of Dust: Settling and Abandoning the Prairie Dry Belt* (Edmonton: University of Alberta Press, 1987), 24, 80, 88, 107, 108.

10.   Ibid., 123, 113, 116, 137, 135, 181, 211.

11.   Roger Boulet, *Vistas on the Canadian Pacific Railway Companion Volume* (Calgary: Glenbow, 2009); R. Douglas Francis, "From Wasteland to Utopia: Changing Images of the Canadian West in the Nineteenth Century," *Great Plains Quarterly* 7, no. 3 (Summer 1987): 178–94.

12.   Malin, *History and Ecology*, 32–35; J.E. Weaver, *Prairie Plants and Their Environment: A Fifty-Year Study in the Midwest* (Lincoln: University of Nebraska Press, 1968), 80–81, 149, 197–201.

13.   Donald Worster, *Dust Bowl: The Southern Plains in the 1930s* (New York: Oxford University Press, 2004).

14.   Ibid.; Leslie Hewes, *The Suitcase Farming Frontier: A Study in the Historical Geography of the Central Great Plains* (Lincoln: University of Nebraska Press, 1973).

15.   Wishart, *Encyclopedia*, 32, 48, 701–2.

11   *Mitigating but Not Rethinking*

1.   George W. Norris, *Fighting Liberal: The Autobiography of George W. Norris* (1945; rept., Lincoln: University Nebraska Press, 1972); Richard Lowitt, *George W. Norris: The Making of a Progressive, 1861–1912* (Syracuse: Syracuse University Press, 1963); Richard Lowitt, *George W. Norris: The Persistence of a Progressive, 1913–1933* (Urbana: University of Illinois Press, 1971); Richard Lowitt, *George W. Norris: The Triumph of a Progressive, 1933–1944* (Urbana: University of Illinois Press, 1978).

2.   Walter Stewart, *The Life and Political Times of Tommy Douglas* (Toronto: McArthur, 2003), 80–81.

3.   Lewis Thomas, ed., *The Making of a Socialist: The Recollections of T.C. Douglas* (Edmonton: University of Alberta Press, 1982), 297, 225, 167, 168.

4. John Richards and Larry Pratt, *Prairie Capitalism: Power and Influence in the New West*, Canada in Transition Series (Toronto: McClelland and Stewart, 1979), 128–29.

5. Thomas, *Making of a Socialist*, 167, 175, 177.

6. Richards and Pratt, *Prairie Capitalism*, 140–43, 156.

7. James M. Pitsula and Ken Rasmussen, *Privatizing a Province: The New Right in Saskatchewan* (Vancouver: New Star Books, 1990), Devine quoted 47, 69, 286.

8. John Feldberg and Warren M. Elofson, "Financing the Palliser Triangle, 1908–1913," *Great Plains Quarterly* 18, no. 3 (1988): 257–68; Kenneth Norrie and Douglas Owram, *A History of the Canadian Economy* (Toronto: HBJ Canada, 1991), 331.

9. Lowitt, *Norris: Making of a Progressive*, 49.

10. Norris, *Fighting Liberal*, 71, 72.

11. See, especially, Turner, *Making of a Socialist*, 174.

12. L.D. Lovick, ed., *Till Power Is Brought to Pooling: Tommy Douglas Speaks* (Lantzville, BC: Oolichan, 1979), 106; Stewart, *Life and Political Times*, 164; Thomas, *Making of a Socialist*, 225; Pitsula and Rassmussen, *Privatizing a Province*, 60.

13. Richards and Pratt, *Prairie Capitalism*, 99–108; Stewart, *Life and Political Times*, 165–66; Henry George, *Progress and Poverty* (San Francisco, 1879; rept., Cambridge: Cambridge University Press, 2009).

14. Lowitt, *Norris: Making of a Progressive*, 97; Lowitt, *Norris: Triumph of a Progressive*, 95.

15. Lowitt, *Norris: Making of a Progressive*, 72; Lowitt, *Norris: Persistence of a Progressive*, 23–25; Norris, *Fighting Liberal*, 172.

16. Walter Stewart, *Uneasy Lies the Head: The Truth About Canada's Crown Corporations* (Toronto: Collins, 1987), 62; Stewart, *Life and Political Times*, 191; Lowitt, *Norris: Triumph of a Progressive*, 129.

17. Richard White, *"It's Your Misfortune and None of My Own": A New History of the American West* (Norman: University of Oklahoma Press, 1991), 153.

18. Lowitt, *Norris: Triumph of a Progressive*, 92–93, 102, 97, 414–15, 366; White, *"It's Your Misfortune,"* 488–89.

19. Richards and Pratt, *Prairie Capitalism*, 113–14; Thomas, *Making of a Socialist*, 202; Lowitt, *Norris: Triumph of a Progressive*, 312; Thomas, *Making of a Socialist*, 202; SaskPower website, accessed 8 August 2004, www.saskpower.com; "History of Lake Diefenbaker," Outlook School Division, 26 February 2002, www.saskschools.ca/~llhs/lakedief/history.

20. Lowitt, *Norris: Triumph of a Progressive*, 408; Lovick, *Till Power Is Brought*, 173–81.

21. F. Laurie Barron, *Walking in Indian Moccasins: The Native Policies of Tommy Douglas and the CCF* (Vancouver: UBC Press, 1997).

22. Douglas R. Parks and Waldo Wedel, "Pawnee Geography: Historical and Sacred," *Great Plains Quarterly* 5, no. 3 (1985): 143–76; Zygmunt J.B. Plater, "Reflected in a River: Agency Accountability and the TVA Tellico Dam Case" (1982), *Boston College Law School Faculty Papers*, Paper 177, http://lawdigitalcommons.bc.edu/lsfp/177 (*Tennessee Law Review* 49 [Summer 1982]: 747–87); Lowitt, *Norris: Triumph of a Progressive*, 454, 462.

23. See Michael Lawson, *Dammed Indians: The Pick-Sloan Plan and the Missouri River Sioux, 1944-1980* (Norman: University of Oklahoma Press, 1982).
24. Norris, *Fighting Liberal*, 407.
25. Lowitt, *Norris: Triumph of a Progressive*, 58–68; Norris, *Fighting Liberal*, 344–56.
26. Norris, *Fighting Liberal*, 193–96; Lowitt, *Norris: Making of a Progressive*, 154; Lovick, *Till Power Is Brought*, 63; Lowitt, *Norris: Triumph of a Progressive*, 257–59.
27. John Herd Thompson, *Forging the Prairie West* (Toronto: Oxford University Press, 1998), 38–40, 292; Jim Harding, "Putting Social Policy Analysis in Its Political Context," in *Social Policy and Social Justice: The NDP Government in Saskatchewan During the Blakeney Years*, ed. Jim Harding (Waterloo: Wilfrid Laurier University Press, 1995), 432.
28. Lowitt, *Norris: Making of a Progressive*, 98; Lowitt, *Norris: Triumph of a Progressive*, 134; White, *"It's Your Misfortune,"* 144, 479.
29. Max Foran, "The Impact of the Depression on Grazing Lease Policy in Alberta," in *Cowboys, Ranchers, and the Cattle Business: Cross-Border Perspectives on Ranching History*, ed. Simon Evans, Sarah Carter, and Bill Yeo (Calgary: University of Calgary Press, and Boulder: University Press of Colorado, 2000), 127–28; Thomas, *Making of a Socialist*, 95; Stewart, *Life and Political Times*, 104–5; Thomas, *Making of a Socialist*, 77.
30. See Angie Debo, *And Still the Waters Run: The Betrayal of the Five Civilized Tribes* (Princeton: Princeton University Press, 1940).
31. Pitsula and Rasmussen, *Privatizing a Province*, 45–46; Murray Mandryk, "Uneasy Neighbours: White-Aboriginal Relations and Agricultural Decline," in *Writing Off the Rural West: Globalization, Governments, and the Transformation of Rural Communities*, ed. Roger Epp and Dave Whitson (Edmonton: University of Alberta Press, 2001), 217–19; "Watershed Moment for Prairie Agriculture: Economist," CBC website, 21 February 2006, www.cbc.ca.
32. Thompson, *Forging the Prairie West*, 135.
33. Richards and Pratt, *Prairie Capitalism*, 143, 186–88, 203, 118–21; Pitsula and Rasmussen, *Privatizing a Province*, 283; Robert Tyre, *Douglas in Saskatchewan: The Story of a Socialist Experiment* (Vancouver: Mitchell Press, 1962), 36.

12 Planning and Economic Theory

1. Marci Barnes Gracey, "Joseph Bruner and the American Indian Federation: An Alternative View of Indian Rights," in *Alternative Oklahoma: Contrarian Views of the Sooner State*, ed. David D. Joyce (Norman: University of Oklahoma Press, 2007), 63–86; David Wishart, ed., *Encyclopedia of the Great Plains* (Lincoln: University of Nebraska Press, 2004), 701.
2. James C. Scott, *Seeing Like a State: How Certain Schemes to Improve the Human Condition Have Failed* (New Haven: Yale University Press, 1998), 49.
3. James M. Pitsula and Ken Rasmussen, *Privatizing a Province: The New Right in Saskatchewan* (Vancouver: New Star Books, 1990), esp. 63.

4. Scott, *Seeing Like a State*, 253.

5. Deborah Popper and Frank Popper, "The Great Plains: From Dust to Dust," *Planning* 53, no. 12 (1987): 12–18.

6. Scott, *Seeing Like a State*, 291.

7 Francis La Flesche, *The Middle Five: Indian School Boys of the Omaha Tribe* (1900; rept., Lincoln, University of Nebraska Press, 1978), esp. 108; Hamlin Garland, *Selected Letters of Hamlin Garland*, ed. Keith Newlin and Joseph B. McCullough (Lincoln: University of Nebraska Press, 1998), 131–32, 162–63.

8. Richard White, *"It's Your Misfortune and None of My Own": A New History of the American West* (Norman: University of Oklahoma Press, 1991), 565, 567.

9. Paul Voisey, *Vulcan: The Making of a Prairie Community* (Toronto: University of Toronto Press, 1988), 107–8.

10. Jennie Abell, "Legal Aid in Saskatchewan: Rhetoric and Reality, 1974–82," in *Social Policy and Social Justice: The NDP Government in Saskatchewan During the Blakeney Years*, ed. Richard Harding (Waterloo: Wilfrid Laurier University Press, 1995), 173–220, quotation on 176.

11. Charles Gore, *Regions in Question: Space, Development Theory and Regional Policy* (London: Methuen, 1984), 244 (emphasis in the original); Joseph Schumpeter, quoted in Benjamin Higgins and Donald J. Savoie, *Regional Development Theories and Their Application* (New Brunswick, NJ: Transaction Publishers, 1997), 36–37.

12. Gunnar Myrdal, *Economic Theory and Underdeveloped Regions* (London: Duckworth, 1957), 34, quoted in Gore, *Regions in Question*, 36.

13. Gore, *Regions in Question*, 37–38, 128.

14. Peter C. Newman, *The Canadian Revolution, 1985–1995: From Deference to Defiance* (Toronto: Viking, 1995), 320–22.

15. Kathleen Ann Pickering, *Lakota Culture, World Economy* (Lincoln: University of Nebraska Press, 2000), 18, 21–22.

16. Wishart, *Encyclopedia*, 587, 419, 556, 567, 592, 607.

17. Higgins and Savoie, *Regional Development Theories*, 147, 49.

18. Scott, *Seeing Like a State*, 331.

19. Higgins and Savoie, *Regional Development Theories*, 50; Scott, *Seeing Like a State*, 324.

20. Gore, *Regions in Question*, 245.

21. Ibid., 256, 254, 253, 250 (emphasis in the original); William Cronon, *Nature's Metropolis: Chicago and the Great West* (New York: W.W. Norton, 1991), 81–96.

22. Janine Brodie, "The Concept of Region in Canadian Politics," in *Federalism and Political Community: Essays in Honour of Donald Smiley*, ed. David P. Shugarman and Reg Whitaker (Peterborough, ON: Broadview Press, 1989), 33–53.

23. James Ferguson, *The Anti-Politics Machine: "Development," Depoliticization, and Bureaucratic Power in Lesotho* (Minneapolis: University of Minnesota Press, 1994); Raj Patel, *Stuffed and Starved: Markets, Power and the Hidden Battle for the World's Food System* (Toronto: HarperCollins, 2007), 5–12, 125, 81–82.

24. Patel, *Stuffed and Starved*, 261, 288.

25. Ibid., 201; see also 173–203.

26. Ibid., 57; see also 114–15, 273, and 53–57.
27. Charles Bowden, *Murder City: Ciudad Juárez and the Global Economy's New Killing Fields* (New York: Nation Books, 2010), 234, 98–99, 137.
28. Voisey, *Vulcan*, 77–86.

13   Mouse Beans and Drowned Rivers

1. W.O. Mitchell, *Who Has Seen the Wind* (Toronto: Macmillan, 1947), 1.
2. See J.R. Miller, *Skyscrapers Hide the Heavens: A History of Indian-White Relations in Canada*, 3rd ed. (Toronto: University of Toronto Press, 2000), esp. 326–27, 203–5; Robert Bensen, *Children of the Dragonfly: Native American Voices on Child Custody and Education* (Tucson: University of Arizona Press, 2001).
3. Patricia Mitchell, Ellie E. Prepas, and J.M. Crosby, *Atlas of Alberta Lakes* (Edmonton: University of Alberta Press, 1990), 564, 558.
4. William Cronon, "Landscapes of Abundance and Scarcity," in *The Oxford History of the American West*, ed. Clyde Milner, Carol O'Connor, and Martha Sandweis (New York: Oxford University Press, 1994), 617.
5. Boyce Richardson, *Strangers Devour the Land: A Chronicle of the Assault upon the Last Coherent Hunting Culture in North America, the Cree Indians of Northern Quebec* (White River Junction, VT: Chelsea Green, 2008); Joy A. Bilharz, *The Allegheny Senecas and Kinzua Dam: Forced Relocation Through Two Generations* (Lincoln: University of Nebraska Press, 1998).
6. Michael Lawson, *Dammed Indians: The Pick-Sloan Plan and the Missouri River Sioux, 1944–1980* (Norman: University of Oklahoma Press, 1994), 75, 56, 125.
7. Jack Glenn, *Once upon an Oldman: Special Interest Politics and the Oldman River Dam* (Vancouver: UBC Press, 1999); *Windspeaker* (Edmonton), 31 August 1990, 14 September 1990, 28 September 1990.
8. Douglas R. Parks and Waldo R. Wedel, "Pawnee Geography: Historical and Sacred," *Great Plains Quarterly* 5, no. 3 (Summer 1985): 143–76.
9. "Badlands National Park," Absolute Astronomy website, accessed 20 October 2008, www.absoluteastronomy.com/topics/badlands_national_park; Indian Claims Commission, Siksika Nation [Castle Mountain claim] (last updated 10 November 2006), http://24583.vws.magma.ca/claimsmap/siksikacastlemed-en.asp, accessed 25 July 2009; James Brooke, "Tsuu T'ina Journal; Indians Stalk a Silent, Deadly Enemy in the Prairie," *New York Times*, 19 June 2000.

14   Oil

1. Richard White, *"It's Your Misfortune and None of My Own": A New History of the American West* (Norman: University of Oklahoma Press, 1991), 398–99; Elizabeth Shogren, "Mining Measure Would Reopen Federal Land Sales," *Weekend*

Edition Sunday, National Public Radio, 6 November 2005; Robert Sherrill, *The Oil Follies of 1970–1980: How the Petroleum Industry Stole the Show (and Much More Besides)* (Garden City: Anchor/Doubleday, 1983), 474; Edmund L. Andrews, "Inspector Finds Broad Failures in Oil Program," *New York Times*, 26 September 2007. Since then the Minerals Management Service of the Department of the Interior has been embroiled in growing scandals; see Charlie Savage, "Sex, Drug Use, and Graft Cited in Interior Department," *New York Times*, 10 September 2008.

2. Sherrill, *Oil Follies*, 33; Christopher T. Rand, *Making Democracy Safe for Oil: Oilmen and the Islamic East* (Boston: Little Brown, 1975); Ed Shaffer, *Canada's Oil and the American Empire* (Edmonton: Hurtig, 1983).

3. Sherrill, *Oil Follies*, 477; John Joseph Mathews, *Life and Death of an Oilman: The Career of E.W. Marland* (Norman: University of Oklahoma Press, 1951); Jim Lyon, *Dome: The Rise and Fall of the House That Jack Built* (Scarborough, ON: Avon Books, 1983).

4. "The Alberta Natural Resources Act" (updated December 2003), http://www. aboriginal.alberta.ca/documents/NRA_Info_Sheet-Dec2003.pdf; Treaty 7 Elders and Tribal Council, with Walter Hildebrandt, Sarah Carter, and Dorothy First Rider, *The True Spirit and Original Intent of Treaty 7* (Montreal and Kingston: McGill-Queen's University Press, 1996), 143–45.

5. Imperial Oil Limited, *The Discovery That Made History: The Legacy of Leduc* (Canadian Petroleum Discovery Centre, 2006); Leduc #1 Energy Discovery Centre, accessed 20 October 2008, www.c-pic.org/mission.htm.

6. John Richards and Larry Pratt, *Prairie Capitalism: Power and Influence in the New West* (Toronto: McClelland and Stewart, 1979), 156.

7. Rand, *Making Democracy Safe*, 330–35; Shaffer, *Canada's Oil*, 159–61; Tammy Nemeth, "1980: Duel of the Decade," in *Alberta Formed, Alberta Transformed*, ed. Michael Payne, Donald Wetherell, and Catherine Cavanaugh (Edmonton: University of Alberta Press, and Calgary: University of Calgary Press, 2005), 2:677–702.

8. G. Bruce Doern and Glen Toner, *The Politics of Energy: The Development and Implementation of the NEP* (Toronto: Methuen, 1985), 89; Shaffer, *Canada's Oil*, 214–15, 221; Nemeth, "1980," 686.

9. Peter Foster, *The Blue-Eyed Sheiks: The Canadian Oil Establishment* (Toronto: HarperCollins, 1980), 38; Sherrill, *Oil Follies*, 33, 51.

10. Doern and Toner, *Politics of Energy*, 145–46.

11. See, for instance, "List of Oil Refineries" (last modified 27 December 2010), www.wikipedia.org/wiki/list_of_oil_refineries#Alberta; Imperial Oil, Operations and Our Neighbours: Strathcona Refinery, http://www.imperialoil.ca/Canada-English/operations_refineries_strathcona.aspx (copyright © 2010 Imperial Oil Limited).

12. Shaffer, *Canada's Oil*, 231–32, 107–8.

13. Nemeth, "1980," 685, 694.

14. Shaffer, *Canada's Oil*, 241; Nemeth, "1980," 683.

15. Nemeth, "1980," 685, 691, 694.
16. Doern and Toner, *Politics of Energy*, 356; Sherrill, *Oil Follies*, 482; Shaffer, *Canada's Oil*, 244–46; Lyon, *Dome*, 19–20, 22–23, 29, 184–85.
17. Roger Gibbins, quoted in Sydney Sharpe, "NEP Remains a Bitter Political Metaphor 25 Years Later," *Calgary Herald*, 26 June 2005.
18. Doern and Toner, *Politics of Energy*, 460; Nemeth, "1980," 687–89.
19. Doern and Toner, *Politics of Energy*, 248–49, 258–59; James Pitsula and Ken Rasmussen, *Privatizing a Province: The New Right in Saskatchewan* (Vancouver: New Star Books, 1990), 181–87.
20. Nemeth, "1980," 684; Shaffer, *Canada's Oil*, 236–37.
21. Shaffer, *Canada's Oil*, 243.
22. White, *"It's Your Misfortune,"* 564–68.
23. Ibid., 567; Sharpe, "NEP Remains."
24. Roger Epp, "The Political De-skilling of Rural Communities," in *Writing Off the Rural West: Globalization, Governments, and the Transformation of Rural Communities*, ed. Roger Epp and Dave Whitson (Edmonton: University of Alberta Press, 2001), 303–4.

15 Arts, Justice, and Hope on the Great Plains

1. Michael Murphy, "Civilization, Self-Determination, and Reconciliation," in *First Nations, First Thoughts: The Impact of Indigenous Thought in Canada*, ed. Anis May Timpson (Vancouver: UBC Press, 2010), 254–60.
2. "The Aboriginal Justice Implementation Commission," accessed 27 July 2009, www.ajic.mb.ca; Frances W. Kaye, *Hiding the Audience: Arts and Arts Institutions on the Prairies* (Edmonton: University of Alberta Press, 2003), 140–43.
3. Jack Glenn, *Once upon an Oldman: Special Interest Politics and the Oldman River Dam* (Vancouver: UBC Press, 1999); Michel F. Girard, "L'aménagement de la forêt d'Oka à la lumière de l'écologie historique," *Journal of Canadian Studies* 27, no. 2 (Summer 1992): 5–21; *Windspeaker* (Edmonton): on Donald Marshall, 9 February 1990; on Wilson Nepoose, 12 October 1990, 23 November 1990, 15 February 1991, 5 July 1991, 2 August 1991, 31 August 1991, 17 January 1992, 16 March 1992, 30 March 1992; on Elijah Harper, 22 June 1990, 6 July 1990; on Oldman Dam, 16 June 1989, 17 August 1990, 31 August 1990, 14 September 1990, 23 November 1990, 7 December 1990, 15 March 1991, 12 April 1991, 7 June 1991, 13 September 1991, 22 November 1991, 30 March 1992, 10 May 1993, March 1994, September 1994.
4. See Richard Wagamese, *The Terrible Summer* (Toronto: Warwick, 1996); *Kanehsatake: 270 Years of Resistance*, dir. Alanis Obomsawin (National Film Board of Canada, 1993); *Windspeaker*, 20 July 1990, 3 August 1990, 17 August 1990, 14 September 1990 (blockade of Louise Bridge).
5. Glenn, *Once upon an Oldman*; *Windspeaker*: on Oldman Dam, 31 August 1990, 14 September 1990; on Daishowa, 12 October 1990, 7 December 1990; on Oka, 28 September 1990.

6. *Windspeaker:* on LaChance and Nerland, 15 February 1991, 26 April 1991, 24 May 1991, 19 July 1991, 7 December 1992, 26 April 1993, 24 May 1993; on sentencing of Born-With-A-Tooth, September 1994; on number of Aboriginal justice inquiries, 7 December 1992. See also Glenn, *Once upon an Oldman,* 123; J.R. Miller, *Skyscrapers Hide the Heavens: A History of Indian-White Relations in Canada,* 3rd ed. (Toronto: University of Toronto Press, 2000), 373–87; "Highlights of Report of Royal Commission on Aboriginal Peoples," Indian and Northern Affairs Canada, www.ainc-inac.gc.ca/ap/pubs/rpt/rpt-eng.asp (last modified 28 April 2010); A.C. Hamilton, *A Feather Not a Gavel* (Winnipeg: Great Plains Publication, 2001).

7. *Windspeaker.* See especially on woman and hunger strike, 2 August 1991; on Tony Thrasher, 28 July 1989; on Ken Ward, 16 February 1990, 16 March 1990, 8 June 1990; on child taking, 17 March 1989, 25 August 1989; on Wilson Nepoose, July 1998.

8. Frederick C. Luebke, ed., *A Harmony of the Arts: The Nebraska State Capitol* (Lincoln: University of Nebraska Press, 1990).

9. Ella Deloria, *Speaking of Indians* (1944; rept., Lincoln: University of Nebraska Press, 1998).

10. Harold Cardinal, *The Unjust Society: The Tragedy of Canada's Indians* (Edmonton: Hurtig, 1969), 1.

11. Vine Deloria, Jr., *Custer Died for Your Sins: An Indian Manifesto* (1969; rept., Norman: University of Oklahoma Press, 1988), 241; Cardinal, *Unjust Society,* 107–15, 120–24.

12. Robert Allen Warrior and Paul Chaat Smith, *Like a Hurricane: The Indian Movement from Alcatraz to Wounded Knee* (New York: New Press, 1996); Jeannette Armstrong, *Slash* (Penticton: Theytus, 1985); Cardinal, *Unjust Society,* 75; Deloria, *Custer Died,* x.

13. Deloria, *Custer Died,* 256, xii, 257; Cardinal, *Unjust Society,* 165, 170.

14. Joane Cardinal Schubert, Gallery Talk, Calgary, Alberta, 16 June 2005; Joane Cardinal Schubert in *Indigena: Contemporary Native Perspectives,* ed. Gerald McMaster and Lee-Ann Martin (Vancouver: Douglas and McIntyre, 1992), 132–33.

15. Schubert, in *Indigena,* 132–33; Hamilton, *Feather.*

16. Jane Ash Poitras, in *Indigena,* 164–69.

17. Wagamese, *Terrible Summer,* 60, 138–39, 164–65.

18. Richard Wagamese, *A Quality of Light* (Toronto: Doubleday Canada, 1997), 279, 319.

19. Wagamese, *Quality of Light,* 272.

20. Thomas Flanagan, *Métis Lands in Manitoba* (Calgary: University of Calgary Press, 1991), 232; Thomas Flanagan, *First Nations? Second Thoughts* (Montreal and Kingston: McGill-Queen's University Press, 2000), 4 (hereafter cited in parentheses in text).

21. John Borrows, *Recovering Canada: The Resurgence of Indigenous Law* (Toronto: University of Toronto Press, 2002), 141.

22. Murphy, "Civilization," 264; see also 253–63.

23. Lenore A. Stiffarm, "The Demography of Native North America: A Question of American Indian Survival," with Phil Lane, Jr., in *The State of Native America,* ed. M. Annette Jaimes (Boston: South End Press, 1992), 23–54.

24. Sarah Carter, *Lost Harvests: Prairie Indian Reserve Farmers and Government Policy* (Montreal and Kingston: McGill-Queen's University Press, 1990), esp. 196–236; Russel Barsh, "The Substitution of Cattle for Bison on the Great Plains," in *The Struggle for the Land: Indigenous Insight and Industrial Empire in the Semiarid World*, ed. Paul A. Olson (Lincoln: University of Nebraska Press, 1990), 103–26; Benjamin Higgins and Donald Savoie, *Regional Development Theories and Their Application* (New Brunswick, NJ: Transaction Publishers, 1997), 47; Angie Debo, *And Still the Waters Run: The Betrayal of the Five Civilized Tribes* (Princeton: Princeton University Press, 1940), 132.

25. Flanagan, *First Nations*, 9.

26. Community Holistic Circle Healing position paper (1993), cited in Royal Commission on Aboriginal Peoples, *Bridging the Cultural Divide: A Report on Aboriginal People and Criminal Justice in Canada* (Ottawa: Canada Communication Group, 1996), 165; Joan Ryan, *Doing Things the Right Way: Dene Traditional Justice in Lac La Martye, N.W.T.* (Calgary: University of Calgary Press, 1995); Rupert Ross, *Dancing with a Ghost: Exploring Indian Reality* (Toronto: Penguin, 2006); Rupert Ross, "Exploring Criminal Justice and the Aboriginal Healing Paradigm," accessed 27 July 2009, www.lsuc.on.ca/media/third_colloquium_rupert_ross.pdf.

27. Reg Crowshoe and Sybille Manneschmidt, *Akat'stiman: A Blackfoot Framework for Decision-Making and Mediation Processes* (Calgary: University of Calgary Press, 1997).

28. Borrows, *Recovering Canada*, 114; see also 86–92, 141–46.

Conclusion

1. Michael Jackson, "Locking Up Natives in Canada," *University of British Columbia Law Review* 23 (1989): 221–83, quoted in Royal Commission on Aboriginal Peoples, *Bridging the Cultural Divide: A Report on Aboriginal People and Criminal Justice in Canada* (Ottawa: Canada Communication Group, 1996), 29–30; Rupert Ross, *Returning to the Teachings: Exploring Aboriginal Justice* (Toronto: Penguin, 1996), 177; Rupert Ross, "Exploring Criminal Justice and the Aboriginal Healing Paradigm," accessed 27 July 2009, www.lsuc.on.ca/media/third_colloquium_rupert_ross.pdf.

2. John Borrows, *Recovering Canada: The Resurgence of Indigenous Law* (Toronto: University of Toronto Press, 2002), 72–76.

3. Murray Mandryk, "Uneasy Neighbours: White-Aboriginal Relations and Agricultural Decline," in *Writing Off the Rural West: Globalization, Governments, and the Transformation of Rural Communities*, ed. Roger Epp and Dave Whitson (Edmonton: University of Alberta Press, 2001), 205–22; Ho-Chunk Inc. website, accessed 28 August 2009, www.hochunkinc.com.

4. See, for instance, Stephen Graham Jones, *The Bird Is Gone: A Manifesto* (n.p.: Fiction Collective, 2003).

5. *The World This Weekend*, CBC Radio, 23 July 2006; Roger Epp, "The Political De-skilling of Rural Communities," in *Writing Off the Rural West: Globalization, Governments, and the Transformation of Rural Communities*, ed. Roger Epp and Dave Whitson (Edmonton: University of Alberta Press, 2001), 301–24; Thomas Frank, *What's the Matter with Kansas? How Conservatives Won the Heart of America* (New York: Henry Holt, 2004).

6. Lorelei L. Hanson, "The Disappearance of the Open West: Individualism in the Midst of Agricultural Restructuring," in *Writing Off the Rural West*, ed. Epp and Whitson, 165–66; Aritha van Herk, *Mavericks: An Incorrigible History of Alberta* (Toronto: Penguin, 2001); "Mavericks: An Incorrigible History of Alberta," permanent exhibition, Glenbow Museum, Calgary, Alberta.

7. Borrows, *Recovering Canada*, 141.

Credits

The map on p. 6 is reproduced courtesy of the Center for Great Plains Studies, University of Nebraska–Lincoln.

Material from *Wah'Kon-Tah*, by John Joseph Mathews, is quoted by permission of the University of Oklahoma Press.

An earlier version of chapter 6, "Intellectual Justification for Conquest," appeared in the *American Review of Canadian Studies* 31, no. 4 (Winter 2001), under the title "An Innis, Not a Turner."

An earlier version of chapter 11, "Mitigating but Not Rethinking," was published in CD-ROM format as part of a compilation of papers presented at the 2006 conference of the Association for Canadian Studies in the United States, in Vancouver, under the title "Nebraskatchewan: George Norris and Tommy Douglas."

Mossman, Manfred, 79, 91, 99
Moul, Maxine, 35
Mourning Dance, 117, 123, 125
Muir, Leilani, 176
Mulroney, Brian, 288, 294
Murdoch, Iris, 182
Murdoch v. Murdoch, 182
Murphy, Emily, 175, 176, 179, 180, 183
Murphy, Michael, 292, 312
Myrdal, Gunnar, 250, 357

Nahathaways, 46, 51, 52. *See also* Crees
National Energy Policy (NEP), 278, 282, 284–90, 329
National Policy, 24, 28, 30, 34, 106, 141, 168, 247, 258
National Recovery Act, 229
Native American Church, 117
Natural Resources Transfer Act, 278
Nature Conservancy, 126, 198, 249
Nebraska, 5, 7, 12–15, 18, 23, 33, 35–37, 41, 89, 109, 146, 154, 173, 179, 193, 207, 248, 261, 264, 266, 270, 298–99, 319, 320, 324, 326; and George Norris, 217–40
Nebraska Public Power, 42
Neihardt, John G., 10, 108, 111, 112, 113, 119–26 passim
Nelson, George, 52
Nelson, Paula, 159
Nemeth, Tammy, 285
Nepoose, Wilson, 294, 297
Nerland, Carney, 295
Neu, Dean, 46
New Deal, 8, 34, 217–18, 222, 226, 228, 229, 233
New Democratic Party (NDP), 30, 219, 221–23, 238, 240, 282
Newman, Peter, 253
New Mexico, 22, 208, 214
Nixon, Richard, 31, 261, 281, 282
Nolin, Charles, 86, 99
Nonpartisan League, 30, 220
Norquay, John, 70, 71
Norrie, Kenneth, 154
Norris, George, 12, 41, 42, 217–19, 221–36, 239, 240–41, 269, 330

Norris, T.C., 180
North American Free Trade Agreement (NAFTA), 127, 261–62
Northcote (steamer), 91
North Dakota, 35, 151, 155, 220, 232, 276
North West Company (Nor'westers), 46, 133–36 passim
North West Mounted Police (NWMP), 72–73, 80, 88, 91
Northwest Rebellion. *See* Northwest Resistance
Northwest Resistance, 28, 73–74, 79, 105, 109, 110, 258, 279
Nor'westers. *See* North West Company

Obomsawin, Alanis, 139
Oglalas, 254. *See also* Lakotas
oil, 13, 33–34, 42, 113, 116–17, 123, 126, 155–56, 165, 200, 202–4, 220, 234, 239, 263, 264, 276–88, 293, 295, 315, 329. *See also* petroleum
Ojibway, 9, 306, 316. *See also* Anishinaabe
Okanagan, 301
Oklahoma, 6, 11–13, 19, 21–23, 38, 39, 40, 43, 112–16 passim, 139, 154, 161, 169, 181–82, 189, 192–93, 196–205 passim, 214, 237, 247, 256, 257, 273, 276–78, 285, 289, 313
Oldman River Dam, 23, 231, 272, 294–96
Oliver, Frank, 209
Omahas, 4, 15, 189, 193, 229, 246, 256–57, 264
One Arrow, 105
Ontario Farmers' Union, 237
Oregon Trail, 47
Orientals, 177. *See also* Chinese
Osages, 8, 10, 12, 21, 36, 40, 103, 111–26, 161, 169, 197, 199, 200, 203, 254, 257, 276–77
Osage Tribal Museum, 112, 123, 126
Osborne, Helen (Betty), 293
Otoes, 15, 54
Ottawas, 21, 144
Otter, William, 90
Owram, Douglas, 52, 68, 72, 133, 154